Mathematical Modelling of Aerospace Dynamic Systems with Practical Applications

Mathematical Modelling of Aerospace Dynamic Systems with Practical Applications provides mathematical models for several aerospace dynamic systems: aircraft, rotorcraft, missiles, unmanned aerial vehicles (UAVs), mini air vehicles (MAVs), autonomous underwater vehicles (AUWVs), and satellite-coordinate systems.

Presenting the use of mathematical models for analysis, prediction, and control of these systems, this book discusses numerous applications in aircraft/helicopter parameter estimation, guidance and navigation of these vehicles, underwater object search, aerial terrain mapping, and satellite orbit determination. It explains path planning with obstacle avoidance, object occlusion detection and tracking, and multisensory target tracking and sensor data fusion.

This book is intended for senior undergraduate mechanical and aerospace engineering students taking courses in aerospace systems and dynamics, flight dynamics and control, and dynamical systems and estimation.

Instructors will be able to utilize a Solutions Manual and Figure Slides for their course.

Mathematical Modelling of Aerospace Dynamic Systems with Practical Applications

Jitendra R. Raol and V.P.S. Naidu

CRC Press
Taylor & Francis Group
Boca Raton London New York

CRC Press is an imprint of the
Taylor & Francis Group, an **informa** business

Designed cover image: Getty Images

First edition published 2025
by CRC Press
2385 NW Executive Center Drive, Suite 320, Boca Raton FL 33431

and by CRC Press
4 Park Square, Milton Park, Abingdon, Oxon, OX14 4RN

CRC Press is an imprint of Taylor & Francis Group, LLC

Library of Congress Cataloging-in-Publication Data
Names: Raol, J. R. (Jitendra R.), 1947- author. | Naidu, V. P. S., author.
Title: Mathematical modelling of aerospace dynamic systems with practical
applications / Jitendra R. Raol and V.P.S. Naidu.
Description: First edition. | Boca Raton, FL : CRC Press, 2025. |
Includes bibliographical references and index.
Identifiers: LCCN 2024040216 (print) | LCCN 2024040217 (ebook) |
ISBN 9781032552750 (hardback) | ISBN 9781032552767 (paperback) |
ISBN 9781003429869 (ebook)
Subjects: LCSH: Aerodynamics—Mathematical models. | Mathematical models. |
Tracking (Engineering) | Parameter estimation.
Classification: LCC TL573 .R35 2025 (print) | LCC TL573 (ebook) |
DDC 629.132/30015118—dc23/eng/20241228
LC record available at https://lccn.loc.gov/2024040216
LC ebook record available at https://lccn.loc.gov/2024040217

ISBN: 978-1-032-55275-0 (hbk)
ISBN: 978-1-032-55276-7 (pbk)
ISBN: 978-1-003-42986-9 (ebk)

DOI: 10.1201/9781003429869

Typeset in Times
by codeMantra

Access the Support Material at: www.routledge.com/9781032552750

Contents

Preface

Engineers and scientists often deal with many dynamic systems, especially the ones in the field of aviation: the atmospheric vehicles, underwater vehicles, and satellites. These vehicles can be considered aerospace dynamic systems. Such vehicles are used for transportation as well as to carry out certain special tasks and missions for improving the lives of people and for their safety and security.

The examples of such usages are aircraft flights, rescue operations by helicopters, missiles for defence needs of a country, unmanned aerial vehicles (UAVs) for surveillance, underwater vehicles for searches in the oceans, and use of some of these for detection and tracking of stationary and/or moving objects, including people on the grounds and in very hazardous places.

These operations require the use of vision or infrared camera, and onboard inertial measurement unit (IMU) for inertial navigation system (INS) data to aid the vehicle navigation process, which is often aided by the information from the global positioning system (GPS).

Since these data would be mostly noisy, and there would be other uncertainties (initial positions), the search and tracking schemes use filtering algorithms which need specification of mathematical models of the vehicles, vision systems, and INS/GPS mechanisms to get accurate navigational results. For many such tasks, the mathematical models are either easily not available or very difficult to formulate.

In this book, we provide several mathematical models for such vehicles, models of such vehicles for the various schemes of operations for the use in the estimation and filtering algorithms to accurately perform the missions for which such systems and devices are to be routinely used. A brief description of some of the estimation algorithms and soft computing paradigms will be given, and often, it would be a part of the presented applications/schemes.

The applications considered are (i) aircraft parameter estimation, (ii) helicopter parameter estimation, (iii) object detection and tracking with UAVs, (iv) terrain-assisted navigation of underwater vehicles (UWVs), (v) multisensory target tracking and sensor data fusion, (vi) two-vehicle coordination, (vii) camera/s (vision-assisted) and INS/GPS-aided navigation, (viii) path planning with obstacle avoidance, (ix) object occlusion detection and tracking, and (x) satellite orbit determination.

The approach would be simple to understand and use in several types of estimation and filtering algorithms that are used for these applications. Where feasible, MATLAB®-based implementation results would be described, and the codes/script written by the authors of this book would be put in the web/URL of the book.

The end users of the material of this book would be practising engineers in several fields such as aviation, aerospace transportation systems, mechanical engineering, air surveillance, telecommunications, and safety & security. Also, it will be very useful for the researchers/teachers who want to build/teach artificial intelligence-based autonomous systems and schemes for the applications mentioned earlier.

Although enough care has been taken in the presentation of the mathematics and the results, the ultimate responsibility of the usage of the material would lie with the readers and the users, and not with the authors and the publisher.

Acknowledgements

We are grateful to several directors of CSIR-NAL for encouraging the research in the areas of mathematical modelling and multisensory data fusion for two to three decades; very notable moral and technical support came from Dr. T. S. Prahlad for nearly 6 years.

Certain colleagues of the Flight Mechanics and Control Division of CSIR-NAL have played an important role in technical discussions and collaborative work for several years; notable ones are Dr. Mrs. Girija, G., Dr. Jatinder Singh, Dr. Sudesh K. Kashyap, and N. Shanthakumar.

The sponsored research from the Integrated Test Range (ITR) and DRDL (both DRDO) and the Aeronautical Development Agency (ADA, DRDO) has been very instrumental in shaping and strengthening our ideas and notions in these areas of practical aerospace problems.

We are grateful to Kyra Lindholm and Sonia Tam of CRC Press for their interest in our work and periodic logistics guidance during writing of this book. The first author (JRR) is very grateful to his wife Virmati, daughter Harshakumari Gohil, son Mayur, and daughter-in-law Amina for their constant moral support and punctual and clinical care for several years.

About the Authors

Jitendra R. Raol received B.E. and M.E. degrees in electrical engineering from M.S. University (MSU) of Baroda, Vadodara, in 1971 and 1973, respectively, and Ph.D. (electrical and computer engineering) from McMaster University, Hamilton, Canada, in 1986, and at both the places, he was also a postgraduate research and teaching assistant. He joined the National Aeronautical Laboratory (NAL, Bangalore) in 1975. At Council of Scientific and Industrial Research (CSIR)-NAL, he was involved in the activities on human pilot modelling in fix- and motion-based research flight simulators. He rejoined NAL in 1986 and retired in July 2007 as Scientist G (and Head, Flight Mechanics and Control Division at CSIR-NAL).

He has visited Syria, Germany, the United Kingdom, Canada, China, the United States of America, and South Africa on deputation/fellowships to (i) work on research problems on system identification, neural networks, parameter estimation, multi-sensor data fusion, and robotics; (ii) present technical papers at international conferences; and/or (iii) deliver guest lectures at some of these places. He has given several guest lectures at many Indian colleges and universities and Honeywell (HTSL, Bangalore).

He was a fellow of the IEE/IET (UK) and a senior member of the IEEE (USA). He is a life fellow of the Aeronautical Society of India and a life member of the Systems Society of India. During his studies at the MSU, he received Dean and Professor Subba Rao memorial prize and M.C. Ghia charitable fellowship. In 1976, he won K. F. Antia Memorial Prize of the Institution of Engineers (India) for his research paper on nonlinear filtering with time delay. He was awarded a certificate of merit by the Institution of Engineers (India) for his paper on parameter estimation of unstable systems. He has received the best poster paper award from the National Conference on Sensor Technology (New Delhi) for a paper on sensor data fusion. He has also received a gold medal and a certificate for a paper related to target tracking (from the Institute of Electronics and Telecommunication Engineers, India). He is also one of the five recipients of the CSIR (India) prestigious technology shield for the year 2003 for the leadership and contributions to the development of integrated flight mechanics and control technology for aerospace vehicles in the country; the shield was associated with a plaque, a certificate, and a project-grant-prize of INR 3,000,000 for the project work. He was one of the five recipients of the Chellaram Foundation Diabetes Research Award 2018 for the best paper (presented at the 2nd International Diabetes Summit, March 2018, Pune, India, which carried a prize of 100,000 INR). He received the Sir Thomas Ward memorial prize of the Institution of Engineers (India) in 2019 (jointly) for the paper on Image Centroid Tracking with Fuzzy Logic…, and it carried a gold medal and a certificate. He is featured in the list of the Stanford University (USA) as one of the top 2% scientists/researchers of the world for the years 2019 and 2021.

He has published nearly 160 research papers and numerous technical reports. He has guest-edited two special issues of *Sadhana* (an engineering journal published by the Indian Academy of Sciences, Bangalore) on (i) advances in modelling, system identification, and parameter estimation (jointly with Late Prof. Dr. Naresh Kumar Sinha) and (ii) multi-source, multi-sensor information fusion. He has also guest-edited two special issues of the *Defense Science Journal* (New Delhi, India) on (i) mobile intelligent autonomous systems (jointly with Dr. Ajith K. Gopal, CSIR-SA), and (ii) aerospace avionics and allied technologies (jointly with Prof. A. Ramachandran, MSRIT).

He has coauthored (i) an IEE/IET (London, UK) Control Series book *Modelling and Parameter Estimation of Dynamic Systems* (2004); and CRC Press (Taylor & Francis, Florida, USA) books, (ii) *Flight Mechanics Modeling and Analysis* (2009, and the revised edition in March 2023), (iii) *Nonlinear Filtering: Concepts and Engineering Applications* (2017), (iv) *Control Systems: Classical, Modern, and AI based Approaches* (2019), and (v) *Situation Assessment in Aviation: Bayesian Network and Fuzzy Logic Approach* (2024). He has also authored CRC Press books (i) *Multi-sensor Data Fusion with MATLAB* (2010) and (ii) *Data Fusion Mathematics–Theory and Practice* (2015). He has edited (with Ajith K. Gopal, CSIR-SA) a CRC Press book titled *Mobile Intelligent Autonomous Systems* (2012).

He has served as a member/chairman of numerous advisory, technical project review, and doctoral examination committees. He has also conducted sponsored research and worked on several projects from industry and other R&D organizations to CSIR-NAL with a substantial budget. Under his formal technical guidance, 11 doctoral and 8 master research scholars have received their degrees successfully; he has also guided jointly/informally one dozen research students for their doctoral studies. He is a reviewer of a dozen national/international journals and has evaluated several M.Tech./doctoral theses (from India and overseas). He had been with (i) MSRIT (M.S. Ramaiah Institute of Technology, Bengaluru) as Emeritus Professor for five years; (ii) the Govt. College of Engineering, Kannur (Kerala) as a senior research advisor; and (iii) the Department of Aerospace Engineering (IISc, Bangalore) as a consultant on modelling and parameter estimation for the Type I diabetes patients' data for a period of three months. He was also a consultant to a private company in Bangalore that provides software and algorithmic solutions to target tracking, navigation, and data fusion problems. He is a member of the governing body of the Centurion University of Technology and Management (CUTM), Andhra Pradesh, India.

His main research interests have been and are data fusion, system identification, state/parameter estimation, flight mechanics-flight data analysis, H-infinity filtering, nonlinear filtering, artificial neural networks, fuzzy logic systems, genetic algorithms, and soft technologies for robotics.

On the literary front, he has also authored a few books as the collection of his 320 (free) verses on various facets closely related to science, philosophy, evolution, and life itself. He has also contributed 80 articles/stories and 1200 'bites' (long quotes) on matrubharti.com (#1 Indian Content Community) and 7 ebooks on Amazon Kindle Direct Publishing (KDP) covering social-, philosophical-, science-, and human life-related aspects and issues. He was featured as one of the top 100 downloaded English authors of the matrubharti.com for the years 2021–2024, nearly four times.

His new area of study and research is data systems analytics (DaSyA).

Dr. V. P. S. Naidu obtained M. Eng. in medical electronics from Anna University, Chennai, and Ph.D. from the University of Mysore, Mysore. He has been working in CSIR-NAL since 2001 as a scientist on multisensory data fusion and target tracking. Before joining CSIR-NAL, he was Professor in the biomedical department of M.S. Ramaiah Institute of Technology, Bangalore. He is a fellow of the Institution of Engineers (India), Chartered Engineer (India), fellow of the Institution of Electronics and Telecommunication Engineering (IETE, India), CEng (IETE), fellow of the Aeronautical Society of India (AeSI), and Faculty of Continuing Education and Training (FCET). He has published several papers on multisensory data fusion and worked on several industry-sponsored projects on data fusion and mathematical modelling. He has earned a gold medal from IETE (India) for his paper on out-of-focus data fusion. He has reviewed several M.Tech. and Ph.D. theses.

He is presently Scientist G in CSIR-NAL and Professor in Academy of Scientific and Innovative Research (AcSIR, India). He is the nodal officer of Jigyasa. His areas of research are data fusion, mathematical modelling, state estimation, image fusion, and enhanced vision systems for civilian aircraft.

1 Introduction

In this chapter, we discuss the preliminary aspects of mathematical model building, and the use of these models for analysis, prediction, and control of a dynamic system is highlighted.

1.1 MATHEMATICAL MODELLING CONCEPTS

Our present day dynamic systems are relatively more complex and sophisticated, and we need to have mathematical models (MM) to understand their behavior, use these models for prediction of the future states and outputs of the underlying systems, and perform analysis (simulation) and design control laws to achieve desired performance of the overall, controlled dynamic system.

The empirical process of model building requires conducting certain specific system identification experiments and collection of input/output (I/O) data from these experiments. Mainly a persistently exciting input signal that has a broader bandwidth (BW) than the system's BW is applied to the dynamic system (to excite its modes), and its I/O data are collected for some time, allowing the system to operate for a sufficient period (to allow the dynamics to settle) such that from the data collected we can perform parameter estimation processes. For this, we need to use a reliable parameter estimation SW (software/algorithm) that uses a postulated mathematical model of the dynamic system under investigation and the I/O data, may be in an iterative manner, after which we obtain the coefficients/parameters of the refined math model. This model needs to be further verified by using a segment of the data that was not used for estimation but belongs to the same set of the data gathered during one experiment; often, many such experiments might need to be conducted for obtaining most reliable math models. This process of verifying the estimated model is called cross-validation. There are several criteria available in the system identification literature for verifying and validating the math models determined from the real test experiments [1].

1.2 AIRCRAFT MATHEMATICAL MODELS

An aircraft is a complex dynamic system. It has basically 6DOF (degrees of freedom) motion, and understanding its behavior is very crucial for the design of new aircraft, control of existing aircraft, prediction of the aircraft maneuvers, and improvement of overall stability and error performance.

1.2.1 For Flight Simulation

Aircraft being a complex and very costly dynamic system, flight simulators are used for various studies related to aircraft performance, and this is more useful during the

DOI: 10.1201/9781003429869-1

design and development of a new aircraft. Also, various faults can be simulated here and pilot's responses can be studied; this would help design reconfigurable control strategies for the aircraft. These simulators are like virtual reality systems and need math model of the dynamics of the aircraft that is being simulated [2]. In fact, this process is that of the analysis of a dynamic system, in which chosen inputs are given to the simulator and its outputs are recorded and then analyzed. The flight simulators are also used for extensive research on pilot-aircraft interactions and pilot training and used for iterative processes for the analysis and design of control laws for artificially stabilized high-performance high-agility fighter aircraft.

1.2.2 For Flight Control

For a new aircraft and even for an existing aircraft, it might be necessary to use some control system (often called control laws or filters) either to stabilize the modes of the aircraft or to improve the performance of the aircraft, especially its mission performance. Many control system design methods need accurate math model of the aircraft. Control aspects for an aircraft are: (i) basic stability augmentation, (ii) autopilot design, and (iii) navigation autopilot [2].

1.3 MISSILE MATHEMATICAL MODELS

In most cases, the missiles are unstable, and they need basic stabilizing and guidance control systems; the former being an inner loop control and the latter being the outer control loop [2]. The main control system required for a missile is to control its lateral acceleration. Again here, the math model of a missile is required. Obtaining these models is a very difficult process; because performing a system identification procedure is a difficult proposition. Often, simple math models are enough because a missile has two axis-symmetries. In such cases, a non-model-based approach can be used.

1.4 ROTORCRAFT MATHEMATICAL MODELS

A rotorcraft (helicopter) can have up to 12 or 15 DOF: basic 6 DOF of an aircraft, tail rotor DOF, interaction modes between main rotor/body, and the tail rotor and aeroelastic modes [2]. Determination of these math models from flight-tests of a rotorcraft is a very difficult task, though not an impossible one. These models are required for flight dynamic simulation and control of a rotorcraft, which is inherently slightly unstable.

1.5 MICRO-MINI-AIR VEHICLES (MAVs)

These vehicles are used for surveillance and are often remotely controlled. The dynamics of such vehicles can be very simple and can be easily determined by conducting flight-test experiments.

1.6 LIGHTER THAN AIR VEHICLES

Lighter than air vehicles (LAVs) have several advantages and find increasing applications in surveillance and advertising. A two-way fiber-optic datalink is used for sensor data and control. The low-pressure differential allows them to sustain for a long time. As for the fixed-wing aircraft, the basic derivation of the equations of motion of an airship is based on rigid body dynamics [2]. One basic difference in developing the equations for an aircraft and an airship is the consideration of buoyancy in the latter case. To account for the apparent mass effect due to the large volume of air mass displaced by an airship, the equations of motion are usually written about the center of buoyancy.

1.7 MANEUVERING TARGET MODELS AND DATA FUSION

The target tracking of ground-based and moving vehicles is a very important task, for which mathematical models are needed. Often, well-known constant velocity and constant acceleration models are used. Such math models are used in Kalman filter, which is used to estimate the states of tracking or the vehicle being tracked. Here, multisensory data fusion methods are very useful [3].

1.8 UNMANNED AERIAL VEHICLES (UAVs)

The UAVs are bigger than MAVs and are used for surveillance; they have their own autopilots and/or are controlled from a ground-based controller. These vehicles need math models for flight dynamics simulation and control.

1.8.1 Vision-Assisted Target Tracking

Due to the advances in MEMs-based inertial sensors, it is possible to build small, inexpensive, and very accurate inertial measurement units (IMUs); these devises are suitable for pose estimation in small-scale systems such as mobile robots and unmanned aerial vehicles (UAVs). These systems are useful where GPS signals are unreliable or not available. Also, low cost, low weight, and low power consumption of cameras make them ideal alternatives for aiding inertial navigation; visual sensing give images that are high-dimensional measurements, with rich information content. An important aspect here is the derivation of a measurement model to express the geometric constraints that arise when a static feature is observed from multiple camera poses. This math model does not require including the 3D feature position in the state vector of the EKF (extended Kalman filter) and is optimal, up to the linearization errors.

1.9 AUTONOMOUS UNDERWATER VEHICLES (AUWVs)

Accurate underwater navigation is crucial in several marine applications. Most AUWV navigation systems are based on INS which drift off with time, even when velocity aiding is used. In order to allow long duration submerged operations, additional position fixes are needed, and since GPS signal is not available underwater, the user is often dependent on acoustic aiding of the vehicle, either from a mother/own ship or by using underwater acoustic transponders. This application requires EKF or related filters which need mathematical models of the system: like dynamic state-model and the measurement model.

1.10 LEADER-FOLLOWER COORDINATION FOR TWO VEHICLES

Such applications are used for training of crew of one vehicle by asking them to follow another vehicle that is called a leader or a lead-vehicle. This strategy is also applicable in case of robots following a lead robot for a given joint mission of tracking in a terrain, for service mission, or mining. This application requires math model of the vehicle/s in order to design suitable control laws and tracking strategies.

1.11 SATELLITE ORBIT DETERMINATION

Satellite orbit determination is very important, especially for search and rescue mission that is aided by a polar orbit man-made satellite. It is also necessary to know where exactly the satellite is at each point in time. This is a very involved process, and mainly, the orbital states of the satellite are determined by using postulated math model/s of the dynamics of the satellite.

1.12 CHAPTER HIGHLIGHTS

In Chapter 2, we discuss mathematical models based on the equations of motion for aircraft, helicopter, missile, and other atmospheric vehicles. Also, aerodynamic models are briefly presented. Simplification aspects of equations of motion are also presented. Illustrative examples' results are presented.

Chapter 3 discusses target motion models used in filtering algorithms that are utilized for tracking the trajectories of the vehicles. Target tracking using imaging sensors, radars, and acoustic array sensors is considered from a mathematical modelling point of view. Illustrative examples' results are presented.

In Chapter 4, we present theory and models of two-vehicle coordination for UAVs and wheeled robots. MATLAB-based implementation results are presented.

Chapter 5 discusses vision-aided inertial navigation (VAIN) and multi-state Kalman filter models.

In Chapter 6, the concepts of terrain-assisted under water navigation (TAUWN) are presented. The point mass and particle filtering approaches are presented with appropriate mathematical models being used in the navigation process.

Chapter 7 discusses object detection and tracking from UAV using thermal camera.

Chapter 8 discusses the aspects of object tracking with UAV using INS, GPS, and vision camera.

Chapter 9 presents the problem of orbit determination of a satellite and discusses various mathematical models used for describing orbital states of a satellite.

In Appendix A, a short description of modelling based on soft computing is given.

In Appendix B, vehicle guidance problem is briefly discussed and the obstacle avoidance aspects are presented with a MATLAB-based illustrative example.

In Appendix C, some aspects of enhanced occlusion detection algorithm for surveillance cameras are presented.

REFERENCES

1. Jitendra, R. R., Girija, G., and Singh, J. *Modelling and Parameter Estimation of Dynamic Systems*, Volume 65 of IEE/IET Control Series book, London, UK, August 2004.
2. Jitendra, R. R., and Singh, J. *Flight Mechanics Modelling and Analysis*, Revised edition. CRC Press, Taylor & Francis, Boca Raton, FL, March 2023.
3. Jitendra, R. R. *Multisensor Data Fusion with MATLAB*. CRC Press, Taylor & Francis, Boca Raton, FL, 2009.

2 Flight Mechanics Models

2.1 INTRODUCTION

To study a system, for example, a flight vehicle, the first thing required is a mathematical model (MM) that is a reality with reduced complexity, and this model closely represents the system under study. This MM helps to understand, simulate, predict, and control the behavior of the system.

A flight vehicle is a complex dynamic system that has several models for the various (dynamic) components that make up the system. Here, we focus on describing the model form that is a reasonable representation of the aircraft dynamics. When an aircraft flies, it experiences the following forces: (i) force arising from the flow of the air over the aircraft frame (and other appendages), (ii) gravity force due to its weight, (iii) engine-propulsive/thrust forces, and (iv) inertial forces, due to momentum (because of the speed of the aircraft). To achieve a steady unaccelerated flight, these forces must balance out: (i) the upward force due to lift (because of the impinging air pockets on the aircraft and velocity of the aircraft) should be in equilibrium with the downward force due to the weight of the aircraft (due to gravity) so it does not experience upward or downward motion (unintended one), and (ii) the forward thrust force should be in equilibrium with the opposing drag force (due to air) so that the aircraft does not accelerate and is in a steady motion. Then, the aircraft is in a state of equilibrium or is flying at trim condition, and the translational and angular accelerations are zero.

The aircraft motion in flight is specified by the (i) equations of motion (EOMs) and (ii) models of the aerodynamic force/moment coefficients. The aircraft EOMs, formulated as differential equations, comprise the aircraft motion variables (aircraft states): (i) airspeed, (ii) flow angles (angle of attack [AOA], and angle of sideslip [AOSS]), (iii) angular velocities/rates (in pitch, roll, and yaw), and (iv) attitude angles (bank, pitch, and heading). The collection of these motion variables is a state vector that helps describe the motion of the flight vehicle at any given point of time. The aircraft EOMs are described in various documents [1–3].

The motion of an aircraft is governed by Newton's laws, provided the aircraft is considered as a rigid body with its mass concentrated at its center of gravity, C.G. There are ways of adding on the aero-elastic effects to the rigid formulation [4]. With this important assumption of the aircraft as a rigid body, one can elaborate on the forces and moments experienced by the aircraft in flight.

2.2 EQUATIONS OF MOTION (EOMs) FOR A RIGID BODY

It is necessary to specify a reference system of the aircraft and its relation to the earth: (i) inertial, (ii) earth, (iii) body, (iv) wind, and (v) stability axes. These are described as follows [4]:

DOI: 10.1201/9781003429869-2

i. **Inertial Axes**: The earth's center is assumed to be the origin of this axis system and is assumed to be stationary with the *Z*-axis pointing to the North pole and the *X*-axis pointing to the zero-degree longitude; the *Y*-axis, perpendicular to the *XZ* plane, completes the right-handed orthogonal axis system.

ii. **Earth Axes**: The origin of the earth axis system lies on the surface of the earth, with the *X*- and *Y*-axes pointing to the North (N) and East (E) direction, respectively; the *Z*-axis points down (D) toward the center of the earth, and the reference frame is fixed to the earth and, at times, called as the NED frame.

iii. **Body Axes**: This reference frame is fixed to the aircraft's body with its origin at the aircraft C.G., and rotates with the aircraft and moves in space. The *X*-axis points forward through the aircraft nose and is called the longitudinal axis. The positive direction of the *Y*-axis is along the right wing and is the lateral axis. Thus, *X*- and *Y*-axes are aligned with some geometric features of the aircraft. The *Z*-axis is perpendicular to the *XY* plane and positive downward. It will point toward the earth when the aircraft is in straight and level flight. In alternative principal body axes, the *X*-axis is aligned with the principal inertial axis. This makes the cross-product of inertia I_{zx} zero. This would simplify the lateral EOMs. Interestingly, the aerodynamic body axes (sometimes known as stability axes in the USA and wind axes in the UK) have the *X*-axis aligned with the projection of the velocity vector, in a datum flight condition, onto the plane of symmetry. The aerodynamic body axes are at an angle α, AOA, relative to the geometric body axis. Since these axes are 'natural' axis system, they are used for defining the basic aerodynamic forces: lift and drag. This system is consistent in the use of the force coefficients in the body axis; however, it has singularity at 90° of pitch angle. Interestingly, the pilot is much aware of the rotations of the aircraft about its C.G., rather than any other axis fixed in space. The pilot 'feels' the accelerating movement of the aircraft because her/his alignment is in the body axis system.

iv. **Wind Axes**: Here, the origin is at the aircraft C.G. with the *X*-axis pointing forward and aligned to the velocity vector; the positive *Y*-axis passes through the right wing, and the *Z*-axis points downward. The aircraft flow angles (AOA, AOSS) define the relation between the wind and the body axis system. Since the direction of the aircraft velocity vector keeps changing during the flight, the wind axis system, unlike the body axis system, is not fixed.

v. **Stability Axes**: This axis system is like that of body axes. The origin is at the aircraft C.G., and the *Y* stability axis is aligned with the *Y* body axis that passes out through the right wing. The *X* stability axis is inclined to the *X* body axis at an angle AOA. The aerodynamic data are generated in a wind tunnel (WT) in body axes or in stability axes.

The axis systems are aligned together by the direction cosine matrix that is a function of the Euler angles ϕ, θ, ψ.

To evaluate aircraft performance (by using the quantities such as range, rate of climb (ROC), take off, and landing distance), the aircraft is generally treated as a point mass model and the lift, drag, and thrust forces acting on it are identified. Although this model is useful for quick check to describe the aircraft dynamics properly, the complete six-degree-of-freedom (6DOF) nonlinear equations would be required. In between these two extremes lie the linearized models that are useful for mathematical modelling and control design applications.

Considering the aircraft to behave like a dynamic system in flight, an input given by the pilot (through control surfaces from actuators, stick, pedals, etc.) will cause the forces and moments acting on the aircraft to interact with its inherent natural characteristics thereby generating responses, which contain the natural dynamic behavior of the aircraft that can be described by a set of EOMs with certain important assumptions:

i. Aircraft is a rigid body; though strictly not valid, it is sufficiently accurate for the purpose of describing aircraft motion in flight. It allows the aircraft motion to be described by the translation of its C.G. and the rotation about it. Various components of aircraft might move in relation to each other: engine, rotorcraft disk, control devices, bending of wings (this would bring in the aspects of aero-elasticity [4]) due to air loads.
ii. The earth axis system can be used as the inertial frame of reference in which Newton's laws are valid, and the earth is assumed to be fixed in the space. The inertial reference frame is fixed or moves at a constant velocity relative to Earth. This description is reasonably accurate for short-term guidance and control.
iii. Aircraft mass is constant, though not always true.
iv. The earth is flat for the period of dynamic analysis. The curvature effects would be required to be considered for long-range flights (international flights, missiles, spacecrafts).
v. Aircraft *XZ* plane is a plane of symmetry.
vi. The steady flight is assumed to be undisturbed; this will simplify several equations. If flow angles are assumed to be small, then sine (~ zero) and cosine (~ one) terms would be simplified. This would render the applicability of the EOM to the conditions where small perturbations could occur.

These basic assumptions are appropriately used in formulating the EOM and their simplification as the case might be. The EOM of the aircraft can be written by considering its linear and angular momentum. Let p be the linear momentum vector (mV, mass times velocity) and H the angular momentum vector (inertia times angular velocity/rate), each measured in the inertial reference frame. Newton's second law states that the summation of all the external forces acting on a body is equal to the time rate of change of momentum of the body:

$$\frac{dp}{dt} = \frac{d(\mathrm{mV})}{dt} = \Sigma F \tag{2.1}$$

The summation of all external moments acting on the body is equal to the time rate of change of the angular momentum (moment of momentum):

$$\frac{dH}{dt} = \Sigma M \tag{2.2}$$

Rewriting the vector form of eqns. (2.1) and (2.2) into the scalar form by resolving the force and angular momentum into components along the *x*-, *y*-, and *z*-axes (of the inertial reference frame that is a right-handed system of the Cartesian axes), the EOM can be written as

$$F_x = \frac{d}{dt}(mu);\ F_y = \frac{d}{dt}(mv);\ F_z = \frac{d}{dt}(mw) \tag{2.3}$$

These equations describe the motion of the airframe C.G. as seen by an observer in the *x*, *y*, *z* frame. The moment equations can be expressed in the scalar form as

$$L = \frac{d}{dt}H_x; M = \frac{d}{dt}H_y; N = \frac{d}{dt}H_z \tag{2.4}$$

The angular momentum equation is given by

$$H = I\omega;\quad I_{x,y,z} = \begin{bmatrix} I_x & -I_{xy} & -I_{xz} \\ -I_{xy} & I_y & -I_{yz} \\ -I_{zx} & -I_{zy} & I_z \end{bmatrix} \tag{2.5}$$

Here, H is the angular momentum, I is the inertia tensor (the rotary analog of the mass), and ω is the angular velocity of the aircraft. The derivative of eqn. (2.5) gives

$$\frac{d}{dt}H = I.\frac{d\omega}{dt} + \frac{dI}{dt}\omega \tag{2.6}$$

Referring to the inertia and angular velocities in the space-fixed inertial frame, the inertia tensor will change as the aircraft rotates about the axes. This would contribute to dH/dt through dI/dt, and the resulting equations would have time-varying parameters. To overcome this, we consider the body axis system that is fixed to the aircraft so that the rotary inertial properties are constant (dI/dts are zero). This implies that the vector quantities V and H in eqns. (2.1) and (2.2) need to be determined in the body axis. A vector in a body axis frame, rotating at an angular speed of ω, can be represented as

$$\left.\frac{d(.)}{dt}\right|_I = \left.\frac{d(.)}{dt}\right|_B + \omega\, x(.) \tag{2.7}$$

Here, subscripts I and B refer to the inertial and body axis frame of references. Using this expression, the forces and moments in the inertial axis system can now be represented in the body axis system as

$$F = m\left.\frac{dV}{dt}\right|_I = m\left.\frac{dV}{dt}\right|_B + m(\omega \times V) \tag{2.8}$$

$$M = \left.\frac{dH}{dt}\right|_I = \left.\frac{dH}{dt}\right|_B + \omega \times H \tag{2.9}$$

The mass is assumed to be constant; however, this would not be true for long-range aircraft flights, missiles, and spacecrafts because of expending of fuels/shedding away certain extended components/stages, etc. The inertial forces and moments can now be resolved in the body axis system.

2.2.1 Resolution of Inertial Forces and Moments

The right-hand side of eqn. (2.8) contains the terms $\dot{V}$ (dV/dt), which represents the time derivative of velocity wrt the body axis system. The cross-product terms $(\omega \times V)$ arise from the centripetal acceleration along any given axis due to angular velocities about the remaining two axes. In the component form, we have

$$\begin{aligned} V &= u\hat{i} + v\hat{j} + w\hat{k} \\ \omega &= p\hat{i} + q\hat{j} + r\hat{k} \\ H &= H_x\hat{i} + H_y\hat{j} + H_z\hat{k} \end{aligned} \tag{2.10}$$

Using eqn. (2.10) in eqn. (2.8), we have

$$F = m\{[\dot{u}\hat{i} + \dot{v}\hat{j} + \dot{w}\hat{k}] + [(p\hat{i} + q\hat{j} + r\hat{k}) \times (u\hat{i} + v\hat{j} + w\hat{k})]\}; \times \text{ is cross product}$$

$$F = m[\dot{u} + qw - rv]\hat{i} + m[\dot{v} + ru - pw]\hat{j} + m[\dot{w} + pv - qu]\hat{k}$$

Thus, the force components acting along the X, Y, and Z body axis are

$$\begin{aligned} F_x &= m(\dot{u} + qw - rv) \\ F_y &= m(\dot{v} + ru - pw) \\ F_z &= m(\dot{w} + pv - qu) \end{aligned} \tag{2.11}$$

Example 2.1

Assume two-dimensional accelerated flight of an aircraft. In this situation, the $U + \Delta U$ and $V + \Delta V$ would have rotated by an incremental angle $R\Delta t$. Of course, the aircraft would have also rotated through the same angle. Here, U and V are the total velocities of the aircraft in forward and lateral directions [4]. Derive the equation for the rectilinear longitudinal acceleration.

Solution 2.1

$$a_X = \lim(\Delta t -> 0)\ \frac{(U+\Delta U)\cos(R\Delta t) - U - (V+\Delta V)\sin(R\Delta t)}{\Delta t}$$

Due to the limit $\frac{\sin(R\Delta t)}{\Delta t} = R$, and for small changes/perturbations, we get $a_X = \dot{U} - VR$.

Example 2.2

Assume two-dimensional accelerated flight of an aircraft. Derive the equation for the corresponding rectilinear lateral acceleration [4].

Solution 2.2

We have for this case the following formula:

$$a_X = \lim(\Delta t -> 0)\ \frac{(U+\Delta U)\sin(R\Delta t) + (V+\Delta V)\cos(R\Delta t) - V}{\Delta t}.$$

With the same argument as the previous one, we get $a_Y = \dot{V} + UR$.

Such small perturbation terms appear in eqn. (2.11) as can be clearly seen. From eqn. (2.9), we have $M = \dot{H} + (\omega \times H)$. The total moment about a given axis is due to both direct angular acceleration about that axis and those arising from linear acceleration gradients due to combined rotations about all axes. The components of angular momentum about the three axes are given by

$$\begin{aligned} H_x &= pI_x - qI_{xy} - rI_{xz} \\ H_y &= -pI_{xy} + qI_y - rI_{yz} \\ H_z &= -pI_{zx} - qI_{zy} + rI_z \end{aligned} \tag{2.12}$$

Even though there could be variation in the mass and its distribution during a flight due to the consumption of the fuel, use of external stores, etc., the assumption that the mass and mass distribution of the air vehicle are constant is reasonable and simplifies the expression for the rate of change of angular momentum. The time derivative of the angular momentum components expressed in the body axis is given by

$$\begin{aligned} \dot{H}_x &= \dot{p}I_x - \dot{q}I_{xy} - \dot{r}I_{xz} \\ \dot{H}_y &= -\dot{p}I_{xy} + \dot{q}I_y - \dot{r}I_{yz} \\ \dot{H}_z &= -\dot{p}I_{zx} - \dot{q}I_{zy} + \dot{r}I_z \end{aligned} \tag{2.13}$$

To get the total time rate of change of angular momentum, and thus the scalar components of the moments, we must add to these the contributions arising due to steady rotations $(\omega \times H)$.

$$\begin{aligned} L &= \dot{H}_x - rH_y + qH_z \\ M &= rH_x + \dot{H}_y - pH_z \\ N &= -qH_x + pH_y + \dot{H}_z \end{aligned} \tag{2.14}$$

Substituting for angular momentum components and regrouping results, the moment equations can now be written (wrt the body axis system) as

$$
\begin{aligned}
L &= \dot{p}I_x - \dot{q}I_{xy} - \dot{r}I_{xz} + qr(I_z - I_y) + (r^2 - q^2)I_{yz} - pqI_{xz} + rpI_{xy} \\
M &= -\dot{p}I_{xy} + \dot{q}I_y - \dot{r}I_{yz} + rp(I_x - I_z) + (p^2 - r^2)I_{xz} - qrI_{xy} + pqI_{yz} \\
N &= -\dot{p}I_{xz} - \dot{q}I_{yz} + \dot{r}I_z + pq(I_y - I_x) + (q^2 - p^2)I_{xy} - rpI_{yz} + qrI_{xz}
\end{aligned}
\quad (2.15)
$$

The total forces and moments acting on the aircraft are made up of contributions from the aerodynamic, gravitational, and aircraft thrust.

2.2.2 Resolution of Aerodynamic, Gravity, and Thrust Forces

The three aerodynamic forces and moments in terms of non-dimensional coefficients are

$$
\begin{aligned}
&F_{xA} = \bar{q}SC_x; F_{yA} = \bar{q}SC_y; F_{zA} = \bar{q}SC_z \\
&L = C_l\,\bar{q}S\,b; M = C_m\,\bar{q}S\bar{c}; N = C_n\,\bar{q}Sb.
\end{aligned}
\quad (2.16)
$$

The coefficients C_X, C_Y, C_Z, C_l, C_m, and C_n are the non-dimensional body axis force and moment coefficients, S is the reference wing area, $\bar{c}$ is the mean aerodynamic chord, b is the wing span, and $\bar{q}$ is the dynamic pressure. The coefficients would be dependent on the Mach number, flow angles, altitude, control surface deflections, and propulsion system. For an example, a deflection of a control surface would mean an effective change of the camber of a wing. This will affect the lift, drag, and moments and hence the aerodynamic coefficients. In fact, these are 'static' coefficients and are generally determined experimentally from various WT tests. The aerodynamic effects when the aircraft does maneuvering flights are captured in the so-called aerodynamic derivatives. Here, a brief on these aerodynamic coefficients [4] is given to connect the EOMs with the aerodynamics.

Lift coefficient: This coefficient, C_L, is related to C_z. Its contributions come from wings, fuselage, and horizontal tail, and vary with the alpha and Mach number. With an increase in lift, an increase in drag is associated. In an aircraft, there would be some mechanisms to increase an effective camber of the wings; these are the additional small surfaces like flaps, the deployment of which would provide required lift at low dynamic pressures. The dependence of the lift coefficient on the AOSS and altitude is usually small and hence neglected. However, the effect of the thrust and the Mach number would be significant on the lift coefficient. With higher thrust, the peak of the lift curve (the lift coefficient plotted v/s alpha) could increase and the peak would also shift to a higher AOA. The effect of the Mach number is such that the slope of the curve would increase up to a certain Mach number and then start decreasing. For supersonic speeds, the peak would decrease.

Drag coefficient: This coefficient, C_D, is related to C_y, C_N (the normal force coefficient), and C_x in general; the latter is called the axial force coefficient. This is an important design parameter for the aircraft. One would always like to have minimal drag, no resistance to the aircraft, and the maximum possible lift (so that the aircraft can carry very large loads, so the lift/drag (L/D) ratio should be large) for an ideal situation; this intuitive requirement defines the L/D ratio, and it is a performance parameter. The drag force is not necessarily a linear and is combination of several components: friction, form, induced, and wave drag. The individual contributions of these effects on the total drag would change with the flight conditions.

Side force coefficient: This force is the result of the aircraft having AOSS different from zero. In general, the sideslip force depends on the alpha, beta, Mach number, rudder control surface deflection, roll rate, and yaw rate.

Pitching moment coefficient: This is very important parameter for the stability of the aircraft, especially in its derivative form, $C_{m\alpha}$, which is supposed to be negative in general. However, when it is positive, the aircraft does not have the required static stability. If the slope of the moment curve is very high, then the aircraft is said to have the pitch 'stiffness'. Thus, a tradeoff is required between the static stability and the maneuverability. The reduced/relaxed static stability airframes are specially designed to obtain high L/D ratio and good maneuverability. The statically unstable configurations for obtaining such benefits, however, need to be artificially stabilized by feedback control filters/transfer functions (TFs)/algorithms. The moment coefficient primarily depends on the alpha, Mach number, altitude, pitch rate, elevator (or elevon for delta-wing aircraft) control surface deflection, inertial damping, aerodynamic damping term $C_{m\dot{\alpha}}$, and flap displacement.

Rolling moment and yawing moment coefficients: These coefficients depend primarily on the alpha, beta, Mach number, roll rate, yaw rate, aileron control surface displacement, and rudder control surface deflection. These coefficients are better understood in the form of their components.

Next, to resolve the gravity force acting on the aircraft into components along the body axes, the relative orientation of the gravity vector to the body axes system is required. If ϕ, θ, and ψ are the angular orientations of the gravity vector wrt body x-, y-, and z-axes, respectively, the components can be written as

$$
\begin{aligned}
F_{xG} &= -\mathrm{mg}\sin\theta \\
F_{yG} &= \mathrm{mg}\sin\phi\cos\theta \\
F_{zG} &= \mathrm{mg}\cos\phi\cos\theta
\end{aligned}
\tag{2.17}
$$

The third force comes from the engine thrust. Assuming the engine tilt angle to be σ_T wrt the body X-axis, the total applied forces, accounting for the contribution from aerodynamic gravity and thrust, can be expressed as

$$
\begin{aligned}
F_x &= \bar{q}SC_x - \mathrm{mg}\sin\theta + T\cos\sigma_T \\
F_y &= \bar{q}SC_y + \mathrm{mg}\sin\phi\cos\theta \\
F_z &= \bar{q}SC_z + \mathrm{mg}\cos\phi\cos\theta - T\sin\sigma_T
\end{aligned}
\tag{2.18}
$$

At times, contribution from the gyroscopic moment from the propulsive system is added to the moment equations. This contribution is, however, neglected in the present model postulates. Although the forces and moments expressed here have been derived for the body axis frame of reference, the orientation of the aircraft must be described wrt the fixed reference frame. Assuming the body axis and the fixed frame to be parallel, three angular rotations in the following specific order are applied to determine the Euler angles:

i. Rotate the fixed frame through yaw angle ψ to attain the frame position x_1, y_1, z_1.
ii. Rotate the x_1, y_1, z_1 frame through pitch angle θ to attain the position x_2, y_2, z_2.
iii. Rotate the x_2, y_2, z_2 frame through roll angle ϕ to attain the position x_3, y_3, z_3.

The frame x_3, y_3, z_3 is the final orientation of the body axis to the fixed axis. The relationship between the body axis angular rates p, q, r and the Euler rates $\dot{\psi}, \dot{\theta}$, and $\dot{\varphi}$ can be expressed as

$$\begin{aligned} \dot{\varphi} &= p + q\tan\theta\sin\phi + r\tan\theta\cos\phi \\ \dot{\theta} &= q\cos\varphi - r\sin\phi \\ \dot{\psi} &= r\cos\varphi\sec\theta + q\sin\phi\sec\theta \end{aligned} \; ; \begin{bmatrix} p \\ q \\ r \end{bmatrix} = \begin{bmatrix} 1 & 0 & -\sin\theta \\ 0 & \cos\phi & \sin\phi\cos\theta \\ 0 & \sin\phi & \cos\phi\cos\theta \end{bmatrix} \begin{bmatrix} \dot{\phi} \\ \dot{\theta} \\ \dot{\psi} \end{bmatrix} \tag{2.19}$$

Example 2.3

Express the inertial velocity components $\dot{x}, \dot{y}, \dot{z}$ in terms of the velocity components u, v, w in the body axis.

Solution 2.3

The rotation of frame through ψ to position x_1, y_1, z_1 gives

$$\dot{x} = u_1\cos\psi - v_1\sin\psi$$

$$\dot{y} = u_1\sin\psi + v_1\cos\psi$$

$\dot{z} = w_1$ (the rotation is about the z-axis)

Next, rotation through θ to the position x_2, y_2, z_2 gives

$$u_1 = u_2\cos\theta + w_2\sin\theta$$

$v_1 = v_2$ (the rotation is about the y_1-axis)

$$w_1 = w_2\cos\theta - u_2\sin\theta$$

Final rotation through ϕ to the position x_3, y_3, z_3 gives

$u_2 = u$ (the rotation is about the y_1-axis)

$$v_2 = v\cos\phi - w\sin\phi$$

$$w_2 = v\sin\phi + w\cos\phi$$

Substituting u_2, v_2, w_2 into equations for u_1, v_1, w_1, we obtain

$$u_1 = u\cos\theta + [v\sin\varphi + w\cos\varphi]\sin\theta$$

$$v_1 = v\cos\varphi - w\sin\varphi$$

$$w_1 = [v\sin\varphi + w\cos\varphi]\cos\theta - u\sin\theta$$

Substituting u_1, v_1, w_1 into equation for $\dot{x}, \dot{y}, \dot{z}$, we obtain

$$\dot{x} = [u\cos\theta + (v\sin\varphi + w\cos\varphi)\sin\theta]\cos\psi - [v\cos\varphi - w\sin\varphi]\sin\psi$$

$$\dot{y} = [u\cos\theta + (v\sin\varphi + w\cos\varphi)\sin\theta]\sin\psi + [v\cos\varphi - w\sin\varphi]\cos\psi$$

$$\dot{z} = [v\sin\varphi + w\cos\varphi]\cos\theta - u\sin\theta$$

In the matrix form

$$\begin{bmatrix} \dot{x} \\ \dot{y} \\ \dot{z}, \dot{H} \end{bmatrix} = \begin{bmatrix} \cos\theta\cos\psi & \sin\theta\sin\phi\cos\psi - \cos\phi\sin\psi & \sin\theta\cos\phi\cos\psi + \sin\phi\sin\psi \\ \cos\theta\sin\psi & \sin\theta\sin\phi\sin\psi + \cos\phi\cos\psi & \sin\theta\cos\phi\sin\psi - \sin\phi\cos\psi \\ -\sin\theta & \cos\theta\sin\phi & \cos\theta\cos\phi \end{bmatrix}$$

$$\times \begin{bmatrix} u \\ v \\ w \end{bmatrix} + \begin{bmatrix} W_x \\ W_y \\ W_{z,v} \end{bmatrix}$$

The last vector contains wind component toward the North, the East, and Vertical, respectively.

A common quantity related to the spatial position is the flight path angle given as

$$\sin\gamma = \frac{\dot{H}}{\sqrt{\dot{x}^2 + \dot{y}^2 + \dot{z}^2}}.$$

2.3 COMPLETE SET OF EOM

Using eqn. (2.11) and eqns. (2.15)–(2.19), the following nine coupled differential equations describe the aircraft motion in flight.

Force equations:

$$\begin{aligned} m(\dot{u} + qw - rv) &= \bar{q}SC_x - \text{mg}\sin\theta + T\cos\sigma_T \\ m(\dot{v} + ru - pw) &= \bar{q}SC_y + \text{mg}\sin\phi\cos\theta \\ m(\dot{w} + pv - qu) &= \bar{q}SC_z + \text{mg}\cos\phi\cos\theta - T\sin\sigma_T \end{aligned} \tag{2.20}$$

Moment equations:

$$
\begin{aligned}
\bar{q}SbC_l &= \dot{p}I_x - \dot{q}I_{xy} - \dot{r}I_{xz} + qr(I_z - I_y) + (r^2 - q^2)I_{yz} - pqI_{xz} + rpI_{xy} \\
\bar{q}ScC_m &= -\dot{p}I_{xy} + \dot{q}I_y - \dot{r}I_{yz} + rp(I_x - I_z) + (p^2 - r^2)I_{xz} - qrI_{xy} + pqI_{yz} \\
\bar{q}SbC_n &= -\dot{p}I_{xz} - \dot{q}I_{yz} + \dot{r}I_z + pq(I_y - I_x) + (q^2 - p^2)I_{xy} - rpI_{yz} + qrI_{xz}
\end{aligned}
\tag{2.21}
$$

Euler angles and body angular velocities:

$$
\begin{aligned}
\dot{\phi} &= p + q\tan\theta\sin\phi + r\tan\theta\cos\phi \\
\dot{\theta} &= q\cos\phi - r\sin\phi \\
\dot{\psi} &= r\cos\phi\sec\theta + q\sin\phi\sec\theta
\end{aligned}
\tag{2.22}
$$

The set of eqns. (2.20)–(2.22) represents the required 6DOF EOMs and can be rearranged and expressed in two equivalent forms.

2.3.1 Rectangular Form

When the velocity is represented in a rectangular coordinate system, we have

$$
\begin{aligned}
\dot{u} &= -qw + rv - g\sin\theta \boxed{+\frac{\bar{q}S}{m}C_x + \frac{T}{m}\cos\sigma_T} \\
\dot{v} &= -ru + pw + g\cos\theta\sin\phi \boxed{+\frac{\bar{q}S}{m}C_y} \\
\dot{w} &= -pv + qu + g\cos\theta\cos\phi \boxed{+\frac{\bar{q}S}{m}C_z - \frac{T}{m}\sin\sigma_T}
\end{aligned}
$$

$$
\begin{aligned}
\dot{p} &= \frac{1}{I_xI_z - I_{xz}^2}\left\{\bar{q}Sb\left(I_zC_l + I_{xz}C_n\right) - qr\left(I_{xz}^2 + I_z^2 - I_yI_z\right) + pqI_{xz}\left(I_x - I_y + I_z\right)\right\} \\
\dot{q} &= \frac{1}{I_y}\left\{\bar{q}ScC_m - \left(p^2 - r^2\right)I_{xz} + pr\left(I_z - I_x\right) + T(l_{tx}\sin\sigma_T + l_{tz}\cos\sigma_T)\right\} \\
\dot{r} &= \frac{1}{I_xI_z - I_{xz}^2}\left\{\bar{q}Sb\left(I_xC_n + I_{xz}C_l\right) - qrI_{xz}\left(I_x - I_y + I_z\right) + pq\left(I_{xz}^2 + I_x^2 - I_xI_y\right)\right\}
\end{aligned}
\tag{2.23}
$$

$$
\begin{aligned}
\dot{\phi} &= p + q\tan\theta\sin\phi + r\tan\theta\cos\phi \\
\dot{\theta} &= q\cos\phi - r\sin\phi \\
\dot{\psi} &= r\cos\phi\sec\theta + q\sin\phi\sec\theta
\end{aligned}
$$

Note that the terms in the rectangular boxes in the expressions for $\dot{u}$, $\dot{v}$, and $\dot{w}$ are the linear accelerations. This gives a set of kinematic equations that can be used to check data consistency and estimate calibration (biases and scale factors) errors in the flight data. The model equations normally used for kinematic consistency check are as follows:

$$
\begin{aligned}
\dot{u} &= -qw + rv - g\sin\theta + a_x \\
\dot{v} &= -ru + pw + g\cos\theta\sin\phi + a_y \\
\dot{w} &= -pv + qu + g\cos\theta\cos\varphi + a_z \\
\dot{\varphi} &= p + q\tan\theta\sin\phi + r\tan\theta\cos\phi \\
\dot{\theta} &= q\cos\phi - r\sin\phi \\
\dot{\psi} &= r\cos\phi\sec\theta + q\sin\phi\sec\theta \\
\dot{h} &= u\sin\theta - v\cos\theta\sin\phi - w\cos\theta\cos\phi
\end{aligned}
\tag{2.24}
$$

2.3.2 Polar Form

At times, it is useful to work with equations expressed in polar coordinates:

a. It is more convenient to use equations in polar coordinates, because the aerodynamic forces and moments are easier to visualize and express in terms of AOA α, AOSS β, and V.
b. The flow angles α and β can be directly measured and hence would yield simple measurements' equations.
c. The primary demerit is that the system becomes singular at zero velocity, since for the hover condition of a rotorcraft, the angles are not defined. Also, singularity occurs for beta = 90°.

V, α, and β in terms of u, v, and w are given by expressions:

$$V = \sqrt{u^2 + v^2 + w^2}; \alpha = \tan^{-1}\frac{w}{u}; \beta = \sin^{-1}\frac{v}{V} \tag{2.24}$$

Differentiating (2.24), we get

$$\dot{V} = \frac{1}{V}(u\dot{u} + v\dot{v} + w\dot{w}); \dot{\alpha} = \frac{u\dot{w} - w\dot{u}}{u^2 + w^2}; \dot{\beta} = \frac{V\dot{v} - v\dot{V}}{V^2\sqrt{(1 - v^2/V^2)}} \tag{2.25}$$

The velocity components u, v, and w can be expressed in terms of V, α, and β as

$$u = V\cos\alpha\cos\beta; v = V\sin\beta; w = V\sin\alpha\cos\beta \tag{2.26}$$

Eqn. (2.26) is obtained from the following transformation:

$$\begin{bmatrix} u \\ v \\ w \end{bmatrix} = \begin{bmatrix} \cos\alpha & 0 & -\sin\alpha \\ 0 & 1 & 0 \\ \sin\alpha & 0 & \cos\alpha \end{bmatrix} \begin{bmatrix} \cos\beta & -\sin\beta & 0 \\ \sin\beta & \cos\beta & 0 \\ 0 & 0 & 1 \end{bmatrix} \begin{bmatrix} V \\ 0 \\ 0 \end{bmatrix}$$

; this is the product of two rotation matrices and the velocity vector in a relative-wind–oriented coordinate system (the wind axes). From eqns. (2.24)–(2.26), the V, α, and β form can be formulated:

$$\dot{V} = g(\cos\phi\cos\theta\sin\alpha\cos\beta + \sin\phi\cos\theta\sin\beta - \sin\theta\cos\alpha\cos\beta) - \frac{\bar{q}S}{m}C_{D\text{wind}} + \frac{T}{m}\cos(\alpha+\sigma_T)\cos\beta$$

$$\dot{\alpha} = \frac{g}{V\cos\beta}(\cos\phi\cos\theta\cos\alpha + \sin\theta\sin\alpha) + q - \tan\beta(p\cos\alpha + r\sin\alpha) - \frac{\bar{q}S}{mV\cos\beta}C_L - \frac{T}{mV\cos\beta}\sin(\alpha+\sigma_T) \quad (2.27)$$

$$\dot{\beta} = \frac{g}{V}(\cos\beta\sin\phi\cos\theta + \sin\beta\cos\alpha\sin\theta - \sin\alpha\cos\phi\cos\theta\sin\beta) + p\sin\alpha - r\cos\alpha + \frac{\bar{q}S}{mV}C_{Y\text{wind}} + \frac{T}{mV}\cos(\alpha+\sigma_T)\sin\beta$$

In (2.27), C_L and C_D are the lift force and drag coefficients in the stability axis. In terms of body axis coefficients, these are given by

$$C_L = -C_Z\cos\alpha + C_X\sin\alpha$$
$$C_D = -C_X\cos\alpha - C_Z\sin\alpha \quad (2.28)$$

Eqn. (2.27) can further be expressed in terms of C_D and C_Y using the following expressions:

$$C_{D\text{wind}} = C_D\cos\beta - C_Y\sin\beta$$
$$C_{Y\text{wind}} = C_Y\cos\beta + C_D\sin\beta \quad (2.29)$$

Note that there are no small angle approximations used, but equations have singularity at $\theta = 90°$. Also, there are several trigonometric and kinematic nonlinearities in the 6DOF EOMs.

Analysis/flight simulation of the full 6DOF equations would require a nonlinear program (algorithm/software). The 6DOF model is useful where nonlinear effects are important and coupling between the longitudinal and lateral modes is strong.

Although it is possible to analyze the 6DOF models, it is always a good idea to work with simplified models.

2.4 MISSILE DYNAMIC MODELS

Like an aircraft, the axis system for missiles for the development of the EOM is assumed to be centered at the C.G. and fixed to the body; the *x*-, *y*-, and *z*-axes are the roll, pitch, and yaw axes, respectively. The *XY* plane represents the yaw plane, while the *XZ* plane is the pitch plane; the angle of incidence in the *XZ* plane is α (like AOA), the AOSS in the *XY* plane is β, and the α_f is defined as the flank angle; with a few simple algebraic calculations, we get

$$\tan\beta = \tan\alpha_f\cos\alpha \quad (2.30)$$

Compared to v and w, the component u is much larger and varies little during the flight. Like the aircraft, the missile too has 6DOF motion which is represented by three force and three moment equations [5]. The standard set of equations normally used to compute missile dynamics are like those in eqns. (2.11) and (2.15):

$$F_x = m(\dot{u} + qw - rv)$$
$$F_y = m(\dot{v} + ru) \tag{2.31}$$
$$F_z = m(\dot{w} - qu)$$

$$L = \dot{p}I_x - \dot{q}I_{xy} - \dot{r}I_{xz} + qr(I_z - I_y) + (r^2 - q^2)I_{yz} - pqI_{xz} + rpI_{xy}$$
$$M = -\dot{p}I_{xy} + \dot{q}I_y - \dot{r}I_{yz} + rp(I_x - I_z) + (p^2 - r^2)I_{xz} - qrI_{xy} + pqI_{yz} \tag{2.32}$$
$$N = -\dot{p}I_{xz} - \dot{q}I_{yz} + \dot{r}I_z + pq(I_y - I_x) + (q^2 - p^2)I_{xy} - rpI_{yz} + qrI_{xz}$$

The F_X equation includes the $\dot{u}$ that represents the change in the forward speed. To compute this, we need to know the drag force and the thrust force acting on the missile. Since w and v are generally small, the contribution of the terms pw and pv in equations for F_Y and F_Z will be small if and only if the roll rate is small. From another perspective, the rolling motion will cause forces in the pitch and yaw plane, a feature that is highly undesirable and should be dealt with at the design stage. The undesirable effects of high roll rate may also be reduced using acceleration feedback from autopilot during the flight (latex). If the roll rate p is assumed to be small and the flight variables q, r, v, and w anyway are not being large, the EOM for a missile can be simplified by neglecting the terms involving the products of these variables, e.g. terms like pv, pw, pq, and pr. Also, assuming the missile to be reasonably symmetrical about the XZ and XY planes, the simplified set of equations can be written as

$$F_x = m(\dot{u} + qw - rv); F_y = m(\dot{v} + ru); F_z = m(\dot{w} - qu) \tag{2.33}$$

$$L = \dot{p}I_x; M = \dot{q}I_y; N = \dot{r}I_z \tag{2.34}$$

The rolling moment, eqn. (2.34), shows no coupling between the roll and pitch motions, and the roll and yaw motions.

2.5 ROTORCRAFT DYNAMIC MODELS

Unlike the fixed-wing aircraft, the lift, propulsion, and control in helicopters are provided by the main rotor. The helicopters can take off vertically, hover, and land even on unprepared terrain. In a helicopter, the upward lift is obtained by keeping the rotor shaft vertical. The main rotor system of the helicopter should be regarded as lift-generating mechanism like the wings of an aircraft; the rotors' airfoils produce the lift. The rotor system could be semi-rigid, rigid, or fully articulated. The engine provides the power to the main rotor and the tail rotor. In a fully articulated rotor, each blade can rotate about the pitch axis that is feathering. This rotation changes the AOA and hence the lift. The blade can move back and forth that is lead/lag movement. The blade can also flap up and down. Generally, such system is used for helicopters with more than two blades. A semi-rigid rotor has two blades, and when one blade flaps up, the other blade will flap down; there is no lead-lag in this system. The rigid rotor does not provide lead-lag or flapping; some of this aspect is achieved by the blending of the blades. Aerodynamics of a rotorcraft can be explained by the momentum theory or the blade element theory; the former provides a basic understanding of the flow through the main rotor and the power required to produce thrust to lift the helicopter – it alone cannot be used to design the rotor blades; and the blade element theory investigates the details of how the thrust is produced by the rotating blades and relates the rotor performance to design parameters.

2.5.1 Momentum Theory

It follows the basic laws of conservation of mass, momentum, and energy [6,7]. The action of the air on the rotor blades produces a reaction of the rotor on the air that manifests in the form of thrust at the rotor disk. It is assumed that the rotor disk is composed of an infinite number of blades; i.e., the rotor is an 'actuator' disk. It is also assumed that the air is incompressible and frictionless and no rotational energy is imparted to the flow outside the slip stream. Velocity in the slip stream increases from zero at upstream infinity to v at the rotor disk, to w at downstream infinity. For the air with density ρ and A being the disk area, the law of conservation of mass gives, from the definition of density, the following:

$$\dot{m} = \rho A v \tag{2.35}$$

If T is the thrust at the rotor disk, it will be equal to the rate of change of momentum:

$$T = \dot{m}(w - 0) = \rho A v w \tag{2.36}$$

From the law of conservation of energy, we have

$$Tv = \frac{1}{2}\dot{m}w^2 \tag{2.37}$$

Substituting for T from the momentum conservation equation:

$$\rho A v w v = \frac{1}{2}\rho A v w^2; \text{or } v = \frac{1}{2}w \tag{2.38}$$

Thus, the velocity at downstream infinity is twice the velocity at the rotor disk. From (2.36) and (2.38), the relation between v and thrust T becomes

$$T = \rho A v 2v; \text{ or } v = \sqrt{\frac{T}{2\rho A}} \tag{2.39}$$

The term '*T/A*' is known as disk loading. The induced power (thrust force times distance/time) for the rotor can be written as

$$P = Tv = T\sqrt{\frac{T}{2\rho A}} = \frac{T^{3/2}}{\sqrt{2\rho A}} \tag{2.40}$$

The non-dimensional form of thrust and power coefficients is given by

$$C_T = \frac{T}{\rho A(\Omega R)^2} \text{ and } C_P = \frac{P}{\rho A(\Omega R)^3} \tag{2.41}$$

Here, Ω is the angular velocity, and R is the radius of the rotor. The inflow ratio λ is obtained by using eqns. (2.39) and (2.41):

$$\lambda = \frac{v}{\Omega R} = \sqrt{\frac{C_T}{2}} \tag{2.42}$$

The hovering efficiency of the rotor is defined by figure of merit (FOM) defined as $\text{FOM} = \frac{\text{minumum power required to hover}}{\text{actual power required to hover}}$, the ideal value of which is equal to 1. However, for most rotors, the value lies between 0.75 and 0.8.

2.5.2 Blade Element Theory

The momentum theory provides estimate of induced power requirements for the rotor, but is not sufficient for designing the rotor blades. The origin of the blade element theory can be traced back to the earlier work on marine propellers by Froude. Subsequently, Drzewiecki took up this study assuming the blade sections to act independently and considering two velocity components ΩR due to rotation and V due to the axial velocity of the rotor. His results indicated correct behavior but were quantitatively in error which was primarily due to the neglecting of the induced velocity at the rotor disk [7]. Several attempts were made later on to account for the induced velocity from the momentum theory into the blade element theory. It was only after Prandtl developed the lifting line theory that the influence of the wake velocity at the rotor disk was incorporated. Thus, it was through the wake vortex theory rather than the momentum theory that the induced velocity at the blade section was finally incorporated into the blade element theory. The blade sees the air coming to it due to rotor rotation, as well as due to downward induced velocity. Today, the blade element theory is the foundation for all analysis of helicopter dynamics and aerodynamics [8].

2.5.3 Rotorcraft Modelling Formulations

Models of different orders may be used depending on what is known about the measurements available, the frequency range of interest, and the degree of coupling between the longitudinal and lateral-directional (LD) motions. Generally, a

6DOF model is the minimum required for a highly coupled rotorcraft. The 6DOF rigid body model for the rotorcraft consists of the usual EOMs derived from Newton's law.

2.5.3.1 Equations for Linear Accelerations

$$\begin{aligned}\dot{u} &= \tilde{X} - g\sin\theta - qw + rv \\ \dot{v} &= \tilde{Y} + g\cos\theta\sin\varphi - ru + pw \\ \dot{w} &= \tilde{Z} + g\cos\theta\cos\varphi - pv + qu\end{aligned} \tag{2.43}$$

2.5.3.2 Equations for Angular Accelerations

$$\begin{aligned}\dot{p} &= \tilde{L} - \frac{I_{zx}}{I_x I_z - I_{zx}^2}\left(I_y - I_x - I_z\right)pq - \frac{1}{I_x I_z - I_{zx}^2}\left(I_{zx}^2 + I_z^2 - I_y I_z\right)qr \\ \dot{q} &= \tilde{M} - \frac{\left(I_x - I_z\right)}{I_y} rp - \frac{I_{zx}}{I_y}\left(p^2 - r^2\right) \\ \dot{r} &= \tilde{N} - \frac{I_{zx}}{I_x I_z - I_{zx}^2}\left(I_z - I_y + I_x\right)qr - \frac{1}{I_x I_z - I_{zx}^2}\left(I_x I_y - I_{zx}^2 - I_x^2\right)pq\end{aligned} \tag{2.44}$$

2.5.3.3 Kinematic Equations for Euler Rates

$$\dot{\varphi} = p + q\sin\varphi\tan\theta + r\cos\varphi\tan\theta;\; \dot{\theta} = q\cos\varphi - r\sin\varphi;\; \dot{\psi} = q\frac{\sin\varphi}{\cos\theta} + r\frac{\cos\varphi}{\cos\theta} \tag{2.45}$$

The equation relating altitude to linear velocities is given by

$$\dot{h} = u\sin\theta - v\cos\theta\sin\phi - w\cos\theta\cos\phi \tag{2.46}$$

In the above equations, the gyroscopic reactions due to rotating elements are neglected and the cross-products of inertia I_{xy} and I_{yz} are assumed zero. The presence of gravitational and rotational related terms in eqn. (2.43), and the product of angular rates in eqn. (2.44), gives these equations a nonlinear structure. The terms $\tilde{X}, \tilde{Y}$, and $\tilde{Z}$ in eqn. (2.43) denote the specific forces along the longitudinal, lateral, and normal body axis, respectively; the terms $\tilde{L}, \tilde{M}$, and $\tilde{N}$ in eqn. (2.44) represent the specific moments about the longitudinal, lateral, and vertical body axis, respectively; these terms basically represent linear and angular acceleration quantities and are related to resultant aerodynamic force and moment coefficients as follows.

2.5.3.4 Specific Forces

$$\tilde{X} = X/m;\; \tilde{Y} = Y/m;\; \tilde{Z} = Z/m \tag{2.47}$$

2.5.3.5 Specific Moments

$$\tilde{L} = \frac{I_z L + I_{zx} N}{I_x I_z - I_{zx}^2}; \tilde{M} = \frac{M}{I_y}; \tilde{N} = \frac{I_{zx} L + I_x N}{I_x I_z - I_{zx}^2} \tag{2.48}$$

Eqn. (2.44) can be further simplified by assuming the product of angular rates to be small and neglecting the corresponding terms in the moment equations; so, we get

$$\dot{p} = \tilde{L}; \quad \dot{q} = \tilde{M}; \quad \dot{r} = \tilde{N} \tag{2.49}$$

The rigid body states described in eqns. (2.43), (2.45), (2.46), and (2.49) are used to characterize the rotorcraft dynamics. Since the forces and moments are presented as specific quantities in the equations, mass and moment of inertia no longer explicitly appear in the MM.

2.5.4 Limitations of the Rigid Body Model

The 6DOF model involving the rigid body states is generally adequate to predict rotorcraft dynamics in the low- and mid-frequency range. In the conventional rigid body model, the main rotor dynamics are omitted and the rotor influence is absorbed by the rigid body derivatives. A better prediction at higher frequencies, however, necessitates the inclusion of rotor dynamics in an estimation model. One approach to include rotor effects in a 6DOF model is by introducing equivalent time delays in control inputs [9,10]. These delays can be determined by correlating the model response with flight data. This is, however, an inappropriate alternative to modelling the rotor dynamics which are highly complex in nature. Model inversion is required for the feedforward controller in the design of the model following control systems, and the time delays become time lead on inversion, which means that the future values of the state variables are needed in advance. This is unrealistic for an on-line real-time process like in-flight simulation. It clearly underlines the need to develop extended models with an explicit representation of rotor dynamic effects. Therefore, the 6DOF rigid body model is generally extended with additional DOFs representing the longitudinal and lateral rotor flapping.

The basic model structure can include two additional DOFs representing the longitudinal and lateral flapping. The state-space matrix defines the submatrices pertaining to the rigid body, rotor, and the cross-coupling matrices for body to rotor and rotor to body. In a simplified rotor model, the longitudinal and lateral flapping angles can be expressed primarily as a function of body angular rates and control inputs. In ref. [10], Kaletka and Grünhagen formulated an 8DOF extended model by redefining the roll and pitch accelerations for rotor/body motion and using them as state variables. Tischler provided a hybrid body/flapping model which coupled the simplified fuselage equations at low frequencies to the simplified rotor equations at high frequencies with equivalent spring terms [11]. The following first-order coupled differential equations for longitudinal flapping a_{1s} and lateral flapping b_{1s} can be appended to a 6DOF model:

$$\dot{a}_{1s} = \tilde{A} \; \dot{b}_{1s} = \tilde{B} \tag{2.50}$$

Here, $\tilde{A}$ and $\tilde{B}$ are the longitudinal and lateral flapping-specific moments.

2.6 AERODYNAMIC COEFFICIENTS AND DERIVATIVES

The forces and moments that act on an aircraft can be distributed and described in terms of the so-called aerodynamic derivatives. For the conventional stable aircraft configurations, these derivatives directly ensure the adequate levels of stability and controllability. For unstable aircraft configurations, these derivatives form the essential input parameters for the design of control laws that in turn ensure the desired stability and controllability characteristics of the aircraft.

The aerodynamic model is an integral part of the EOM that in totality represents the aerodynamic and dynamic behavior of the vehicle. In general, the aerodynamic derivatives are applicable to small perturbations of the motion about an equilibrium and this leads to quasi-static aerodynamic derivatives. Large amplitude motions/maneuvers require sophisticated and more complex aerodynamic modelling. We specifically deal with aerodynamic modelling that basically consists of the aerodynamic coefficients that in turn expand to (or encompass) the aerodynamic derivatives. Certain stability and control aspects (of aircraft) are captured in certain aerodynamic derivatives.

2.6.1 Basic Aerodynamic Forces and Moments

If we specify certain geometric parameters of the aircraft like reference surface area of the wings as S, the mean aerodynamic chord $\bar{c}$, the wing span b, and the so-called dynamic pressure $\bar{q}$, then we have the following straightforward aerodynamic force and moment equations in terms of the component forces in the x, y, and z direction:

Force = Dynamic Pressure × Area →

$$F_x = C_x \bar{q}S; F_y = C_y \bar{q}S; F_z = C_z \bar{q}S \tag{2.51}$$

Moment = Force × lever arms (respective ones) →

$$L = C_l \bar{q}S\, b; M = C_m \bar{q}S\bar{c}; N = C_n \bar{q}Sb \tag{2.52}$$

These expressions are in terms of the component moments: rolling moment L about the x-axis, pitching moments M about the y-axis, and the yawing moment N about the z-axis. Here, $C_x, C_y, C_z, C_l, C_m,$ and C_n are the non-dimensional body axis force and moment coefficients, called the aerodynamic coefficients; these are the proportionality constants (that do vary with certain variables). The total velocity V (in the formula for the dynamic pressure) and the total angle can be resolved in terms of component velocities: u, v, and w and AOA (in the vertical xz plane) and AOSS (in the horizontal xy plane). From the geometry of the flow velocity, the following expressions of the so-called flow angles and component velocities emerge very naturally:

$$\text{AOA} \Rightarrow \alpha = \tan^{-1}\frac{w}{u}; \text{ AOSS} \Rightarrow \beta = \sin^{-1}\frac{v}{V} \tag{2.53}$$

By trigonometric transformation, we get equivalent expressions as

$$u = V\cos\alpha\cos\beta;\ v = V\sin\beta;\ w = V\sin\alpha\cos\beta \tag{2.54}$$

The total velocity V of the airplane is expressed as

$$V = \sqrt{u^2 + v^2 + w^2} \tag{2.55}$$

If we know the values of the aerodynamic coefficients, then we can easily determine the forces and moments acting on the aircraft, since we know V and the geometric parameters like area and lever arms. Alternatively, if we know the forces (like accelerations) by some measurements on the aircraft, then we can determine the aerodynamic coefficients. The latter is true of the WT experiments conducted on a scaled model of the actual aircraft. In a WT, the model is mounted and compressed air is released thereby putting the model in the flow field. The experiments are conducted at various AOA and AOSS settings, and forces and moments are measured/calculated from these measurements. Since the flow velocity and the dynamic pressure are known, simple computations would lead to the determination of the aerodynamic coefficients. One can see that these coefficients can be obtained as a function of the flow angles, Mach number, etc. Similar analysis can be done using CFD and other analytical methods. To determine the aerodynamic coefficients from experimental/flight data, we need to know V and measure the force, i.e. force = mass x acceleration. Since mass, m, is known, we can get acceleration as $\frac{\bar{q}SC_x}{m}$. So, if we measure the acceleration of an airplane in motion, we can get the aerodynamic coefficient C_x. Since the coefficients depend on aerodynamic derivatives and the dynamic variables, and if we measure these responses, then we can also get these derivatives. This calls for establishing a formal relationship between these dynamic responses and the aerodynamic derivatives thus leading to aerodynamic modelling.

What we have not said so far is the fact that the pilot will maneuver an aircraft by using pitch-stick, roll stick, and rudder pedals besides using throttle lever arm. The movements of these devices forward or backward or sideways are transmitted via the (servo motors and) control actuators to the moving/hinged surfaces of the aircraft. These surfaces are extensions of the main surfaces, like wing and rudder fin, and are called control surfaces. The movements of the control surfaces interact with the flow field while aircraft is in motion and alter the force and moment balance (equations), thereby imparting the 'changed' motion to the aircraft until the new balance is achieved. During this changing motion (that can be called perturbation), the aircraft attains new altitude/orientation, etc., depending upon the energy exchanges and balance between kinetic, potential, and propulsive (thermal) energies. At the microscopic level (or scale), this behavior can be studied by exploring the effect of aerodynamic parameters. At the macroscopic level, the behavior can be studied using the simplified EOMs and resultant TF analysis.

2.6.2 Aerodynamic Parameters

The aerodynamic coefficients would, in general, be function of not only flow angles but also the Mach number, angular rates, control surface deflections, and changes in the thrust force due to the movement of the throttle lever arm. For example, C_m from eqn. (2.52) can be expressed in terms of these independent response variables in a straightforward manner as

$$C_m = \text{function of } (\alpha, q, \delta_e, M, F_T)$$

$$C_m = C_{m0} + C_{m\alpha}\alpha + C_{m_q}\frac{q\bar{c}}{2V} + C_{m\delta_e}\delta_e + C_{mM}M + C_{mF_T}F_T \tag{2.56}$$

Here, the pitching moment aerodynamic coefficient is considered as a function of several independent variables: AOA, pitch rate q, elevator control surface deflection, Mach number, and thrust. The effect on the pitching moment, due to small changes in these response variables, is captured in the terms that are called aerodynamic derivatives (e.g. $C_{m\alpha}$), which can be considered as proportionality constants. However, these constants vary with certain variables. The quantity $\frac{q\bar{c}}{2V}$ is the non-dimensional angular rate (velocity) as can be seen from (rad/s) (m/2) (s/m) => rad. The small perturbation theory when applied to the simplification of the EOM leads to the fact that the aerodynamic forces and moments depend on some constants that are called the stability derivatives. The parametric terms of eqn. (2.56) are called aerodynamic derivatives, and often stability and control derivatives, since many of these derivatives govern the stability and control of the aircraft (dynamics). For example, $C_{m\alpha}$ signifies ${}^{\Delta C_m}/_{\Delta\alpha}$ when other variables are set to zero.

2.6.3 Definition of Aerodynamic Derivatives

Essentially, these derivatives are the sensitivities to changes in flight variables. If there is a change in AOA, there is a change in the pitching moment coefficient, and appropriate change in the pitching moment M (see eqn. (2.52)). How this change in turn alters the aircraft responses is the subject of the EOMs. There could be cross-coupling terms, i.e. a change in an aerodynamic coefficient of one axis due to a change in the variable of the other axis. Several possibilities exist. The analysis is limited to straightforward effects. The force and moment coefficients of eqns. (2.51) and (2.52) are expressed in the term of the aerodynamic derivatives as seen in eqn. (2.56). These are often called stability and control derivatives. This is because they directly or implicitly govern the stability and control behavior of the vehicle. They are called derivatives, since they specify the variation of aerodynamic forces and moments wrt a small change in the perturbed variable. A particular derivative could vary with velocity or the Mach number, altitude, and AOA. And in turn the coefficients that depend on these derivatives also vary with these and several other variables. However, the variation of these coefficients soon gets translated to variations with time due to the composition of the coefficients in terms of motion variables that vary with time.

There is a matrix of relationships in terms of the sensitivities between the forces/moments and the response variables u, v, w, p, q, r, and control surface deflections in terms of the aerodynamic derivatives; also, there is a matrix of the major functional (effect) relationships between forces/moments and the aircraft response variables. In other words, we have the classification: (i) speed derivatives, (ii) static derivatives, (iii) dynamic derivatives, and (iv) control derivatives. The static derivatives are mainly with respect to AOA and AOSS. They govern the static stability of the vehicle. Dynamic derivatives are with respect to the rotational motion of the vehicle and mainly specify the damping (of certain modes of the vehicle) in the respective axis. The speed derivatives are with respect to linear velocities of the vehicle. The control derivatives specify the control effectiveness of the control surface movements in changing the forces and moments acting on the vehicle. The static and the dynamic derivatives as they are termed imply something more in terms of the behavior of the aircraft from the control theory point of view. What is important to know is that these names themselves relate to certain stability and damping properties of an aircraft which is considered as a dynamic system.

The derivative effects are considered in isolation assuming that the perturbations occur in isolation. The aerodynamic derivatives are evaluated at steady-state equilibrium condition/s with small perturbation about it. Hence, these derivatives are called quasi-static aerodynamic derivatives. These derivatives are used in dynamically varying conditions also, mainly for small amplitude maneuver analysis, thereby assuring the linear domain operations and analysis. For large amplitude maneuvers where AOAs are high, the unsteady and nonlinear effects need to be incorporated into the analysis for which additional aerodynamic derivatives must be included. For definitions of the dimensional aerodynamic derivatives (DADs) and non-dimensional aerodynamic derivatives (NDADs), we closely follow ref. [1] since the definitions and procedures therein are consistent and straightforward. The present treatment is also enhanced by ref. [3]. Although there might be some differences in the formulae of certain derivatives between various sources [1–3,12], we feel that the presentation given in this book is more uniform and highly standardized as in ref. [1]. The DADs are defined as

$$M_w = \left(1/I_y\right)\frac{\partial M}{\partial w}; I_y \text{ is moment of inertia } (\text{MOI}); Z_w = (1/m)\frac{\partial Z}{\partial w} \tag{2.57}$$

The derivatives specify the change in pitching moment and vertical force due to a small change in vertical speed (w) of the aircraft. Next, we have the NDAD defined as

$$C_{m\alpha} = \frac{\partial\, C_m}{\partial\, \alpha} \tag{2.58}$$

Using $C_m = M/(\bar{q}S\bar{c})$ and $\alpha = w/U$ (for small AOAs), we get

$$C_{m\alpha} = \frac{\partial M}{\partial(w/U)}\frac{1}{\bar{q}S\bar{c}} = \frac{U}{\bar{q}S\bar{c}}\frac{\partial M}{\partial w} = \frac{UI_y}{\bar{q}S\bar{c}}M_w;$$

$$M_w = \bar{q}S\bar{c}/(I_yU)C_{m\alpha}; M_w = \rho US\bar{c}/(2I_y)C_{m\alpha} \tag{2.59}$$

Similar procedure can be applied to all the DADs, and the equivalent NDADs can easily be obtained.

2.6.3.1 For Aircraft

The decoupling between longitudinal and LD derivatives is presumed. We also assume that during a small perturbation maneuver, the Mach number, Reynolds number, dynamic pressure, velocity, and engine parameters do not change much so that their effects can be neglected.

2.6.3.1.1 Longitudinal Derivatives

These derivatives are related to the sensitivities of the forces in axial and vertical (normal) directions and the pitching moment wrt changes in forward speed, normal speed, pitch angle rate, and elevator/elevon control surface deflections.

Effect of forward/axial speed u along the x-axis: (X_u, Z_u, M_u)

When there is an increase in the axial velocity, the drag (D) and lift (L) forces will generally increase. Also, the pitching moment (M) will increase.

X_u is the change in the forward force (often known as the axial force) due to a change in the forward (also known as the axial) speed. It is called the speed damping derivative. Following eqn. (2.57) (for force derivatives, mass m is used for normalization, and u, as a perturbation, and U are interchangeably used for convenience, and appropriately, U should be taken as V, the airspeed):

$X_u = \frac{1}{m}\frac{\partial X}{\partial u}$ or equivalently $X_u = -\frac{1}{m}\frac{\partial D}{\partial u}$, since the drag force acts in the negative direction, it contributes to a negative force X. Using the expression of the drag force D, we get

$$X_u = -\frac{1}{m}\frac{\partial(1/2\,\rho U^2 S C_D)}{\partial u};\; X_u = -\frac{\rho S}{2m}\frac{\partial(U^2 C_D)}{\partial u}$$

$$X_u = -\frac{\rho S}{2m}\left(U^2\frac{\partial C_D}{\partial\, u} + 2UC_D\right);\; X_u = -\frac{\rho SU}{m}\left(U/2\frac{\partial C_D}{\partial\, u} + C_D\right),$$

Also, by definition

$$C_{D_u} = U/2\frac{\partial C_D}{\partial\, u};\; X_u = -\frac{\rho SU}{m}(C_{D_u} + C_D) + \frac{\partial T_x}{m\,\partial U};$$
$$T_x = \text{Thrust provided by the engine} \tag{2.60}$$

An increase in the thrust T contributes to a positive X force. The drag is measured along the direction of the relative wind, and the drag coefficient is measured along the negative equilibrium x-axis (in the stability axis system). The drag coefficient is always positive in sign, and the derivative X_u is always negative in sign. The C_{D_u} derivative can arise from the Mach number (Mn) and aero-elastic effects (usually small and negligible). The Mach effect will be significant in the range 0.8 < Mach number < 1.2 (transonic region). The derivative C_{D_u} can be determined from the plot of the drag coefficient wrt the Mach number.

Z_u is the change in the normal force due to a change in the forward speed. Since the lift acts along the negative z-axis (the lift acts upward and that is the –ve z-axis;

Z_u is always measured along the +ve z-axis and is positive downward), one gets $Z_u = -\frac{1}{m}\ \frac{\partial L}{\partial \mathrm{u}}$; by following the previous procedure, one gets

$$Z_u = -\frac{\rho SU}{m}\ (C_L + C_{L_u}) \tag{2.61}$$

The low values of C_L are associated with low AOA and high speeds, and high values are associated with high AOA and low speeds. The C_{L_u} can arise from the Mach number (Mn) and aero-elastic effects. For the low Mach number, it is negligibly small and it could be high at the critical Mach number. The magnitude of C_{L_u} can vary considerably, and its sign can change, depending not only on airframe geometry and its elastic properties, but also on the Mach number and the dynamic pressure at which the aircraft is flying.

The derivative wrt the Mach number (here denoted as M) is derived next.

$$Z_{\mathrm{Mn}} = -\frac{1}{m}\frac{\partial L}{\partial \mathrm{Mn}}; Z_{\mathrm{Mn}} = -\frac{1}{m}\frac{\partial(1/2\ \rho U^2 S C_L)}{\partial \mathrm{Mn}}; Z_{\mathrm{Mn}} = -\frac{\rho S}{2m}\frac{\partial(U^2 C_L)}{\partial \mathrm{Mn}}$$

$$Z_{\mathrm{Mn}} = -\frac{\rho S}{2m}\left(U^2\frac{\partial C_L}{\partial \mathrm{Mn}} + 2a^2 \mathrm{Mn} C_L\right); \text{ since } \mathrm{Mn} = U/a = \text{airspeed/speed of sound;}$$

$$Z_{\mathrm{Mn}} = -\frac{\rho S}{2m}\ (U^2\frac{\partial C_L}{\partial \mathrm{Mn}} + 2U^2 C_L/\mathrm{Mn}); Z_{\mathrm{Mn}} = -\frac{\rho S}{m}\ (C_{L\mathrm{Mn}} + C_L / \mathrm{Mn})$$

Alternatively, it can be expressed as

$Z_{\mathrm{Mn}} = -\frac{\rho SUa}{m}\ (C_L + C_{L_u})$, since $Z_{\mathrm{Mn}} = Z_u a$, and $\frac{\partial Z}{\partial \mathrm{Mn}} = \frac{\partial Z}{\partial \mathrm{u}}\frac{u}{\mathrm{Mn}}$

M_u is the change in the pitching moment (M) due to a change in the forward speed. By definition, one gets from the basic equation of the pitching moment

$$M_u = 1/I_y\left[\frac{\rho SU^2\bar{c}}{2}\ \frac{\partial C_m}{\partial u} + \rho US\bar{c}C_m\right]$$

$$M_u = \frac{\rho SU\bar{c}}{I_y}\left(U/2\frac{\partial C_m}{\partial u} + C_m\right); C_{M_u} = U/2\left(\frac{\partial C_M}{\partial U}\right); C_M = \frac{M\,(\text{moment})}{qS\bar{c}}$$

Again by definition, one gets

$$M_u = \frac{\rho SU\bar{c}}{I_y}\ (C_{m_u} + C_m) \tag{2.62}$$

The C_m is the aerodynamic portion of the total trimmed pitching moment. In the presence of thrust asymmetry, C_m can change due to (change in) the airframes' geometry and the elastic properties. Also, the sign can change due to the Mach number and dynamic pressure variation. These derivatives originate from the power, Mach number, and aero-elastic effects, the latter two are more important. For subsonic flights, often M_u is considered negligible. However, at high speeds, bending due to aero-elastic effects can induce large changes in C_{m_u}.

Effect of change in vertical speed w (equivalently AOA) along the vertical z-axis: (X_w, Z_w, M_w)

According to the definition of stability axes, the relative wind during the steady flight condition is parallel to the x-axis. The only component of the linear velocity is U_0; when the aircraft is disturbed from the steady flight so that it has a component of velocity w along the z-axis, as well as a forward velocity u, the relative wind shifts to a new position resulting in an increase in AOA. X_w is the change in the forward (axial) force due to the change in the vertical (normal) speed. We have

$$X_w = \frac{1}{m}\frac{\partial X}{\partial w} \text{ or } X_w = \frac{1}{mU}\frac{\partial X}{\partial \alpha}; X_w = \frac{1}{mU}\frac{\partial(1/2\,\rho U^2 SC_x)}{\partial \alpha}$$

We also have

$$C_x = C_L \sin\alpha - C_D \cos\alpha; C_L = \frac{L}{\bar{q}S}; C_{D\alpha} \triangleq \frac{\partial C_D}{\partial \alpha}$$

Using the above expression, we obtain

$$X_w = \frac{\rho U^2 S}{2mU}\frac{\partial(C_L \sin\alpha - C_D \cos\alpha)}{\partial \alpha}$$

By evaluating the partials, we get

$$X_w = \frac{\rho S U^2}{2mU}(C_L \cos\alpha + C_{L\alpha} \sin\alpha + C_D \sin\alpha - \cos\alpha\, C_{D\alpha})$$

Assuming that the AOA is very small, after simplification we get

$$X_w = \frac{\rho U S}{2m}(C_L - C_{D\alpha}) \tag{2.63}$$

When the AOA (or vertical speed w) increases, the total drag increases; hence, $C_{D\alpha}$ is +ve in sign; the main contribution is from the wing, and other from tail and fuselage, and is a nonlinear function of the AOA.

Z_w is the change in the normal force due to the change in the vertical (normal) speed. We have

$$Z_w = \frac{1}{m}\frac{\partial Z}{\partial \mathrm{w}} \text{ or } Z_w = \frac{1}{mU}\frac{\partial Z}{\partial \alpha}; Z_w = \frac{1}{mU}\frac{\partial(1/2\,\rho U^2 SC_z)}{\partial \alpha}$$

Also, we have

$$C_z = -(C_L \cos\alpha + C_D \sin\alpha)$$

By using the above equation, we obtain

$$Z_w = \frac{1}{mU} \frac{\partial(-1/2\,\rho U^2 S(C_L \cos\alpha + C_D \sin\alpha))}{\partial\alpha}$$

By evaluating the derivatives and simplifying (for small AOA), we get

$$Z_w = -\frac{\rho SU}{2m}(-C_L \sin\alpha + C_{L\alpha}\cos\alpha + C_D \cos\alpha + \sin\alpha\; C_{D\alpha});$$
$$Z_w = -\frac{\rho SU}{2m}(C_{L\alpha} + C_D) \tag{2.64}$$

$C_{L\alpha}$ (known as lift curve slope) is +ve below 'stall-α', and 80%–90% contribution comes from wing, fuselage, and tail.

M_w is the change in the pitching moment due to a change in the vertical (normal) speed. A change in the vertical speed causes a change in AOA that in turn causes a change in lift and drag of the corresponding surfaces (like wing and horizontal tail). We have

$$M_w = \frac{1}{I_y}\frac{\partial M}{\partial w} \text{ also } M_w = \frac{1}{I_y U}\frac{\partial M}{\partial\alpha}; M_w = \frac{\rho SU^2\bar{c}}{2I_y U}\frac{\partial C_m}{\partial\alpha}; M_w = \frac{\rho SU\,\bar{c}}{2I_y} C_{m\alpha} \tag{2.65}$$

Effect of change in pitch rate q: (X_q, Z_q, M_q)

X_q is the change in axial force due to a change in pitch (angle) rate. This is mainly due to the change of AOA of the horizontal tail ('T') that contributes to the total drag. The contribution from the wing is usually negligible, and the resulting drag increase is neglected in the first approximation, and the derivative is also assumed to be zero.

$$X_q = 1/m\left[-\frac{\rho U^2 S_T}{2}\frac{\partial C_{D_T}}{\partial q}\right]; \text{ by definition } C_{D_{T_q}} = \frac{\partial C_{D_T}}{\partial(q\bar{c}/2U)},$$
$$\text{we obtain } X_q = -\frac{\rho S_T U\bar{c}}{4m} C_{D_{T_q}} \tag{2.66}$$

Z_q is the change in the normal force due to a change in pitch rate. The pitching causes a change in the AOA of the horizontal tail, thereby resulting in the change of the lift at the tail. Following the previous development for X_q ($X = -L$), we obtain

$$Z_q = -\frac{\rho SU\bar{c}}{4m} C_{L_q} \tag{2.67}$$

We can obtain C_{Z_q} (equivalently C_{L_q}) from the known variables and constants as follows realizing the change of tail-lift:

$$\Delta L_t = \frac{\rho U_t^2 S_t}{2} C_{L\alpha}\Delta\alpha;\; \Delta Z = -\Delta L_t = -\frac{\rho U_t^2 S_t}{2} C_{L\alpha}\left(\frac{q l_t}{U_0}\right)$$

The last term is the change of the AOA of the tail due to the pitch rate–induced vertical speed of the tail. Next, we have

$$C_Z = \frac{2Z}{\rho U^2 S}; \Delta C_Z = -\frac{2\rho S_t U_t^2}{2\rho U^2 S} C_{L\alpha}\left(\frac{ql_t}{U_0}\right)$$

Defining $C_{Z_q} = \dfrac{\partial C_Z}{\partial(q\bar{c}/2U_0)}$ and substituting from the above equation, we get

$$C_{Z_q} = -2\left(\frac{2\rho U_t^2}{2\rho U^2}\right) C_{L\alpha}\left(\frac{l_t S_t}{S\bar{c}}\right); C_{Z_q} = -2\left(\frac{U_t^2}{U^2}\right) C_{L\alpha}\left(\frac{l_t S_t}{S\bar{c}}\right) \Rightarrow C_{Z_q} = -2C_{L\alpha}\eta V_H \tag{2.68}$$

Here, η is the ratio of the dynamic pressure at the tail and the main dynamic pressure and is often called the tail efficiency. V_H is horizontal tail volume ratio, the term in the second parenthesis. As is with $C_{L\dot{\alpha}}$, the effect of C_{L_q} on longitudinal stability is usually very small and it is neglected in the dynamic analysis.

M_q is the change in the pitching moment due to a change in the pitch rate. We get

$$M_q = \frac{1}{I_y}\frac{\partial M}{\partial q}; M_q = \frac{\rho S U^2 \bar{c}}{2I_y}\frac{\partial C_m}{\partial q}; M_q = \frac{\rho S U \bar{c}^2}{4I_y}\frac{\partial C_m}{\partial(q\bar{c}/2U)}; M_q = \frac{\rho S U \bar{c}^2}{4I_y} C_{m_q} \tag{2.69}$$

Following the steps as for C_{Z_q}, we can obtain the formula for C_{m_q} realizing the change in the pitching moment at the C.G. due to a change in the lift force at the tail and the lever arm as follows:

$$\Delta M = -\Delta L_t l_t = -\frac{\rho S_t U_t^2}{2} C_{L\alpha}\Delta\alpha\, l_t; C_{m_q} = \frac{\partial C_m}{\partial(q\bar{c}/2U)} = \frac{2U}{\bar{c}}\frac{\partial C_m}{\partial q} = -2C_{L\alpha}\eta V_H \frac{q}{\Delta q}\frac{l_t}{\bar{c}}$$

Since we are dealing with only small perturbations, q and Δq are identical; hence, we cancel out them resulting in

$$C_{m_q} = -2C_{L\alpha}\eta V_H \frac{l_t}{\bar{c}} \tag{2.70}$$

The formula (2.70) gives the damping derivative in terms of the known or measurable constants. It is very important because it contributes a large portion of the damping of the short period (SP) mode for a conventional aircraft. This damping effect comes from the AOA changes at the horizontal tail, which are proportional to the tail length; that is also a lever arm converting the tail length into the moment.

Next, we consider the effects of aircraft control surfaces which include (i) elevators, (ii) stabilizers, (iii) flaps, (iv) slats, and (v) dive brakes which pertain to longitudinal control. The +ve direction is taken as that giving +ve lift for convenience.

Effect of change in elevator control surface deflection: $(X_{\delta_e}, Z_{\delta_e}, M_{\delta_e})$

These are control effectiveness derivatives. However, $C_{m\delta_e}$ derivative is called the primary control derivative or often control power. If the absolute value of the derivative is higher, then for a given deflection, more moment is generated. This can be regarded as higher control sensitivity for a given moment of inertia. The Z_{δ_e}, M_{δ_e} derivatives are closely related via the moment arm.

X_{δ_e} derivative is defined/given as (since the axial force X is in the negative direction of the drag force D):

$$X_{\delta_e} = \frac{1}{m}\frac{\partial X}{\partial \delta_e}; X_{\delta_e} = \frac{\rho S U^2}{2m}(-C_{D\delta_e}); C_{D\delta_e} = \frac{\partial C_D}{\partial \delta_e}$$

Similarly, since the lift force is in the opposite direction to the z-normal force, we get

$$Z_{\delta_e} = \frac{1}{m}\frac{\partial Z}{\partial \delta_e} = \frac{\rho S U^2}{2m}(-C_{L\delta_e}); C_{L\delta_e} = \frac{\partial C_L}{\partial \delta_e} \tag{2.71}$$

M_{δ_e} is the change in the pitching moment due to a small change in the elevator/elevon control surface deflection.

$$M_{\delta_e} = \frac{1}{I_y}\frac{\partial M}{\partial \delta_e} = \frac{\rho S U^2 \bar{c}}{2I_y}\frac{\partial C_m}{\partial \delta_e} = \frac{\rho S U^2 \bar{c}}{2I_y} C_{m\delta_e}; C_{m\delta_e} = \frac{\partial C_m}{\partial \delta_e} \tag{2.72}$$

The $C_{m\delta_e}$ derivative can be considered as a primary design parameter, and it specifies longitudinal axis control power.

2.6.3.1.2 *Most Important Longitudinal Derivatives Are Explained*

i. $C_{L\alpha}$ is defined as a change in the lift coefficient for a unit change in AOA. The lift force can be easily given by $L = C_L \bar{q} S$; $C_{L\alpha}$ represents the lift curve slope wrt to AOA. This is very important derivative because it almost directly determines the contribution to the lift force. It also signifies the fact that as AOA increases, the lift force increases proportionally, in the linear region up to a certain AOA. Beyond this AOA, the lift would remain somewhat constant and would even decrease further. This condition is called 'wing stall'; this means that the aircraft losses some lift after it stalls.
ii. $C_{m\alpha}$ is the basic (longitudinal) static stability derivative and is referred to as pitch stiffness parameter. A negative value of $C_{m\alpha}$ indicates that the aircraft is statically stable; i.e., if the AOA increases, then the pitching moment becomes more negative thereby decreasing the AOA and hence restoring the stability. The positive change in the AOA increases the lift on the tail inducing −ve pitching moment about the C.G.; and increases the lift of the wing inducing the +ve moment if the lift acts fore of the C.G., and −ve moment if the lift acts aft of the C.G. Thus, for a stable equilibrium, C_{m0} must be +ve, and $C_{m\alpha}$ −ve; and for an airfoil with +ve camber, C_{m0} is −ve; with zero camber, it is 0; and with –ve camber, it is +ve. The $C_{m\alpha}$ derivative is proportional to the distance between the aerodynamic center (better referred to and used as neutral point, NP) of the aircraft and the C.G. This distance is related

to the static margin as static margin= (NP distance, $h_{NP}-h_{CG}$)/MAC, i.e. $C_{m\alpha} = C_{L\alpha}$ ($h_{CG}-h_{NP}$). The distances (of NP and C.G.) are from some reference point in the front of the aircraft on the x-axis. If the static margin is positive, then the aircraft is said to have static stability; this means that the NP distance is larger than that of the C.G. distance from the reference point. If the C.G. is continuously moved (by some means) rearward, then at one point the aircraft will become neutrally stable (neutrally unstable! here, $C_{m_\alpha} = 0$). This point is called the neutral point of the aircraft, or this C.G. position or location is called the neutral point. The NP must be always behind the C.G. location for the guaranteed static stability (C_{m_α} is −ve). A slightly more rearward movement of the C.G. will render the aircraft statically unstable (the C.G. is aft and C_{m_α} is +ve). If the aircraft is dynamically stable, then it must have been statically stable. For dynamic stability, the static stability is a must, but the converse is not true, meaning that if the aircraft is statically stable, it could be dynamically unstable. This derivative is of primary importance for the longitudinal stability of the atmospheric vehicles. If the aircraft is inherently statically unstable (by design or so for gaining certain benefits in case of a relaxed static stability aircraft like FBW and many modern-day high-performance fighter aircraft), then artificial stabilization is required in order that the aircraft can fly. The level of instability is dictated by (i) available state-of-the-art technology (computers, actuators, sensors, control law design tools), (ii) possible reduction in the size of the aircraft and subsequent decrease in weight, (iii) increase in weight due to additional hardware (redundancy, computer, etc.), (iv) obtainable performance, (v) maneuverability, and (vi) agility. This is the subject of design, development, and analysis of flight control laws.

iii. C_{m_q} signifies a change in the pitching moment coefficient due to a small change in the pitch rate q. Being (angular) rate-related derivative, it implies, from the control theory point of view, that it must have something to do with damping; as such, it contributes to the damping in pitch. Usually, more negative values of C_{m_q} signify increased damping. The sign is generally −ve for both stable and unstable configurations. Since the statically unstable configurations will have stability augmentation control laws, the lower values of this derivative (less -ve values) are acceptable. The aircraft flight control system (AFCS) also provides some artificial damping. This derivative often has higher prediction uncertainty levels (scatter in estimates); however, this is not a problem since the AFCS generally tolerates these uncertainties.

iv. Longitudinal control effectiveness derivative, $C_{m\delta_e}$, is the elevator control effectiveness. Its function is through the applied pitching moment to control the AOA of the airframe in equilibrium and maneuvering flight, and this function is the most important. In conventional sense, more −ve value means more control effectiveness. It also helps determine the sizing/design of the control surface. More elevator (elevons for FBW delta-wing aircraft configuration) control power means more effective control in generating control moment for the aircraft. This also applies to horizontal tails, canard, or some combination of these surface movements. The knowledge of the available elevator/elevon power at all the flight conditions in the flight envelope is very important. The design value of this derivative is determined by the anticipated fore and aft travel of C.G. The larger C.G. travel and higher lift coefficient demand larger value of this derivative. Supersonic flight may impose additional requirements because of the attendant aft movement of the Aerodynamic center (a.c). The sign of this derivative depends on the location of the elevator, fore or aft of the C.G.; for the aft locations and the elevator sign convention used here, the derivative has a negative value. The unstable configurations demand higher control power compared to the stable ones.

v. The $C_{L\delta_e}$ is always positive in accordance with the convention used here. For the conventional aircraft with the horizontal tail mounted at an appreciable distance aft of the C.G., this derivative is usually small and its effect is relatively unimportant. The $C_{D\delta_e}$ is invariably smaller than $C_{L\delta_e}$ (because of the usual of drag with lift), and it is normally neglected. Its sign could be +ve or −ve, depending on the trim position of the elevator and the trim AOA.

2.6.3.1.3 *Lateral-Directional Derivatives*

These derivatives are defined as the sensitivities of the side force (Y), the rolling moment (L), and the yawing moment (N) wrt small changes in side speed, roll rate, yaw rate, and aileron and rudder control surface deflections. These derivatives are, in general, difficult to estimate with a good degree of confidence.

Effect of change in side speed v (equivalently the AOSS): (Y_v, L_β, N_β)

Y_v is the change in the side force due to a small change in the side velocity v:

$$Y = \frac{\rho U^2 S}{2} C_y; \partial Y/\partial v = \frac{\rho U^2 S}{2} \frac{\partial C_y}{\partial v};$$

Using $v = \beta U$, we obtain

$$Y_v = (1/m)\partial Y/\partial v = \frac{\rho US}{2m} C_{Y\beta}; C_{y\beta} = \frac{\partial C_y}{\partial \beta} \tag{2.73}$$

This derivative is difficult to estimate. It is usually negative, since the side slipping causes the side force that opposes the sideward motion. In general, several contributory forces are generated: (i) the one, say F_1, that arises from the change of the AOA of the vertical tail, (ii) side force acting on the fuselage, F_2, and (iii) the two forces (F_3, F_4) due to differential lift forces acting on each semi-span of the wing due to its 'effective dihedral'. Small or +ve values of the derivative are undesirable because the resulting small (or reversal) side forces make the detection of SS difficult; accordingly, coordination of banked turns becomes a piloting problem. Also, such values of the derivative contribute little to the damping of the Dutch roll (DR), whereas normal (−ve) values of $C_{y\beta}$ can contribute substantially to the total damping.

L_β is the change in the rolling moment due to a change in sideslip. The rolling moment L about the x-axis is caused mainly by the components (F_3, F_4) that act normal to the wing, and by F_1 at the fin center of pressure which can be either above or below the x-axis. We have

$$L_v = \frac{1}{I_x}\frac{\partial L}{\partial v} = \frac{\rho U^2 Sb}{2I_x}\frac{\partial C_l}{\partial v}; L_v = \frac{\rho USb}{2I_x}\left(\frac{\partial C_l}{\partial \beta}\right); L_\beta = \frac{\rho U^2 Sb}{2I_x} C_{l\beta} \quad (2.74)$$

$C_{l\beta}$ is the 'effective' dihedral, and it is negative for the positive dihedral. It is very important for lateral stability and control and hence considered as a preliminary design parameter.

N_β is the change in the yawing moment N due to a change in the side velocity, v. It is mainly caused by the side force F_1 on the vertical tail. Following the previous development for L_v, we obtain

$$N_v = \frac{\rho USb}{2I_z} C_{n\beta}; N_\beta = \frac{\rho U^2 Sb}{2I_z} C_{n\beta}; C_{n\beta} = \frac{\partial C_{N(or\,n)}}{\partial \beta} \quad (2.75)$$

The vertical tail and related lever arm contribute to $C_{n\beta}$; the contribution is positive and signifies the fact that due to the positive sideslip, this force creates a positive yawing moment that in turn tries to reduce the sideslip thereby maintaining the static stability.

Effect of change in the time rate change in the side velocity ($\dot{v}$): ($Y_{\dot{v}}$, $L_{\dot{v}}$, $N_{\dot{v}}$)

In fact, nothing much is known of these derivatives; hence generally, these are neglected, but their expressions are given here for the sake of completion.

$$Y_{\dot{v}} = \frac{\rho Sb}{4m} C_{Y_{\dot{\beta}}};\ C_{y_{\dot{\beta}}} = \frac{\partial C_y}{\partial(\dot{\beta}b/2U)}$$

$$L_{\dot{v}} = \frac{\rho Sb^2}{4I_x} C_{l_{\dot{\beta}}};\ C_{l_{\dot{\beta}}} = \frac{\partial C_l}{\partial(\dot{\beta}b/2U)}$$

$$N_{\dot{v}} = \frac{\rho Sb^2}{4I_z} C_{n_{\dot{\beta}}};\ C_{n_{\dot{\beta}}} = \frac{\partial C_n}{\partial(\dot{\beta}b/2U)}$$

Effect of change in roll rate p: (Y_p, L_p, N_p)

These derivatives arise due to the change in rolling angular velocity and roll rate p; the latter creates a linear velocity of the vertical, horizontal, and wing surfaces causing a local change in the AOA of these surfaces and the lift distribution which in turn causes the moment about the C.G.

Y_p is the change in the side force due to a change in the roll rate. We have

$$Y_p = (1/m)\partial Y/\partial p = \frac{\rho USb}{4m}\frac{\partial C_y}{\partial(pb/2U)} = \frac{\rho USb}{4m} C_{Yp} \quad (2.76)$$

This derivative is not significant and is often neglected.

L_p is the change in the rolling moment due to a change in the roll rate p, known as a roll damping derivative. The induced rolling moment is such that it opposes the rolling moment thereby providing the dynamic stability to the aircraft. We have

$$L_p = (1/I_x)\partial L/\partial p = \frac{\rho USb^2}{4I_x}\frac{\partial C_l}{\partial(pb/2U)}; L_p = \frac{\rho USb^2}{4I_x}C_{l_p} \tag{2.77}$$

The derivative C_{l_p} (−ve in sign) is composed of contributions from the wing and the horizontal/vertical tails, the major being from the wing, and the variations in the derivative with the Mach number and AOA closely follow the variations in wing lift curve slope $C_{L\alpha}$.

N_p is a change in the yawing moment due to a change in the roll rate. The lift forces acting on the down-going and up-going semi-spans are rotated forward and backward, respectively; the change in the direction of these forces results in a −ve yawing moment about the z-axis. For the flight near the stall, the drag forces may become important and would result in a yawing moment of the opposite sign:

$$N_p = \frac{\rho USb^2}{4I_z}C_{n_p} \tag{2.78}$$

The derivative C_{n_p} arises mainly from the wing and the vertical tail (+ve or −ve), depending on its geometry.

Effect of change in yaw rate r: (Y_r, L_r, N_r)

The yawing angular velocity, yaw rate, causes these derivatives. The yaw (angle) rate causes the change in the side force F_1 that acts on the vertical tail surface; and the fact that effective AOA of the vertical tail increases.

Y_r is the change in the side force due to a change in the yaw rate:

$$Y_r = (1/m)\partial Y/\partial r = \frac{\rho USb}{4m}.C_{y_r};\ C_{y_r} = \frac{\partial C_y}{\partial(rb/2U)} \tag{2.79}$$

This derivative is of little importance, and it is generally neglected.

L_r is the change in the rolling moment due to a change in the yaw rate r. The forward speed of a station/point distance d_1 toward the +ve y-axis, from the X, Z plane on the semi-span of the wing, is decreased by an amount d_1r, resulting in a decrease in lift at this point, and similarly, the forward speed of a point distance d_2 normal to the X, Z plane on the semi-span of the wing is increased (because of the yawing moment and the rotation of the airframe clockwise about the z-axis) by an amount d_2r, resulting in an increase in lift at this station; and this differential lift yields a rolling moment (usually +ve). The corresponding derivative is

$$L_r = (1/I_x)\partial L/\partial r = \frac{\rho USb^2}{4I_x}C_{l_r}; C_{l_r} = \frac{\partial C_l}{\partial(rb/2U)} \tag{2.80}$$

N_r is a change in the yawing moment due to a change in the yaw rate. The side force F_1 also causes a moment about the z-axis, since the vertical tail is some distance aft of the C.G.; this moment is usually –ve. The derivative is given as with the usual definition of the dimensional/non-dimensional derivatives:

$$N_r = \frac{\rho USb^2}{4I_z} C_{n_r} \tag{2.81}$$

Effect of change in aileron and rudder control surface deflection: (Y_{δ_a}, L_{δ_a}, N_{δ_a}, Y_{δ_r}, L_{δ_r}, N_{δ_r})

The L_{δ_a} and N_{δ_r} are the primary control derivatives or control power. If the absolute magnitude of these derivatives is higher, then for a given deflection, more rolling and yawing moments are generated. This is regarded as higher control sensitivity for a given moment of inertia.

$Y_{\delta_a} = (1/m)\partial Y/\partial\delta_a = \frac{\rho U^2 S}{2m} C_{y\delta_a}$, Y-axis force created by deflections ailerons.

$$Y_{\delta_r} = (1/m)\partial Y/\partial\delta_r = \frac{\rho U^2 S}{2m} C_{y\delta_r}, \tag{2.82}$$

Y-axis force created by deflections of rudder pedals.

The side force due to +ve δ_r (rudder deflection) is +ve $C_{y\delta_r}$. Its effects are relatively unimportant in lateral stability and control, except when considering lateral acceleration feedback to an autopilot.

N_{δ_a} and L_{δ_r} are cross-derivatives (the change in yawing moment due to ailerons' deflection; and the change in the rolling moment due to the deflections of the rudder pedals, respectively) and will be useful if there is rudder-aileron interconnect for certain high-performance fighter aircraft or rotorcrafts.

$$N_{\delta_a} = (1/I_z)\partial N/\partial\delta_a = \frac{\rho U^2 Sb}{2I_z} C_{n\delta_a}; N_{\delta_r} = (1/I_z)\partial N/\partial\delta_r = \frac{\rho U^2 Sb}{2I_z} C_{n\delta_r} \tag{2.83}$$

$$L_{\delta_a} = (1/I_x)\partial N/\partial\delta_a = \frac{\rho U^2 Sb}{2I_x} C_{l\delta_a}; L_{\delta_r} = (1/I_x)\partial N/\partial\delta_r = \frac{\rho U^2 Sb}{2I_x} C_{l\delta_r} \tag{2.84}$$

2.6.3.1.4 Some Most Important L-D Derivatives Are Explained Now

i. $C_{n\beta}$ represents the directional static stability; if $C_{n\beta} > 0$, then the aircraft possesses static directional stability; the aircraft will always point into the relative wind direction, and hence, this stability is called the 'weathercock' stability, synonymous with the concept of the weathercock at airports, etc. At high angles of attack (HAOA), the $C_{n\beta}$ will be affected significantly, since the fin will get submerged in the wing-body wake; because the major portion of this derivative comes from the vertical tail area and the lever arm that stabilize the airframe just as the tail feathers of an arrow stabilize the arrow shaft. The contribution to $C_{n\beta}$ from the tail is +ve (stabilizing), and that from the body is –ve (signifying instability); the one from the wing is usually +ve but very small. Also, since the major portions depend on the dimensions of the body, the C.G. variation does not affect the derivative much, since this variation would be a small fraction of the body length. This derivative is significant in dynamic lateral stability and control. It governs the natural frequency of the DR oscillatory mode of the aircraft. It also contributes to the spiral stability of the aircraft. The high values of $C_{n\beta}$ aid the pilot in affecting coordinated turns and prevent excessive sideslip and yawing motion in extreme flight maneuver.

ii. $C_{l\beta}$ is the rolling moment derivative (due to a small change in AOSS). This provides the dihedral stability for $C_{l\beta} < 0$; for the +ve dihedral. The contributions to this derivative come from (i) wing (as a function of geometric dihedral, sweep, aspect ratio, and AOA); (ii) wing location on the fuselage, if it is high, then it is –ve increment to the derivative, and if it lower, then there is a +ve increment; and (iii) direct force on the vertical tail that contributes decreasingly –ve increments as the fin moves down with respect to the x-axis with increasing trim AOA. The dihedral instability will contribute to the spiral divergence. A stable dihedral will tend to decrease the AOSS by rolling into the direction of yaw. Both $C_{l\beta}$ and $C_{n\beta}$ affect the aircraft DR mode and spiral mode as will be discussed in Chapter 5. It is also involved in maneuvering characteristics of an aircraft especially about lateral control with the rudder alone near stall. To improve the DR damping of an aircraft, small –ve values of the derivative are sought.

iii. C_{l_p} is the damping-in-roll derivative and determines roll subsidence. The change in rolling velocity causes the change in rolling moment. The +ve roll rate induces a restoring/opposing moment, and hence, the derivative has generally –ve magnitude. It can also be considered as a design criterion, since it directly affects the design of ailerons. Since C_{l_p} in conjunction with the $C_{l\delta_a}$ establishes the airframe's maximum available rolling velocity, this is an important criterion of flying qualities. The +ve values occur only when the wing or portions thereof are stalled, and such flight situations (including the separated flows) are avoided.

iv. C_{n_p} is a cross-derivative, usually negative, that influences the frequency of DR damping, The larger its –ve value, the greater the reduction in the DR damping; also, this causes higher (uncoordinated) side-slipping motions accompanying turn entry or exit; hence, +ve values of C_{n_p} are to be desired.

v. C_{n_r} is the 'yaw damping derivative', and it is made up of contributions (all of –ve sign) from the wing, the fuselage, and the vertical tail. It is the main contributor to the damping of the DR oscillatory mode, and it is important to the spiral mode; for each mode, large –ve values of the derivative are sought.

vi. C_{l_r} affects mainly the aircraft's spiral mode, and for the stability, it is desirable that its +ve value be as small as possible. For a conventional airframe configuration, the changes in C_{l_r} of reasonable magnitude show only slight effects on the DR damping.

vii. $C_{l\delta_a}$, $C_{n\delta_a}$, $C_{n\delta_r}$ are important control derivatives: $C_{l\delta_a}$ is the aileron control effectiveness, and in conjunction with damping in roll (C_{lp}) establishes the maximum available roll rate of an airframe which is a very important consideration in fighter tactics at high speeds. $C_{n\delta_r}$ is the 'rudder effectiveness' derivative or the rudder power. $C_{n\delta_a}$ arises in part from the difference in drag due to the down aileron compared to the drag of the up aileron. If it is –ve, then it is called the adverse yaw, because it causes the airframe to yaw initially in a direction opposite to that desired by the pilot when she deflects the ailerons for turn; if it is +ve, then it produces favorable or 'proverse yaw' in the turning maneuver. In an overall sense, for the proverse or stable control of an aircraft (including a spacecraft), the following should be true:

$$C_{n\beta} > \frac{C_{n\delta_a} C_{l\beta}}{C_{l\delta_a}}.$$

The DADs are directly related to the forces and moments and to the aircraft response variables. The DADs are useful since they directly appear in the simplified EOM/differential equations that describe the vehicle dynamic responses. That is to say that the DADs appear in the TF models that can be used to describe the dynamic responses/Bode diagrams for the aircraft in a particular axis in an approximate way. The NDADs are directly related to the aerodynamic coefficients and non-dimensional velocities. The merits of the non-dimensional forms are as follows: (i) easy correlation between the results from different test methods like WT, flight tests, theoretical predictions (CFD, DATCOM), (ii) correlation between the results from different flight test conditions, and (iii) possible comparison/correlation of the results of different configurations of the same aircraft or the different aircraft of similar kind (or even the other kinds!). The DADs and NDADs defined and studied here can be very conveniently used in the aerodynamic model of the atmospheric flight vehicles.

2.6.3.2 For Missiles

Due to the axis symmetry of a missile, pitch derivatives are identical to yaw derivatives in magnitude with some sign changes due to the asymmetry of an orthogonal right-handed axis system [5]. As in the case of aircraft, the force derivatives are divided by mass m and the moment derivatives by respective moment of inertia. Interestingly, these derivatives are called acceleration derivatives. The reason will be clear from the following development:

$$\ddot{\theta} = \frac{\text{pitching moment}}{\text{pitch inertia}} = \frac{C_m \bar{q} S \bar{c}}{I_y} \rightarrow \frac{\delta \ddot{\theta}}{\delta \alpha} = \frac{\rho U^2 S \bar{c}}{2 I_y} C_{m\alpha} \tag{2.85}$$

We know that

$$M_\alpha = \frac{1}{I_y} \frac{\partial M}{\partial \alpha} = \frac{\rho U^2 S \bar{c}}{2 I_y} \frac{\partial C_m}{\partial \alpha} = \frac{\delta \ddot{\theta}}{\delta \alpha} \tag{2.86}$$

Hence, $\frac{\delta\ddot{\theta}}{\delta\alpha} = M_\alpha$, and we see that the corresponding derivative is also called pitch angular acceleration derivative.

2.6.3.2.1 Longitudinal Derivatives

The normal force is given by $Z = 1/2\ \rho U^2 SC_z$

Here, the area S is the maximum body cross-section reference area. C_z is the normal force coefficient and is function of AOA (called incidence in pitch) for a given altitude and the Mach number.

Z_α is defined as

$$Z_\alpha = \frac{1}{2m}\ \rho U^2 S\ \frac{\partial C_z}{\partial \alpha};\ Z_\alpha = \frac{1}{2m}\ \rho U^2 SC_{z\alpha}\ \text{or}\ Z_w = \frac{1}{2m}\ \rho USC_{z\alpha} \tag{2.87}$$

The actual values of the normal force coefficient depend on the shape of the missile and the reference area S. The normal force coefficient usually for a missile is expressed as

$$C_N = C_{N_0} + C_{N\alpha}\alpha + C_{N_q}\frac{qc}{2V} + C_{N\dot{\alpha}}\left(\frac{\dot{\alpha}c}{2V}\right) + \cdots \tag{2.88}$$

Here, V is the resultant velocity or the cruising speed of the missile. The parameter c is the total body length of the missile. In case of the missile, N denotes the normal force unlike for the aircraft where Z denotes the vertical force. We will use only Z for the normal force uniformly. The meanings of the derivatives as such do not change.

M_w and M_q have the same significance as in the case of aircraft aerodynamics. M_{δ_e} will have an opposite sign for a missile with canard control surface (in the front). M_w will also change sign in case of an unstable missile configuration. Z_{δ_e} is related to M_{δ_e} in a usual way, and both are elevator control surface effectiveness derivatives. The derivative Z_q is of no practical significance and hence neglected in missile aerodynamics.

2.6.3.2.2 Lateral-Directional Derivatives: There Are Only a Few of Such Important Derivatives

i. **Roll derivatives:** There are only two important derivatives in roll DOF.

   L_p is the usual damping derivative in the roll, change in the rolling moment due to a change in the roll rate. As usual, it is negative in sign. It is often considered to have a constant value for a given Mach number and altitude. Due to positive roll rate, the change in the rolling moment is such that it will oppose this change in the roll rate and hence provide (increased) damping in the roll axis. It is usually defined as

$$L_p = \frac{1}{I_x}\ \frac{\partial L}{\partial p} \tag{2.89}$$

In an aircraft, the major contribution to this derivative is from wing; however, in the case of a missile, this is not the case.

L_{δ_a} is the aileron control surface effectiveness derivative. It is defined as

$$L_{\delta_a} = \frac{1}{I_x}\frac{\partial L}{\partial \delta_a} \tag{2.90}$$

ii. **Yaw derivatives**

Effect of change in side speed v/AOSS:(Y_v, N_β); Y_v is given as

$$Y_v = (1/m)\partial Y/\partial v = \frac{\rho US}{2m} C_{Y_\beta}; Y_\beta = \frac{\rho U^2 S}{2m} C_{Y_\beta} \tag{2.91}$$

It must be noted here that for a symmetrical missile, the normal force coefficient for yaw has the same value as the C_z. This lateral force derivative should be calculated from the total force from the wings, body, and the control surfaces. It is presumed that the control surfaces are in a central position. N_β is the force derivative Y_v times the distance of center of pressure from the C.G. As in the case of an aircraft, this distance is known as the static margin. It is a measure of the static stability of the missile. In usual manner, the derivative is defined as

$$N_v = \frac{\rho USb}{2I_z} C_{n\beta}; N_\beta = \frac{\rho U^2 Sb}{2I_z} C_{n\beta} \tag{2.92}$$

It will change sign for an unstable missile configuration.

Effect of change in yaw rate r:(Y_r, N_r)

Y_r is a difficult derivative to calculate and measure, and hence, it is usually neglected.

$$N_r \text{ given as}: N_r = \frac{\rho USb^2}{4I_z} C_{n_r} \tag{2.93}$$

It is a damping derivative in yaw. It has a small value.

Effect of change of rudder control surface deflection: $(Y_{\delta_r}, N_{\delta_r})$

Y_{δ_r} is the side force derivative due to deflection of the rudder control surface.

N_{δ_r} is the yawing moment derivative due to deflection of the rudder control surface. It has an opposite sign for a missile with canard control surface. It is given as Y_{δ_r} times the distance of the center of pressure of the rudder from the C.G. The formats of these derivatives are usually like the ones for the aircraft.

A missile has very few derivatives that are most important to describe its dynamics, especially from the control point of view. An interesting equivalence is illustrated, since acceleration = force/mass:

$$\frac{\partial a_y}{\partial \beta} = \frac{\partial F_y}{m\partial \beta} = \frac{\partial(\bar{q}SC_y)}{m\partial \beta} = \frac{\rho U^2 S\,\partial C_y}{2m\partial \beta} = \frac{\rho U^2 SC_{y\beta}}{2m} = Y_\beta \tag{2.94}$$

2.6.3.3 For Rotorcraft

In rotorcraft flight dynamic analysis, the same stability and control derivative expressions as used in the aircraft analysis are used. The asymmetric cross-coupling derivatives are usually neglected in aircraft flight mechanics analysis; however, in rotorcraft analysis, these are also used. For a rotorcraft, we have

$$M_w = \frac{1}{I_y}\frac{\partial M}{\partial w}$$

In this case, the controls are held at the trim values. The higher order rotor DOFs are allowed to reach a new equilibrium. The conventional 6DOF derivative model is a poor approximation for rotorcrafts, since the dynamics of the rotor response are not well separated from the airframe dynamic responses [6–8]. The use of the 6DOF quasi-static derivatives is found to give an impression of greater aircraft stability. This is so because the regressing flapping mode (of the rotor) is neglected. In flight, the rotor is continuously being excited by control and turbulence inputs and the estimates from the flights might differ considerably from the analytical perturbation derivatives (e.g. WT derivatives). The main reason is that the linear rotor plus body models are not easily derived by analytical means. These models should be determined from more comprehensive nonlinear models. Also, because of the lack of physical interpretation of the higher order rotor plus body models, the use of the quasi-static derivatives persists in rotorcraft analysis [4]. In general, force/moment equations are given as follows, by representing the changes about the trim conditions as the linear functions of the rotorcraft's states:

$$\begin{bmatrix} X_{\delta_0} & X_{\delta_x} & X_{\delta_y} & X_{\delta_p} \\ Y_{\delta_0} & Y_{\delta_x} & Y_{\delta_y} & Y_{\delta_p} \\ Z_{\delta_0} & Z_{\delta_x} & Z_{\delta_y} & Z_{\delta_p} \\ L_{\delta_0} & L_{\delta_x} & L_{\delta_y} & L_{\delta_p} \\ M_{\delta_0} & M_{\delta_x} & M_{\delta_y} & M_{\delta_p} \\ N_{\delta_0} & N_{\delta_x} & N_{\delta_y} & N_{\delta_p} \end{bmatrix} \begin{bmatrix} \delta_{\text{lon}} \\ \delta_{\text{lat}} \\ \delta_{\text{ped}} \\ \delta_{\text{col}} \end{bmatrix} + \text{biases} + \text{gravity terms}$$

$$+ \text{centrifugal specific (forces) terms} \tag{2.95}$$

If the coupling between various axes is strong, then the rotorcraft has many more derivatives (say up to 36 force and moment derivatives and 24 control derivatives) than an aircraft. For example, if the coupling between roll and pitch axes is quite strong, then the L_q and M_p will have significant values. Also, X_u is the drag damping derivative.

It is emphasized again here that the meanings of these aerodynamic derivatives across aircraft, missile, and rotorcraft should remain the same, with some specific interpretations depending on the special interactions between various axes of the airplane. This will depend on the extra DOFs entering the EOM, e.g. for a rotorcraft.

The longitudinal modes for helicopters with zero or low forward speeds are very different from the SP and phugoid modes. This is because certain derivatives disappear for the hovering flight. The derivatives X_w, M_w, and $M_{\dot{w}}$ are negligible. Also, in the hovering condition $U_0 = N_p = L_r = N_v = Y_p = Y_r = 0$, under certain conditions.

2.7 AERODYNAMIC MODELS

As we have seen earlier, aerodynamic forces and moments are expressed in a Taylor's series expansion of the instantaneous (numerical) values of the aircraft variables. Specifically, the aerodynamic forces and moments (in fact the aerodynamic coefficients) can be expressed as functions of the Mach number Mn, engine thrust F_T, and other aircraft motion and control variables $\alpha, \beta, p, q, r, \varphi, \theta, \delta_e, \delta_a$, and δ_r. A complete set of the aerodynamic coefficients can be represented as [13]

$$C_x = C_{x0} + C_{x\alpha}\alpha + C_{xq}\frac{q\bar{c}}{2V} + C_{x\delta_e}\delta_e + C_{xMn}Mn + C_{xF_T}F_T$$

$$C_y = C_{y0} + C_{y\beta}\beta + C_{y_p}\frac{pb}{2V} + C_{y_r}\frac{rb}{2V} + C_{y\delta_a}\delta_a + C_{y\delta_r}\delta_r$$

$$C_z = C_{z0} + C_{z\alpha}\alpha + C_{z_q}\frac{q\bar{c}}{2V} + C_{z\delta_e}\delta_e + C_{zMn}Mn + C_{zF_T}F_T$$

$$C_m = C_{m0} + C_{m\alpha}\alpha + C_{m_q}\frac{q\bar{c}}{2V} + C_{m\delta_e}\delta_e + C_{mMn}Mn + C_{mF_T}F_T \tag{2.96}$$

$$C_l = C_{l0} + C_{l\beta}\beta + C_{l_p}\frac{pb}{2V} + C_{l_r}\frac{rb}{2V} + C_{l\delta_a}\delta_a + C_{l\delta_r}\delta_r$$

$$C_n = C_{n0} + C_{n\beta}\beta + C_{n_p}\frac{pb}{2V} + C_{n_r}\frac{rb}{2V} + C_{n\delta_a}\delta_a + C_{n\delta_r}\delta_r$$

The non-dimensional coefficients of lift and drag are denoted by C_L and C_D. Due to disposition of forces in x and z directions, we need to resolve the relevant coefficients in the following way:

$$C_x = C_L \sin\alpha - C_D \cos\alpha$$

$$C_z = -(C_L \cos\alpha + C_D \sin\alpha) \tag{2.97}$$

Here, the effect of AOSS is neglected. If AOA is small, we can further simplify (2.97) as follows:

$$C_x = C_L\alpha - C_D, \qquad \text{If AOA is zero or very small } C_x = -C_D, \text{and } C_z = -C_L$$

$$C_z = -(C_L + C_D\alpha),$$

This shows the straightforward relations between the axial force coefficients and the drag and lift coefficients. Further, we get

$$C_L = -C_z \cos\alpha + C_X \sin\alpha,\ C_D = -C_X \cos\alpha - C_z \sin\alpha \tag{2.98}$$

In terms of several independent variables, we have

$$\begin{aligned} C_D &= C_{D_0} + C_{D_\alpha}\alpha + C_{D_q}\frac{q\bar{c}}{2V} + C_{D\delta_e}\delta_e + C_{DM}M + C_{D_{FT}}F_T \\ C_L &= C_{L_0} + C_{L_\alpha}\alpha + C_{L_q}\frac{q\bar{c}}{2V} + C_{L\delta_e}\delta_e + C_{LM}M + C_{L_{FT}}F_T \end{aligned} \tag{2.99}$$

To account for nonlinear and unsteady aerodynamic effects, the higher order derivatives and down-wash effects should be included. Also, some additional derivatives can be included to account for aerodynamic cross-coupling effects:

$$\begin{aligned} C_m &= C_{m0} + C_{m\alpha}\alpha + C_{m_q}\frac{q\bar{c}}{2V} + C_{m\delta_e}\delta_e + C_{mM}M + C_{m_{FT}}F_T + C_{m_{\beta^2}}\beta^2 + C_{m_{\beta^2\alpha}}\beta^2\alpha \\ &+ C_{m_{\delta^2_r}}\beta^2\delta_r{}^2 + C_{m_p}\left|pb/(2V)\right| + C_{m_r}\left|rb/(2V)\right| + \ldots + \end{aligned} \tag{2.100}$$

However, the model of eqn. (2.100) is quite complex and it will be difficult to determine the additional aerodynamic derivatives from flight test data of an aircraft. When the change in airspeed is not significant during the flight maneuver, the forces X, Y, and Z and the moments L, M, and N can be expanded in terms of the dimensional derivatives rather than non-dimensional derivatives for parameter estimation. Let us derive these expressions. Starting from eqn. (2.100), considering only three important terms, and substituting the formula for the NDADs (in terms of the dimensional derivatives), we obtain

$$\frac{M}{\bar{q}S\bar{c}} = \frac{2I_y}{\rho SU\bar{c}}M_w\alpha + \frac{4I_y}{\rho SU\bar{c}^2}M_q q\bar{c}/(2U) + \frac{2I_y}{\rho SU^2\bar{c}}M_{\delta e}\delta_e \tag{2.101}$$

Canceling out the common terms on both the sides, we get

$$M = I_y(M_w w + M_q q + M_{\delta e}\delta_e)$$

Substituting the defining terms for the dimensional derivatives, we obtain

$$M = \frac{\partial M}{\partial w}w + \frac{\partial M}{\partial q}q + \frac{\partial M}{\partial \delta_e}\delta_e$$

We see that the pitching moment is expressible in terms of the dimensional derivatives in a very straightforward manner. Similarly, other moments and forces can be expressed in terms of the corresponding dimensional derivatives (with certain simple assumptions) as follows:

$$
\begin{aligned}
X &= m(X_u u + X_w w + X_q q \; + X_{\delta e}\delta_e) \\
Y &= m(Y_v v + Y_p p + Y_q q + Y_r r \; + Y_{\delta_a}\delta_a + Y_{\delta_r}\delta_r) \\
Z &= m(Z_u u + Z_w w + Z_q q \; + Z_{\delta e}\delta_e) \\
L &= I_x(L_v v + L_p p + L_q q + L_r r \; + L_{\delta_a}\delta_a + L_{\delta_r}\delta_r) \\
M &= I_y(M_u u + M_w w + M_q q \; + M_{\delta e}\delta_e) \\
N &= I_z(N_v v + N_p p + N_q q + N_r r \; + N_{\delta_a}\delta_a + N_{\delta_r}\delta_r)
\end{aligned}
\tag{2.102}
$$

Starting from the natural phenomenon that an airplane experiences aerodynamic (airflow/interactions) forces and moments in flight, we have come to gradually appreciate the concept of the aerodynamic coefficients/derivatives that form the key elements in understanding the flight dynamic responses of an aircraft. The aerodynamic coefficients can (i) also be expressed in terms of derivatives due to the change in forward speed, e.g. $C_{L_u}, C_{D_u}, C_{z_u}$, and C_{m_u}, (ii) use higher order derivatives (e.g. $C_{x_{\alpha^2}}$, $C_{z_{\alpha^2}}$, and $C_{m_{\alpha^2}}$) to account for nonlinear effects, and (iii) use $C_{L\dot{\alpha}}$ and $C_{m\dot{\alpha}}$ derivatives to account for unsteady aerodynamic effects. The choice is problem-specific.

2.8 SIMPLIFIED STATE-SPACE AND TRANSFER FUNCTION (TF) MODELS

The MMs are used in research for interpretation of acquired knowledge/measurements that provide clues for further investigations; e.g., first a simplified model of the system is used, and then, based on the analysis of experimental data, this model is revised. A MM for the purpose of design of a new system or design of a controller for the same system is expressed in a manner that is compatible with the design criteria: guaranteeing the basic static/dynamic stability, overall performance, tracking error minimization, safety, and cost. For example, a time-domain model will be more suitable for use in the control-optimization methods based on time-domain methods, and hence, the knowledge of components and subsystems should be in such usable forms. Finally, the control actions on a system depend on the knowledge about the system.

2.8.1 Mathematical Model Structures

The EOMs form the basis of MMs of such a flight vehicle and are backbone for flight control, flight simulation, parameter estimation from flight data, and handling qualities analysis. In most of the aircraft identification and parameter estimation applications [13], the model can be represented by a set of rigid body equations in the body axis system [4]. The basic EOMs, derived from the Newtonian mechanics, define the aircraft characteristic motion. They involve the fundamental assumption that the forces and moments acting on the aircraft can be synthesized. In many practical cases, the forces and moments, which include the aerodynamic, inertial, gravitational, and propulsive forces, are approximated by terms in the Taylor's series expansion. This invariably leads to a model that is linear in parameters. A more complete model can be justified for the correct description of the aircraft dynamics. The degree of relationship between the model complexity and measurement information needs to be ascertained. To arrive at a complex and complete model, one would need lot of repeated experimentation under varieties of environmental conditions and lot of data. This will ensure the statistical consistency of the modelling results in face of the uncertainties of measurement noise/errors. But, how much of this additional effort/data will be useful in building the complete model will depend on the methods of reduction of the data and engineering judgment.

2.8.2 Choice of Coordinate Systems

The selection of the appropriate coordinate system for EOMs plays an important role in simplification of the EOM. Some important points to be considered are as follows [14]: (i) body axis system is most suitable for defining rotational DOFs. It is a normal practice to include the angular rates and the Euler angles in the aircraft state model, and the equations in body axes, under certain assumptions, can be considerably simplified; (ii) either of the polar or rectangular coordinate system can be used for translational DOFs; (iii) the polar axis system uses the (α,β,V) form which involves wind axis coefficients C_L and C_D; the wind axis coefficients introduce nonlinearities because of the way they are related to the body axis force coefficients C_X, C_Y, and C_Z. However, the α,β,V are commonly measured quantities which encourages the use of the polar coordinate system. The expressions for α,β,V in observation equations are simple if polar coordinates are used for data analysis; the linear accelerations are normally defined in terms of C_X, C_Y, C_Z which suggests the use of u,v,w form of state equations. This form is particularly useful in case of the rotorcraft where the (α,β,V) form may lead to singularities. With the differential equations for u,v,w in the state model, the (α,β,V) expressions in the observation model will be nonlinear.

2.8.3 Linearization of Model Equations

The nonlinear coupled 6DOF EOMs can be linearized using small disturbance theory. In using this approach, we assume that the excursions about the reference flight condition are small. Thus, we assume small values of the variables u,v,w,p,q,r,ϕ,θ. The approach cannot be applied to situations with large change in motion variables. All the variables in the 6DOF model are expressed as sum of a reference value (steady-state value) and a small perturbation, e.g.

$$\begin{aligned} &U = u_0 + u;\; V = v_0 + v;\; W = w_0 + w \\ &P = p_0 + p;\; Q = q_0 + q;\; R = r_0 + r \end{aligned} \tag{2.103}$$

The nonlinear translational accelerations from the full set of 6DOF equations are expressed as

$$\begin{aligned} &\dot{U} = -qw + rv - g\sin\phi + a_y \\ &\dot{V} = -ru + pw + g\cos\theta\sin\phi + a_y \\ &\dot{W} = -pv + qu + g\cos\theta\cos\phi + a_z \end{aligned} \tag{2.104}$$

The $\dot{u}$ equation can be expressed in terms of perturbed variables as follows:

$$\dot{u} = -(q+q_0)(w+w_0)+(r+r_0)(v+v_0)-g\sin(\theta+\theta_0)+(a_x+a_{x0}) \tag{2.105}$$

Eqn. (2.105) can be simplified by neglecting products of the perturbations and using small angle approximation; also, assuming wing level, symmetric reference flight, we have

$$w_0 = v_0 = p_0 = q_0 = r_0 = \phi_0 = \psi_0 = 0 \tag{2.106}$$

Setting all disturbances in the perturbed $\dot{U}, \dot{V}, \dot{W}$ equations to zero gives the reference flight conditions as follows:

$$\begin{aligned} a_{x0} &= q_0 w_0 - r_0 v_0 - g\sin\theta_0 \\ a_{y0} &= r_0 u_0 - p_0 w_0 - g\cos\theta_0 \sin\phi_0 \\ a_{z0} &= p_0 v_0 - q_0 u_0 - g\cos\theta_0 \cos\phi_0 \end{aligned} \tag{2.107}$$

Thus, the set of simplified linearized perturbation EOMs can be expressed as

$$\begin{aligned} \dot{u} &= -g\theta\cos\theta_0 + a_x \\ \dot{v} &= -ru_0 + g\phi\cos\theta_0 + a_y \\ \dot{w} &= +qu_0 - g\theta\sin\theta_0 + a_z \\ \dot{p} &= \frac{1}{I_x I_z - I_{xz}^2}(I_z L + I_{xz} N) \\ \dot{q} &= \frac{1}{I_y} M \\ \dot{r} &= \frac{1}{I_x I_z - I_{xz}^2}(I_x N + I_{xz} L) \\ \dot{\phi} &= p \\ \dot{\theta} &= q \end{aligned} \tag{2.108}$$

The linearized implementation is frequently used (i) for routine batch analysis of large amount of data in cruise flight regimes, (ii) for the analysis of small amplitude maneuvers, and (iii) where computational efficiency is important and nonlinear effects are minimal.

2.8.4 Simplification Using Measured Data

In this approach, we do away with some of the differential equations from the state model and work with a reduced set of equations. However, since the eliminated variables will mostly appear in other equations, measured values of these variables from flight data are used. A very good example is that of the model equations used to analyze data generated from longitudinal SP maneuver. It is well known that during a SP pitch-stick maneuver, there is only marginal change in velocity, implying that the longitudinal SP mode is independent of velocity. Therefore, in the α, β, V form of equations, the $\dot{V}$ equation can be omitted from SP data analysis; instead, the measured value of V can be used in the model equations without any loss of accuracy. The measured data for the variables used in this manner are also sometimes referred to as 'pseudo-control inputs'. It is clear from this discussion that the use of this approach would require measurements of the motion variables omitted from analysis; the disadvantage of this approach is that it is sensitive to the noise in the measurements/data. Further simplification is possible, since most aircraft can be assumed to be symmetric about the XZ plane and fly at small AOSS, by separating the 6DOF coupled equations into nearly independent sets: longitudinal and LD. Each set has nearly half the number of differential equations in the state model compared to the full coupled nonlinear 6DOF state equations. The simplified non-dimensional and dimensional models normally used for longitudinal and LD flight data analysis are discussed next.

2.8.4.1 Aircraft Models

2.8.4.1.1 Longitudinal Models and Modes

Assuming that the aircraft is not performing maneuvers with large excursion, it is possible to characterize the aircraft response to pitch-stick inputs by considering only the lift force, drag force, and pitching moment equations. Assuming the aircraft to be symmetric about the XZ plane, i.e. I_{xy} and I_{yz} are zero, the general free longitudinal motion can be described by eliminating the $\dot{\beta}, \dot{p}, \dot{r}$, and $\dot{\phi}$ equations from the 6DOF state model; thus, we have

$$\begin{aligned}
\dot{V} &= g(\cos\phi\cos\theta\sin\alpha\cos\beta + \sin\phi\cos\theta\sin\beta - \sin\theta\cos\alpha\cos\beta) \\
&\quad - \frac{\bar{q}S}{m}C_{Dwind} + \frac{T}{m}\cos(\alpha+\sigma_T)\cos\beta \\
\dot{\alpha} &= \frac{g}{V\cos\beta}(\cos\phi\cos\theta\cos\alpha + \sin\theta\sin\alpha) + q - \tan\beta(p\cos\alpha + r\sin\alpha) \\
&\quad - \frac{\bar{q}S}{mV\cos\beta}C_L - \frac{T}{mV\cos\beta}\sin(\alpha+\sigma_T) \\
\dot{q} &= \frac{1}{I_y}\left\{\bar{q}ScC_m - \left(p^2 - r^2\right)I_{xz} + pr\left(I_z - I_x\right) + T(l_{tx}\sin\sigma_T + l_{tz}\cos\sigma_T)\right\} \\
\dot{\theta} &= q\cos\phi - r\sin\phi
\end{aligned} \tag{2.109}$$

In eqn. (2.109), although the differential equations for $\dot{\beta}, \dot{p}, \dot{r}$, and $\dot{\phi}$ are omitted from analysis, these terms still appear in the right-hand side of the equations, and their measured values can be used to solve eqn. (2.109), which can be further simplified by assuming these quantities to be small during longitudinal maneuvers and neglecting them altogether, and we obtain

$$\dot{V} = g(\cos\theta\sin\alpha + \cos\theta - \sin\theta\cos\alpha) - \frac{\overline{q}S}{m}C_{Dwind} + \frac{T}{m}\cos(\alpha+\sigma_T)\cos\beta$$

$$\dot{\alpha} = \frac{g}{V}(\cos\theta\cos\alpha + \sin\theta\sin\alpha) + q - \frac{\overline{q}S}{mV}C_L - \frac{T}{mV}\sin(\alpha+\sigma_T) \qquad (2.110)$$

$$\dot{q} = \frac{1}{I_y}\{\overline{q}ScC_m + T(l_{tx}\sin\sigma_T + l_{tz}\cos\sigma_T)\}$$

$$\dot{\theta} = q$$

We also know that

$$C_{\text{Dwind}} = C_D\cos\beta - C_Y\sin\beta;\ \text{For small values of } \beta C_{\text{Dwind}} = C_D \qquad (2.111)$$

Substituting eqn. (2.111) back in eqn. (2.110) and assuming the thrust to be acting at C.G., i.e. $l_{tx} = l_{tz} = 0$, the differential equations for longitudinal motion now take the following form:

$$\dot{V} = g\sin(\theta-\alpha) - q\frac{\overline{q}S}{mV}C_D + \frac{T}{m}\cos(\alpha+\sigma_T)$$

$$\dot{\alpha} = \frac{g}{V}(\cos\theta - \alpha) + q - \frac{\overline{q}S}{mV}C_L - \frac{T}{mV}\sin(\alpha+\sigma_T) \qquad (2.112)$$

$$\dot{q} = \frac{1}{I_y}\overline{q}ScC_m$$

$$\dot{\theta} = q$$

In the u, v, w form, the longitudinal motion of the aircraft in flight can be described by the following fourth-order model:

$$\dot{u} = \frac{\overline{q}S}{m}C_X - qw - g\sin\theta$$

$$\dot{w} = \frac{\overline{q}S}{m}C_Z + qu + g\cos\theta \qquad (2.113)$$

$$\dot{q} = \frac{\overline{q}S\overline{c}}{I_y}C_m$$

$$\dot{\theta} = q$$

Here, C_X, C_Z, and C_m are the non-dimensional aerodynamic coefficients. The set (5.75) is commonly used to analyze the aircraft longitudinal motion. Certain aerodynamic derivatives rarely appear in the technical literature concerning with the aircraft dynamics, since these are of secondary importance because either they are very small (or their effects are negligible) or they are very difficult to determine: $X_{\dot{w}} = X_q = Z_{\dot{w}} = Z_q = 0$.

Also, in the steady-state flight conditions, the flight path of the airplane is assumed to be horizontal, $\gamma_0 = 0$. Using the foregoing two conditions, the longitudinal EOMs (in the stability axes) become as follows, in the form of complex frequency 's', useful for the TF forms:

$$
\begin{aligned}
&(s - X_u)u - X_w w + g\theta = X_{\delta_e}\delta_e - X_u u_g - X_w w_g \\
&-Z_u u + (s - Z_w)w - U_o s\theta = Z_{\delta_e}\delta_e - Z_u u_g - Z_w w_g \\
&-M_u u - (M_{\dot{w}} s + M_w)w + s(s - M_q)\theta = M_{\delta_e}\delta_e - M_u u_g - [M_{\dot{w}} - M_q/U_0)s + M_w]w_g \\
&s\theta = q
\end{aligned}
\tag{2.114}
$$

This set can be further simplified, under certain assumptions, to describe the two basic oscillatory longitudinal modes: (i) the heavily damped SP mode and (ii) the lightly damped long period or phugoid (fleeing away) mode.

Example 2.4

The state-space equations in matrix form for LTV A-7A Corsair aircraft [3] are given as

$$
\begin{bmatrix} \dot{u} \\ \dot{w} \\ \dot{q} \\ \dot{\theta} \end{bmatrix} = \begin{bmatrix} 0.005 & 0.00464 & -73 & -31.34 \\ -0.086 & -0.545 & 309 & -7.4 \\ 0.00185 & -0.00767 & -0.395 & 0.00132 \\ 0 & 0 & 1 & 0 \end{bmatrix} \begin{bmatrix} u \\ w \\ q \\ \theta \end{bmatrix} + \begin{bmatrix} 5.63 \\ -23.8 \\ -4.52 \\ 0 \end{bmatrix} \delta
$$

At the flight condition: 4.57 km (15 kft.), Mn = 0.3, obtain the frequency responses for all the variables and unit step responses for pitch rate and pitch attitude, and see the discerning effects.

Solution 2.4

The TFs are obtained by [num,den] = ss2tf(a,b,c,d,1); sysu = tf(num(1,:),den); sysw = tf(num(2,:),den); sysq = tf(num(3,:),den); and systheta = tf(num(4,:),den). The Bode diagrams are shown in Figure 2.1. We notice from these frequency responses that the aircraft dynamics has two distinct modes: one at a lower frequency and another at a relatively higher frequency. This shows the feasibility of modelling these two modes separately as they are relatively well separated. One can also discern from these plots that the low-frequency mode is lightly damped compared to the higher frequency mode. We see from Figure 2.2 that the pitch rate response

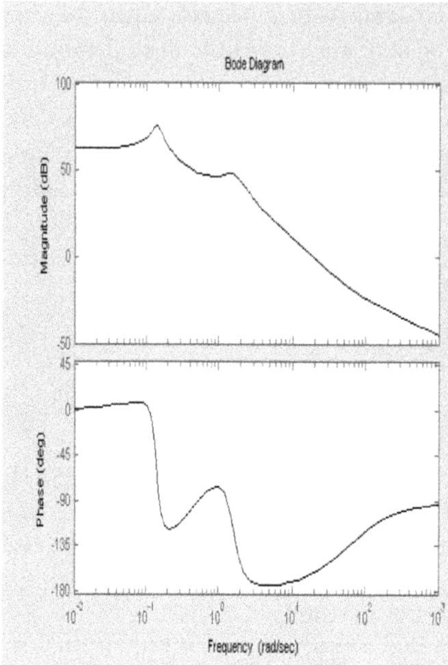

(a) Forward velocity

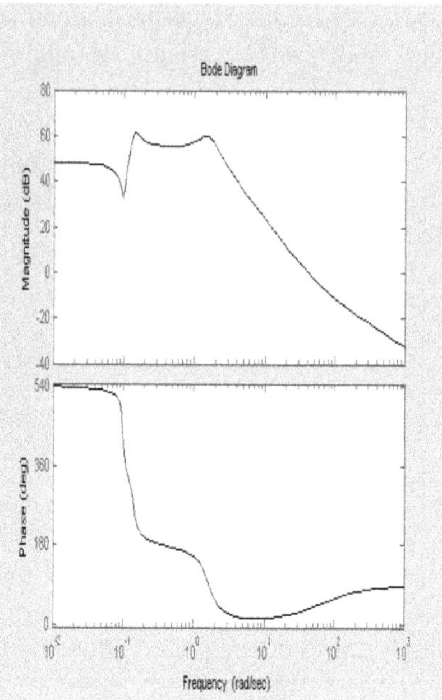

(b) vertical velocity

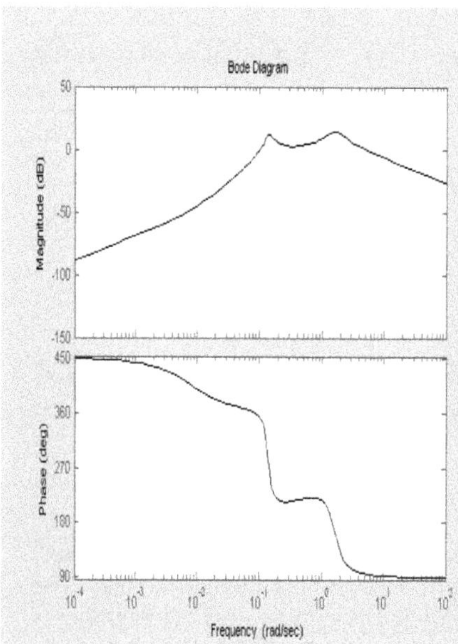

(c) pitch rate

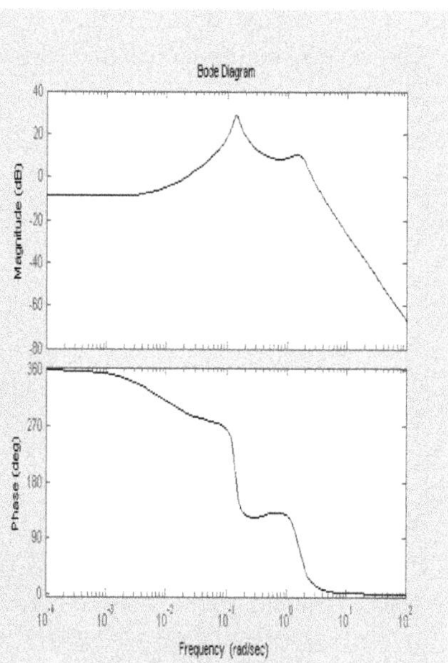

(d) pitch attitude

FIGURE 2.1 Frequency responses/Bode diagrams of the combined longitude modes of the LTV A-7A Corsair aircraft (a) forward velocity; (b) verticle velocity; (c) pitch rate; (d) pitch attitude.

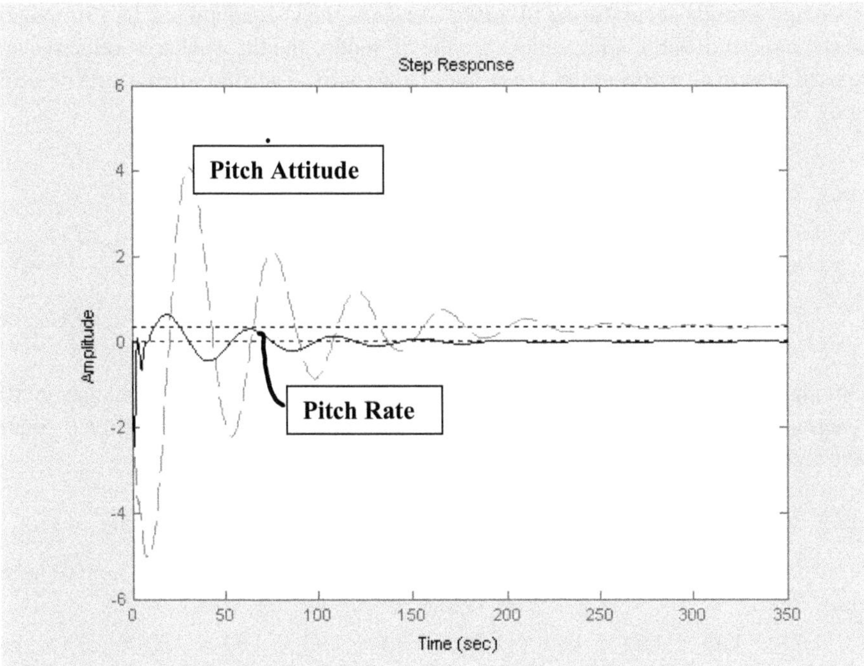

FIGURE 2.2 Unit step responses for pitch rate (-) and pitch attitude (–) of the longitude modes of the LTV A-7A Corsair aircraft.

is well damped and the pitch attitude response takes more time to settle. This leads to the two distinct modes discussed in the next section. Some observations about the possible TFs arising from eqn. (2.114) are as follows: (i) $|u/\delta_e| \rightarrow$ the amplitude ratio is much smaller at the natural frequency of the SP than at that of the phugoid, (ii) $|\alpha/\delta_e| \rightarrow$since the numerator cancels nearly the phugoid denominator, there is no change in the AOA during the phugoid mode, (iii) $|\theta/\delta_e| \rightarrow$ the values of the amplitude ratios at the SP and the phugoid modes are nearly the same as that of $|u/\delta_e|$ at the same frequencies, and (iv) $|a_z/\delta_e| \rightarrow$ the vertical (normal) acceleration amplitude at phugoid is only somewhat higher than that at the SP; however, considering the large differences in the frequencies, a given oscillatory acceleration will involve much higher excursions in h and (h-dot) at phugoid frequency than at SP frequency. Successive integration of the normal acceleration to obtain h-dot and then h involves successive clockwise rotations of the $|\alpha/\delta_e|$ TF/Bode diagram each of 20 dB/decade, and such rotation progressively suppresses the SP hump relative to the phugoid peak.

2.8.4.1.2 Short Period Mode

The SP is a relatively well-damped, high-frequency oscillation mode. The amplitude ratio V/δ_e is generally much smaller at the natural frequency of the SP mode. In other words, the velocity is weakly related to the SP mode and only small changes in speed and altitude occur during SP mode. As such, the $\dot{V}$ equation can be eliminated from the set of differential equations for the SP mode, and the measured velocity can be used wherever it appears in the $\dot{\alpha}$ and $\dot{q}$ equations. The simplified set for the SP mode is written as

$$\dot{\alpha} = \frac{g}{V}\cos(\theta-\alpha)+q-\frac{\bar{q}S}{mV}C_L-\frac{T}{mV}\sin(\alpha+\sigma_T)$$

$$\dot{q} = \frac{1}{I_y}\bar{q}ScC_m \tag{2.115}$$

$$\dot{\theta} = q$$

Although pitch angle θ changes during a SP maneuver, the effect of the change on SP dynamics is small. Approximating the gravity term in $\dot{\alpha}$ equation by setting θ equal to α and neglecting $\dot{\theta}$ equation and thrust, we get

$$\dot{\alpha} = \frac{g}{V}+q-\frac{\bar{q}S}{mV}C_L$$
$$\dot{q} = \frac{1}{I_y}\bar{q}ScC_m \tag{2.116}$$

Equation (2.116) is the simplest usable form for characterizing the SP mode. In the u, v, w form of state equations and in terms of dimensional stability and control derivatives, the following simplified model is generally used for longitudinal SP motion (including in the TF form):

$$\begin{array}{l} \dot{w} = Z_w w+(u_0+Z_q)q+Z_{\delta_e}\delta_e \\ \dot{q} = M_w w+M_q q+M_{\delta_e}\delta_e \end{array} ; \quad \begin{array}{l} (s-Z_w)w-u_0 s\theta = Z_{\delta_e}\delta_e \\ -(sM_{\dot{w}}+M_w)w+(s-M_q)s\theta = M_{\delta_e}\delta_e \end{array} \tag{2.117}$$

Here, w and q are the vertical velocity and pitch rate, respectively, and u_0 is the forward speed under the steady-state condition. The stability and control derivatives are defined as [1]

$$Z_w = \frac{1}{m}\frac{\partial Z}{\partial w}; Z_q = \frac{1}{m}\frac{\partial Z}{\partial q}; Z_{\delta_e} = \frac{1}{m}\frac{\partial Z}{\partial \delta_e}; M_w = \frac{1}{I_y}\frac{\partial M}{\partial w}; M_q = \frac{1}{I_y}\frac{\partial M}{\partial q}; M_{\delta_e} = \frac{1}{I_y}\frac{\partial M}{\partial \delta_e} \tag{2.118}$$

Since $\alpha \approx \frac{w}{u_0}$, eqn. (5.80) can also be written in terms of α instead of w:

$$\dot{\alpha} = \frac{Z_\alpha}{u_0}\alpha+(1+\frac{Z_q}{u_0})q+\frac{Z_{\delta_e}}{u_0}\delta_e; \dot{q} = M_\alpha\alpha+M_q q+M_{\delta_e}\delta_e \tag{2.119}$$

$$Z_w = \frac{Z_\alpha}{u_0}; M_w = \frac{M_\alpha}{u_0} \tag{2.120}$$

From (2.119), we have the following TFs:

$$\frac{\alpha(s)}{\delta_e(s)} = \frac{1}{U_0}\frac{Z_{\delta_e}s + (U_0 M_{\delta_e} - Z_{\delta_e}M_q)}{(s^2 - (U_0 M_{\dot{w}} + Z_w + M_q)s + (M_q Z_w - U_0 M_w)}$$

$$\frac{\theta(s)}{\delta_e(s)} = \frac{(Z_{\delta_e}M_{\dot{w}} + M_{\delta_e})s + (Z_{\delta_e}M_w - M_{\delta_e}Z_w)}{s(s^2 - (U_0 M_{\dot{w}} + Z_w + M_q)s + (M_q Z_w - U_0 M_w)}$$

Putting the SP 2DOF model, from (2.119) in the state-space form $\dot{x} = Ax + Bu$, and neglecting Z_q, we get

$$\begin{bmatrix} \dot{\alpha} \\ \dot{q} \end{bmatrix} = \begin{bmatrix} \dfrac{Z_\alpha}{u_0} & 1 \\ M_\alpha & M_q \end{bmatrix} \begin{bmatrix} \alpha \\ q \end{bmatrix} + \begin{bmatrix} \dfrac{Z_{\delta_e}}{u_0} \\ M_{\delta_e} \end{bmatrix} \delta_e \tag{2.121}$$

The characteristic equation of the form $(\lambda I - A) = (sI - A)$ will be

$$\lambda^2 - \left(M_q + \frac{Z_\alpha}{u_0}\right)\lambda + \left(M_q \frac{Z_\alpha}{u_0} - M_\alpha\right) = 0 \tag{2.122}$$

Solving for the eigenvalues of the characteristic equation yields the following frequency and damping ratio for the SP mode:

$$\text{Frequency } \omega_{n_{sp}} = \sqrt{\frac{Z_\alpha M_q}{u_0} - M_\alpha} \tag{2.123}$$

$$\text{Damping ratio } \zeta_{sp} = -\frac{M_q + \dfrac{Z_\alpha}{u_0}}{2\omega_{n_{sp}}} \tag{2.124}$$

The SP occurs before there is any appreciable change in the forward speed.

Example 2.5

The aerodynamic derivatives determined from flight test conducted with 3-2-1-1 pilot-stick command input for a transport aircraft are given as

$Z_0 = -0.0085,\ Z_\alpha = -0.480,\ Z_q = 0.102,\ Z_{\delta_e} = 0.652,\ M_0 = 0.472,$

$M_\alpha = -4.916,\ M_q = -1.946,\ M_{\delta_e} = -7.011$

The Z-force derivatives are assumed to be normalized with longitudinal velocity component u_0. Write an appropriate MM and obtain TF and frequency response of the SP model.

Solution 2.5

Based on the derivatives given, we can build the following longitudinal SP model:

$$\dot{\alpha} = Z_0 + Z_\alpha \alpha + (1 + Z_q)q + Z_{\delta_e}\delta_e$$
$$\dot{q} = M_0 + M_\alpha \alpha + M_q q + M_{\delta_e}\delta_e \tag{2.125}$$

The state-space form is given as

$$\begin{bmatrix} \dot{\alpha} \\ \dot{q} \end{bmatrix} = \begin{bmatrix} -0.482 & (1+0.102) \\ -4.916 & -1.946 \end{bmatrix} \begin{bmatrix} \alpha \\ q \end{bmatrix} + \begin{bmatrix} 0.652 \\ -7.011 \end{bmatrix} \delta_e + \begin{bmatrix} -0.0085 \\ 0.472 \end{bmatrix} \tag{2.126}$$

The TFs are obtained as [numsp,densp] = ss2tf(a,b,c,d,1) and sysalpha = tf(numsp(1,:),densp) and sysq = tf(numsp(2,:),densp). The bias/ 'nought' derivatives are ignored. The two TFs are

$$\frac{\alpha(s)}{\delta_e(s)} = \frac{0.652s - 6.457}{s^2 + 2.428s + 6.355} \text{ and } \frac{q(s)}{\delta_e(s)} = \frac{-7.011s - 6.585}{s^2 + 2.428s + 6.355} \tag{2.127}$$

The SP frequency and damping ratio are obtained as [spfreq, spzeta] = damp(sysalpha) = [2.521 rad/s, 0.4816]. The frequency/Bode diagrams are shown in Figure 2.3. Thus, we see the distinct SP mode, though for a different aircraft.

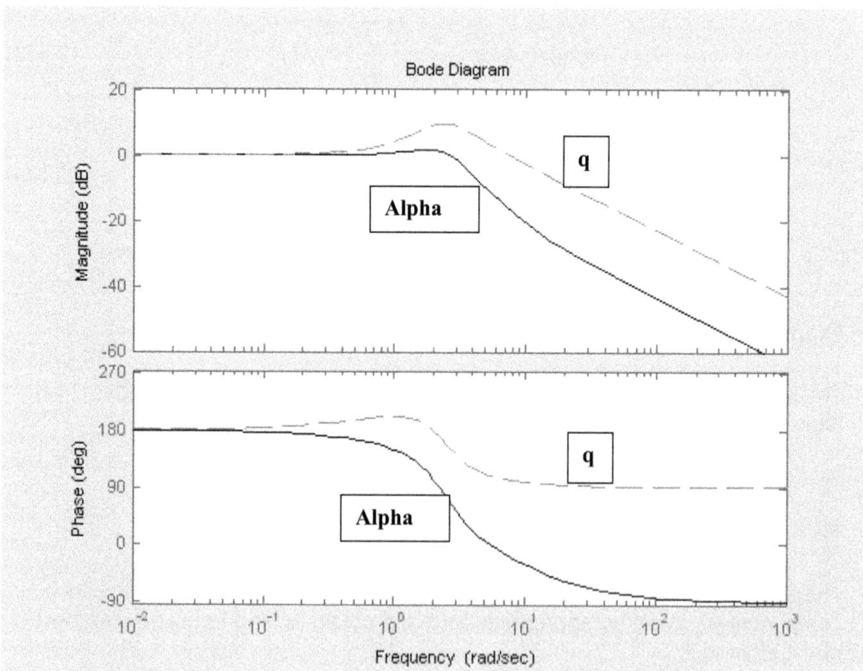

FIGURE 2.3 Bode diagrams/frequency responses for the SP mode of a transport aircraft alpha (-) and q (–) TFs.

Example 2.6

The aerodynamic derivatives for an Air Force medium transport aircraft are given as:

$$Z_\alpha/u_0 = -0.66,\ Z_{\delta_e}/u_0 = 0.01,\ M_\alpha = -1.74,\ M_q = -0.67,\ M_{\delta_e} = -5.33.$$

Write an appropriate MM and obtain TFs and frequency responses of the SP model. Also obtain the unit step responses.

Solution 2.6

Based on the derivatives given, we can build the following longitudinal SP model:

$$\begin{aligned}\dot{\alpha} &= (Z_\alpha/u_0)\alpha + q + (Z_{\delta_e}/u_0)\delta_e \\ \dot{q} &= M_\alpha\alpha + M_q q + M_{\delta_e}\delta_e\end{aligned} \tag{2.128}$$

The state-space form is given as

$$\begin{bmatrix}\dot{\alpha} \\ \dot{q}\end{bmatrix} = \begin{bmatrix}-0.66 & 1 \\ -1.74 & -0.67\end{bmatrix}\begin{bmatrix}\alpha \\ q\end{bmatrix} + \begin{bmatrix}0.01 \\ -5.33\end{bmatrix}\delta_e \tag{2.129}$$

The TFs are obtained as [num,den] = ss2tf(a,b,c,d,1) and sysalpha = tf(num(1,:),den) and sysq = tf(num(2,:),den). The two TFs are as follows:

$$\frac{\alpha(s)}{\delta_e(s)} = \frac{0.01s - 5.323}{s^2 + 1.33s + 2.182} \text{ and } \frac{q(s)}{\delta_e(s)} = \frac{-5.33s - 3.535}{s^2 + 1.22s + 2.182} \tag{2.130}$$

The SP frequency and damping ratio are obtained as [spfreq, spzeta] = damp(sys alpha) = [1.48 rad/s, 0.45]. The frequency/Bode diagrams are shown in Figure 2.4. The unit step responses are shown in Figure 2.5.

2.8.4.1.3 *Phugoid*

The phugoid mode is a lightly damped low-frequency oscillation. The aircraft responses in phugoid are very slow compared to the changes in the motion parameters in the SP mode. This mode describes the long-term translatory motions of the vehicle center of mass with practically no change in the AOA. This mode involves large oscillatory changes in forward speed, pitch angle θ, and altitude h at constant AOA. An approximation to the phugoid mode can be made by omitting the pitching moment equation and having only the 2DOF phugoid mode state equations:

$$(s - X_u)u = -g\theta + X_{\delta_e}\delta_e - U_o s\theta = Z_u u + Z_{\delta_e}\delta_e \text{ (The TF form)}; \rightarrow \begin{bmatrix}\dot{u} \\ \dot{\theta}\end{bmatrix}$$
$$= \begin{bmatrix}X_u & -g \\ -\dfrac{Z_u}{U_0} & 0\end{bmatrix}\begin{bmatrix}u \\ \theta\end{bmatrix} + \begin{bmatrix}(X_{\delta_e} =)0 \\ -Z_{\delta_e}/U_0\end{bmatrix}\delta_e; \tag{2.131}$$

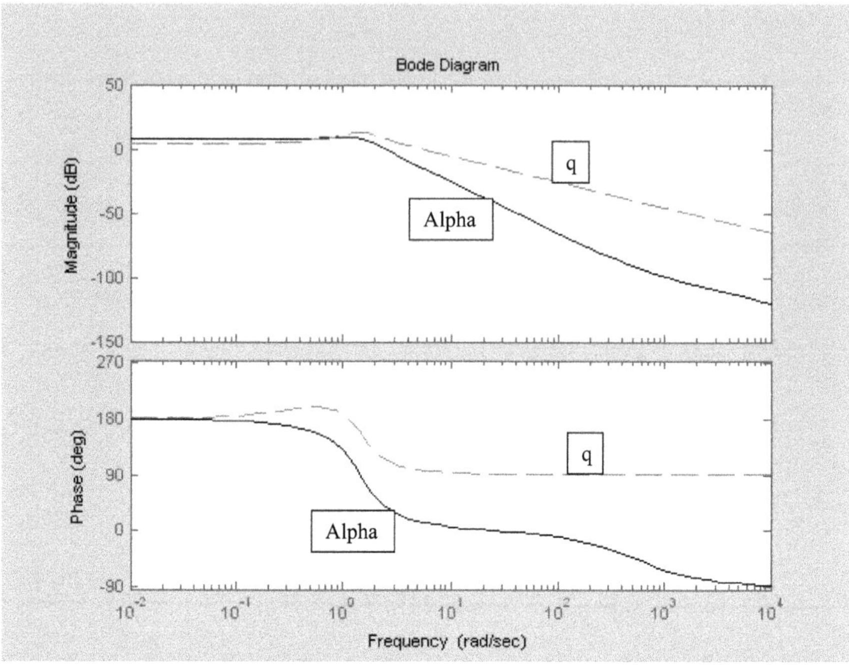

FIGURE 2.4 Bode diagrams/frequency responses for the SP mode of the medium transport aircraft: alpha (-) and q (–) TFs.

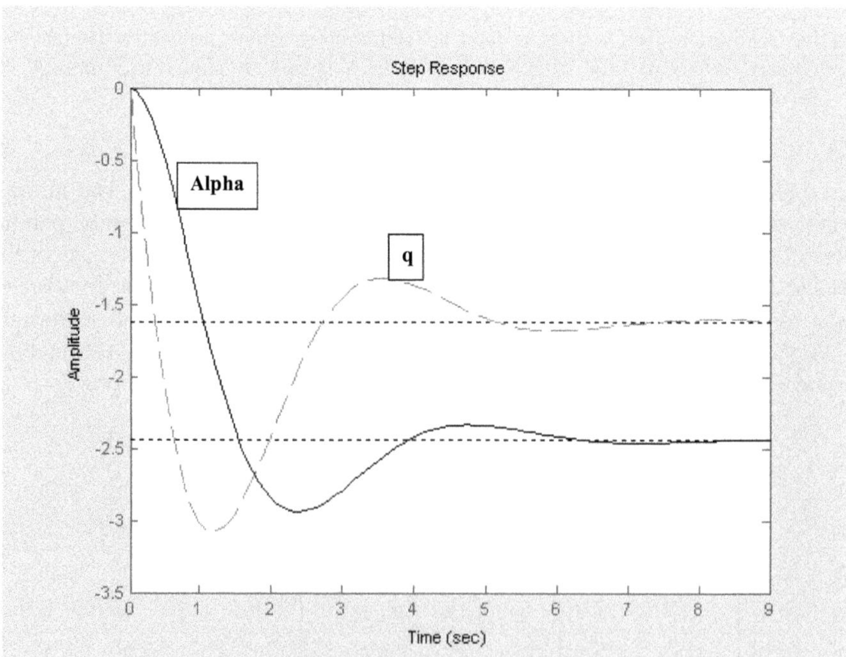

FIGURE 2.5 Unit step responses of the SP mode of the medium transport aircraft: alpha (-) and q (–).

Forming the characteristic equation and solving for the eigenvalues yield the following expressions for the phugoid natural frequency and damping ratio:

$$\text{Frequency } \omega_{n_{ph}} = \sqrt{-\frac{Z_u g}{u_0}}\text{; With Mu = 0.0; With Mu = 0.0} \tag{2.132}$$

Damping ratio

$$\zeta_{ph} = -\frac{X_u}{2\omega_{n_{ph}}} \tag{2.133}$$

Since $Z_u = -\frac{\rho SU}{m}(C_L + C_{L_u})$; for $C_{Lu}=0$, we have $Z_u=-2L/(mU_0)=-2g/U_0$ (since $L=mg$). Thus, $\omega_{n_{ph}} \simeq \sqrt{2}\left(g/U_0\right)$.

Example 2.7

The aerodynamic derivatives of a transport aircraft in the phugoid mode are given as

$$X_u = -0.015,\ Z_u = -0.1,\ u_0 = 60\,\text{m/s} \tag{2.134}$$

Use the phugoid model of eqn. (2.131) and obtain the frequency responses and other characteristics.

Solution 2.7

The state-space model is written as

$$\begin{bmatrix} \dot{u} \\ \dot{\theta} \end{bmatrix} = \begin{bmatrix} -0.015 & -g \\ \frac{0.1}{60} & 0 \end{bmatrix} \begin{bmatrix} u \\ \theta \end{bmatrix} + \begin{bmatrix} 0.01 \\ 0 \end{bmatrix} \delta_e \tag{2.135}$$

Based on the procedure of the previous examples, the TFs and the frequency responses are obtained for this phugoid model. The natural frequency of this mode is 0.128 rad/s, and the damping ratio is 0.0587. The frequency and unit step responses are shown in Figure 2.6. Since the damping ratio of this mode is very small, the oscillations last for several seconds. Since the amplitudes are small and oscillations are of low frequency, the pilot would be able to manage these oscillations very well. There is possibility of the 3DOF phugoid approximation model. This can be done by retaining the derivatives related to the vertical speed:

$$\begin{aligned} &(s - X_u)u - X_w w + g\theta = X_{\delta_e}\delta_e \\ &-Z_u u + (s - Z_w)w - U_o s\theta = Z_{\delta_e}\delta_e \\ &-M_u u - M_w w = M_{\delta_e}\delta_e \end{aligned} \tag{2.136}$$

$$\omega_{n_{ph}} = \sqrt{-\frac{g\left(Z_u - \frac{M_u}{M_w}Z_w\right)}{u_0}} \tag{2.137}$$

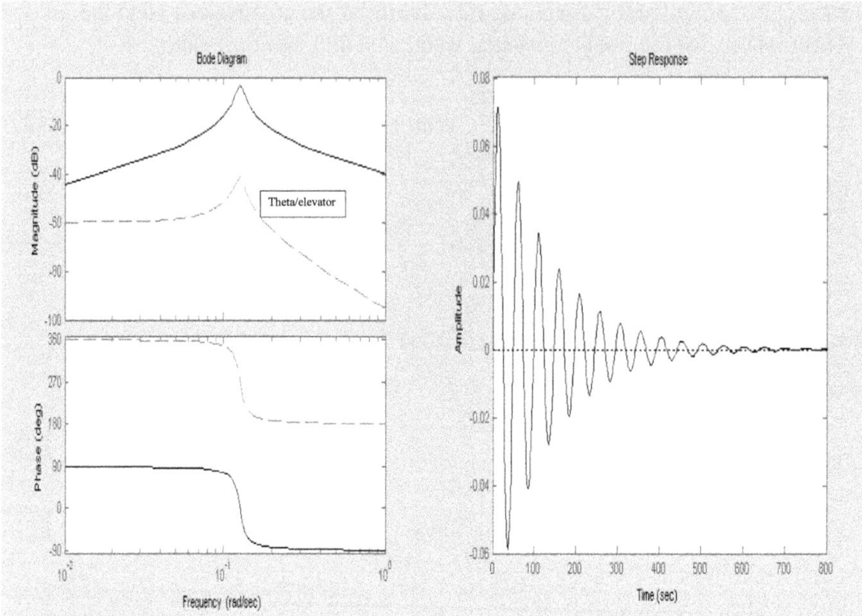

(a) Frequency responses for forward velocity/elevator (-); theta/elevator (--)

(b) unit step response for u

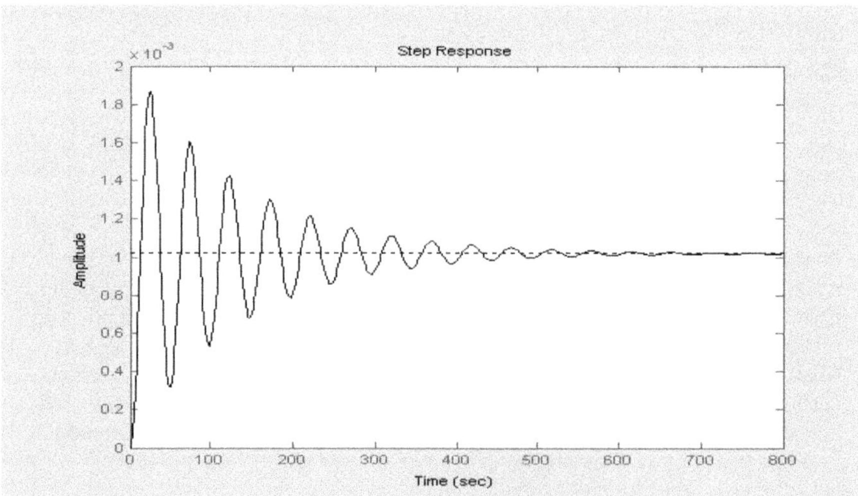

(c) unit step response for theta

FIGURE 2.6 Phugoid mode characteristics for a transport aircraft.

Generally, the phugoid frequency and damping increase in proportion to M_u. In summary, we see that free longitudinal motions of the aircraft comprise two oscillatory mode characteristics. This means that the forward velocity to elevator amplitude/magnitude (of the TF) will be much smaller at the natural frequency of the SP compared to that at the phugoid frequency. In the phugoid mode, there will not be any change in the AOA. The higher vertical acceleration has a pronounced effect on the altitude and its rate of change. In an approximate sense, the phugoid mode describes the long-term motion of the vehicle center of mass (C.G.), whereas the SP describes rotation about the C.G.; for practical purposes, in the phugoid motion, the 'dynamic' pitching, inertial, and damping moments are small compared to the 'static' pitching moment changes with speed and AOA. Conventionally, with $M_u = 0$, the phugoid was/is considered to involve fairly large oscillatory changes in forward speed u, pitch attitude/angle, and the altitude, with approximately constant AOA, and the static stability, M_α, being sufficiently large to decouple the phugoid and SP modes, and to maintain AOA perturbations small.

For the normally positive values of (Z_w/M_w), the phugoid frequency and damping increase proportionally to M_u; for sufficiently –ve values of M_u, ω_p^2 becomes –ve, and the phugoid mode is then characterized by two first-order TF (factors): one convergent, and the other a divergent mode, i.e. 'tuck' mode, so-called because as speed increases, the aircraft's nose has a tendency to 'tuck' under –ve M_u. The phugoid modal response is such that roughly, the sum of kinetic energy and the potential energy remains constant, mainly the change in the altitude, h, and the forward speed, u; similar energy play would be happening in all the dynamic systems, including say SP of an aircraft; however, here in the phugoid mode, it is more apparent.

In general, irrespective of the size and weight of an aircraft, its basic characteristics can be described in a very simple manner in terms of only few parameters, say two or four aerodynamic derivatives. This is made possible due to the fact the linear system theory is applicable to simplified dynamics and the related TF/control system concepts.

2.8.4.1.4 Lateral and Lateral-Directional Models and Modes

A general LD motion of an aircraft can be described by eliminating (more of the longitudinal state variable) $\dot{V}, \dot{\alpha}, \dot{q}$, and $\dot{\theta}$ equations from the 6DOF state model:

$$\dot{\beta} = \frac{g}{v}(\cos\beta\sin\phi\cos\theta + \sin\beta\cos\alpha\sin\theta - \sin\alpha\cos\phi\cos\theta\sin\beta)$$

$$+ p\sin\alpha - r\cos\alpha + q\frac{\bar{q}S}{mV}C_{Ywind} + \frac{T}{mV}\cos(\alpha + \sigma_T)\sin\beta$$

$$\dot{p} = \frac{1}{I_xI_z - I_{xz}^2}\left\{\bar{q}Sb(I_zC_l + I_{xz}C_n) - qr(I_{xz}^2 + I_z^2 - I_yI_z) + pqI_{xz}(I_x - I_y + I_z)\right\}$$

$$\dot{r} = \frac{1}{I_xI_z - I_{xz}^2}\left\{\bar{q}Sb(I_xC_n + I_{xz}C_l) - qrI_{xz}(I_x - I_y + I_z) + pq(I_{xz}^2 + I_x^2 + I_xI_y)\right\}$$

$$\dot{\phi} = p + q\tan\theta\sin\phi + r\tan\theta\cos\phi$$

$$\dot{\psi} = r\cos\phi\sec\theta + q\sin\phi\sec\theta$$

(2.138)

If the excursions (of the major longitudinal axis state variable) are small, one can neglect the longitudinal motion altogether. Also, we have

$$C_{Y\text{wind}} = C_Y \cos\beta + C_D \sin\beta \tag{2.139}$$

For small values of β, we obtain

$$C_{Y\text{wind}} = C_Y \tag{2.140}$$

Since the variable ψ does not appear in any of the equations for $\dot{\beta}, \dot{p}, \dot{r}$, and $\dot{\phi}$, the $\dot{\psi}$ equation is generally omitted from the analysis; using these assumptions in $\dot{\beta}$ equation, a simplified set to describe the LD motion can be written as

$$\begin{aligned}
\dot{\beta} &= \frac{g}{V}(\sin\phi\cos\theta + p\sin\alpha - r\cos\alpha + \frac{\bar{q}S}{mV}C_Y \\
\dot{p} &= \frac{1}{I_x I_z - I_{xz}^2}\left\{\bar{q}Sb(I_z C_l + I_{xz}C_n) - qr(I_{xz}^2 + I_z^2 - I_y I_z) + pqI_{xz}(I_x - I_y + I_z)\right\} \\
\dot{r} &= \frac{1}{I_x I_z - I_{xz}^2}\left\{\bar{q}Sb(I_x C_n + I_{xz}C_l) - qrI_{xz}(I_x - I_y + I_z) + pq(I_{xz}^2 + I_x^2 + I_x I_y)\right\} \\
\dot{\phi} &= p + q\tan\theta\sin\phi + r\tan\theta\cos\phi
\end{aligned} \tag{2.141}$$

One can find several approximate forms of the EOMs in literature [1–4]. The coefficients C_Y, C_l and C_n can be expressed in terms of stability and control derivatives using the Taylor series expansion. Expressing the side force, rolling, and yawing moments in terms of dimensional derivatives, the following state-space model can be used to describe aircraft LD motion for most applications:

$$\begin{bmatrix} \dot{\beta} \\ \dot{p} \\ \dot{r} \\ \dot{\varphi} \end{bmatrix} = \begin{bmatrix} \frac{Y_\beta}{u_0} & \frac{Y_p}{u_0} & \frac{Y_r}{u_0} - 1 & \frac{g\cos\theta_0}{u_0} \\ L_\beta & L_p & L_r & 0 \\ N_\beta & N_p & N_r & 0 \\ 0 & 1 & 0 & 0 \end{bmatrix} \begin{bmatrix} \beta \\ p \\ r \\ \varphi \end{bmatrix} + \begin{bmatrix} \frac{Y_{\delta_a}}{u_0} & \frac{Y_{\delta_r}}{u_0} \\ L_{\delta_a} & L_{\delta_r} \\ N_{\delta_a} & N_{\delta_r} \\ 0 & 0 \end{bmatrix} \begin{bmatrix} \delta_a \\ \delta_r \end{bmatrix} \tag{2.142}$$

Here, u_0 and θ_0 are the forward speed and pitch angle, respectively, under the steady-state condition. This state-space model can be used to obtain the LD TF characteristic equation, solving which for the eigenvalues will yield two real roots corresponding to the spiral mode and roll subsidence, while a pair of complex roots defines the DR mode. These three/four distinct LD modes are explained using the following example:

Example 2.8

The Douglas DC-8 aircraft state-space LD model [4] is given as

$$\begin{bmatrix}\dot{v}\\ \dot{p}\\ \dot{r}\\ \dot{\varphi}\end{bmatrix}=\begin{bmatrix}-0.1 & 0 & -468 & 32\\ -0.0058 & -1.232 & 0.397 & 0\\ 0.0028 & -0.0346 & -0.257 & 0\\ 0 & 1 & 0 & 0\end{bmatrix}\begin{bmatrix}v\\ p\\ r\\ \varphi\end{bmatrix}+\begin{bmatrix}0 & 13.48\\ -1.62 & 0.392\\ -0.0188 & -0.864\\ 0 & 0\end{bmatrix}\begin{bmatrix}\delta_a\\ \delta_r\end{bmatrix} \tag{2.143}$$

Obtain the all the eight TFs using MATLAB. Obtain the characteristic modes of the aircraft dynamics.

Solution 2.8

The TFs are obtained by [numail, denail] = ss2tf(a,b,c,d,1); [numrud,denrud] = ss2tf (a,b,c,d,2); for aileron input: sysva = tf(numail(1,:),denail); syspa = tf(numail(2,:), denail); sysr = tf(numail(3,:), denail); and dampsystpha = tf(numail(4,:),denail). For rudder input: sysvrud = tf(numrud(1,:), denail);sysprud = tf(numrud(2,:),denrud); sysrrud = tf(numrud(3,:),denrud); and systphrud = tf(numrud(4,:),denrud). These TFs are given in Table 2.1.

TABLE 2.1
Lateral Mode Transfer Functions for the Chosen Aircraft

	Aileron Control Input	Rudder Control Input
Side velocity	$-5.107\times10^{-15}s^3+8.798s^2-67.23s-13.56$ --- --------- $s^4+1.589s^3+1.78s^2+1.915s+0.01238$	$13.48s^3+424.4s^2+521.5s-7.752$ --- --------- $s^4+1.589s^3+1.78s^2+1.915s+0.01238$
Roll rate	$-1.62s^3-0.5858s^2-2.201s-3.123\times10^{-17}$ --- --------- $s^4+1.589s^3+1.78s^2+1.915s+0.01238$	$0.392s^3-0.2813s^2-1.865s-1.301\times10^{-16}$ --- --------- $s^4+1.589s^3+1.78s^2+1.915s+0.01238$
Yaw rate	$-0.0188s^3+0.03101s^2+0.003289s-0.1476$ --- ----------- $s^4+1.589s^3+1.78s^2+1.915s+0.01238$	$-0.864s^3-1.127s^2-0.05891s-0.1255$ --- --------- $s^4+1.589s^3+1.78s^2+1.915s+0.01238$
Roll angle	$4.441\times10^{-16}s^3-1.62s^2-0.5858s-2.201$ --- --------- $s^4+1.589s^3+1.78s^2+1.915s+0.01238$	$8.882\times10^{-16}s^3+0.392s^2-0.2813s-1.865$ --- -------- $s^4+1.589s^3+1.78s^2+1.915s+0.01238$

Next, we discuss the lateral EOMs referred to stability axes and written in terms of AOSS, with assumptions $Y_{\dot{v}} = Y_p = Y_r = L_{\dot{v}} = N_{\dot{v}} = 0$:

$$(s - Y_v)\beta - \frac{g}{U_0}\frac{p}{s} + r = Y_\delta^*\delta; \quad Y_\delta^* = Y_\delta / U_0$$

$$-L_\beta\beta + (s - L_p)p - (\frac{I_{xz}}{I_x}s + L_r)r = L_\delta\delta$$

$$-N_\beta\beta - (\frac{I_{xz}}{I_x}s + N_p)p + (s - N_r)r = N_\delta\delta$$

Also, $p = s\phi, r = s\varphi$; $a_{y_{cg}} = U_0\dot{\beta} - g(p/s) + U_0 r = \dot{v} - g\phi + U_0 r$, and the possible TFs are as follows: $\frac{\beta(s)}{\delta(s)}, \frac{\phi(s)}{\delta(s)}, \frac{r(s)}{\delta(s)}$, and $\frac{a_y(s)}{\delta(s)}$ for aileron and rudder inputs : δ_a, δ_r

2.8.4.1.5 Dutch Roll Mode

2DOF DR: It is a relatively lightly damped oscillatory mode that consists of primarily the sideslip and yawing motions. Neglecting equations for $\dot{p}$ and $\dot{\phi}$ in eqn. (2.142) and with no control inputs, the simplified form of the state-space model for the DR oscillatory mode can be expressed as [2]

$$\begin{bmatrix} \dot{\beta} \\ \dot{r} \end{bmatrix} = \begin{bmatrix} \frac{Y_\beta}{u_0} & \frac{Y_r}{u_0} - 1 \\ N_\beta & N_r \end{bmatrix} \begin{bmatrix} \beta \\ r \end{bmatrix}$$; Sometimes Y_r is neglected.; Sometimes Y_r is neglected. (2.144)

The characteristic equation for the above model will be

$$\lambda^2 - \left(\frac{Y_\beta + u_0 N_r}{u_0}\right)\lambda + \left(\frac{Y_\beta N_r - N_\beta Y_r + u_0 N_\beta}{u_0}\right) = 0 \qquad (2.145)$$

Solving equation (2.144) for the eigenvalues of the characteristic equation yields the following expressions for the natural frequency and damping ratio for this oscillatory mode:

$$\text{Frequency } \omega_{n_{DR}} = \sqrt{\frac{Y_\beta N_r - N_\beta Y_r + N_\beta u_0}{u_0}} \qquad (2.146)$$

$$\text{Damping ratio } \zeta_{DR} = -\left(\frac{Y_\beta + N_r u_0}{u_0}\right)\frac{1}{2\,\omega_{n_{DR}}} \qquad (2.147)$$

Even though $N_\beta + Y_v N_r \rightarrow 0$, it is still possible to have +ve stiffness and a finite DR frequency if I_{xz} and L_β are of the same sign (i.e. for normally −ve L_β and nose-up inclination of the principal axis of inertia to the direction of the flight). Such situations are not uncommon for HAOA conditions, where a negative N_β, e.g. due to fin immersion in the wing/body wake, can be stabilized by overriding negative values of I_{xz} and L_β. The DR damping is almost always +ve because usually both the derivatives are negative (see the numerator of the bracketed term of (5.109)). Thus, the possible motion of 2DOF simplified equations must therefore be either a damped (or overdamped) oscillation or a subsidence/divergence combination.

Example 2.9

The state-space model of the DR mode of a supersonic fighter aircraft is given as

$$\begin{bmatrix} \dot{\beta} \\ \dot{r} \end{bmatrix} = \begin{bmatrix} \dfrac{Y_\beta}{u_0} & \dfrac{Y_r}{u_0} - 1 \\ N_\beta & N_r \end{bmatrix} \begin{bmatrix} \beta \\ r \end{bmatrix} + \begin{bmatrix} Y_{\delta_a}/u_0 & Y_{\delta_r}/u_o \\ N_{\delta_a} & N_{\delta_r} \end{bmatrix} \begin{bmatrix} \delta_a \\ \delta_r \end{bmatrix} \tag{2.148}$$

With the derivatives incorporated into this model, we obtain

$$\begin{bmatrix} \dot{\beta} \\ \dot{r} \end{bmatrix} = \begin{bmatrix} -0.139 & 2.218/75 - 1 \\ 1.125 & -0.571 \end{bmatrix} \begin{bmatrix} \beta \\ r \end{bmatrix} + \begin{bmatrix} 0.056/75 & 1.134/75 \\ 0.083 & -0.343 \end{bmatrix} \begin{bmatrix} \delta_a \\ \delta_r \end{bmatrix} \tag{2.149}$$

Obtain the corresponding TFs.

Solution 2.9

Following the procedure outlined in previous examples, four TFs are obtained (see Table 2.2).

The frequency responses are shown in Figure 2.7. The unit step responses for sideslip and roll rate are shown in Figure 2.8; the lightness of the damping is evident from these responses.

TABLE 2.2
DR TFs for the Supersonic Fighter Aircraft

	Aileron Control Input	Rudder Control Input	DR Frequency and Damping Ratio
Beta TF	0.0007467 s − 0.08012	0.01512 s + 0.3415	1.08 rad/s
	------------------------------	---------------------------	0.328
	s² + 0.71 s + 1.171	s² + 0.71 s + 1.171	
Roll rate TF	0.083 s + 0.01238	−0.343 s − 0.03067	
	---------------------------	---------------------------	
	s² + 0.71 s + 1.171	s² + 0.71 s + 1.171	

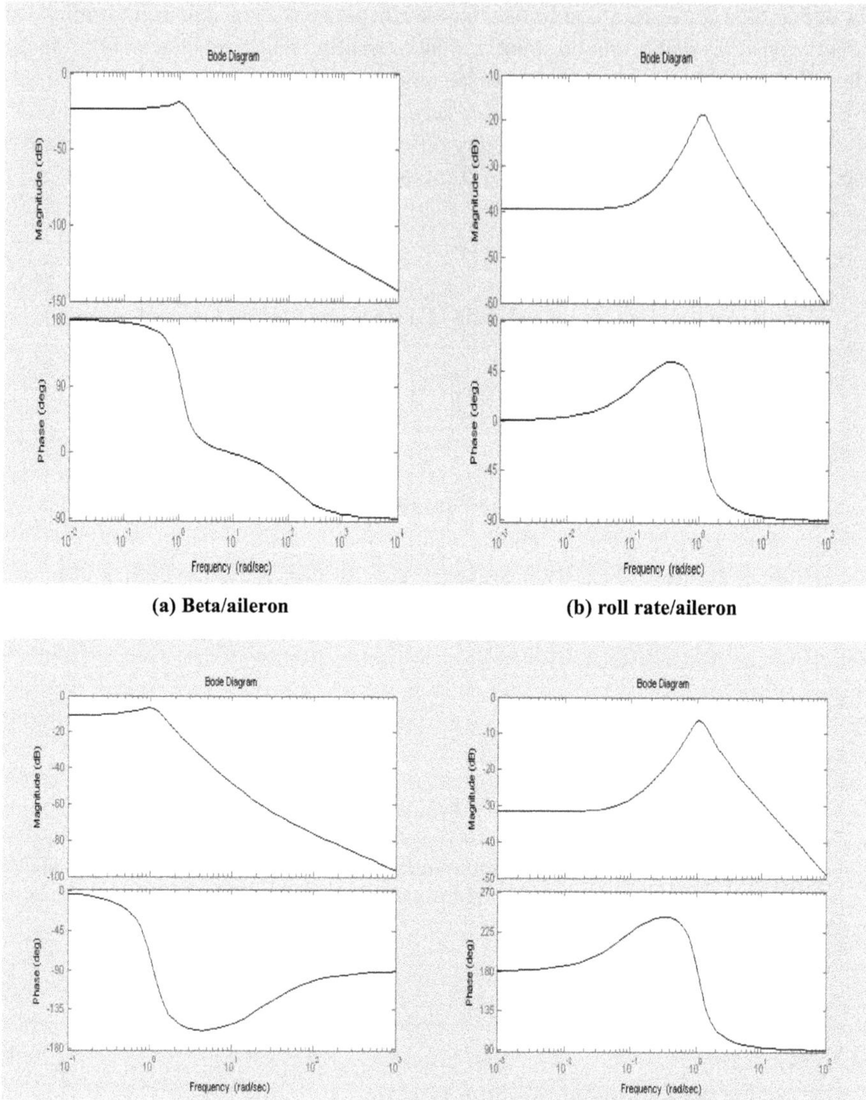

(a) Beta/aileron **(b) roll rate/aileron**

(c) Beta/Rudder **(d) roll rate/rudder**

FIGURE 2.7 DR mode frequency responses of a fighter aircraft.

3DOF DR: Although the rudder TFs are well matched, the 2DOF approximation may depart reasonably from the complete 3DOF case, which is given with its characteristic equation as follows:

$$
\begin{aligned}
&(s - Y_v)\beta + r = Y_\delta^* \delta; \quad Y_\delta^* = Y_\delta / U_0 \\
&-L_\beta \beta + s(s - L_p)\phi = L_\delta \delta \qquad ; s(s - L_p)[s^2 + (-Y_v - N_r)s + (N_\beta + Y_v N_r)] \\
&-N_\beta \beta + (s - N_r)r = N_\delta \delta
\end{aligned}
$$

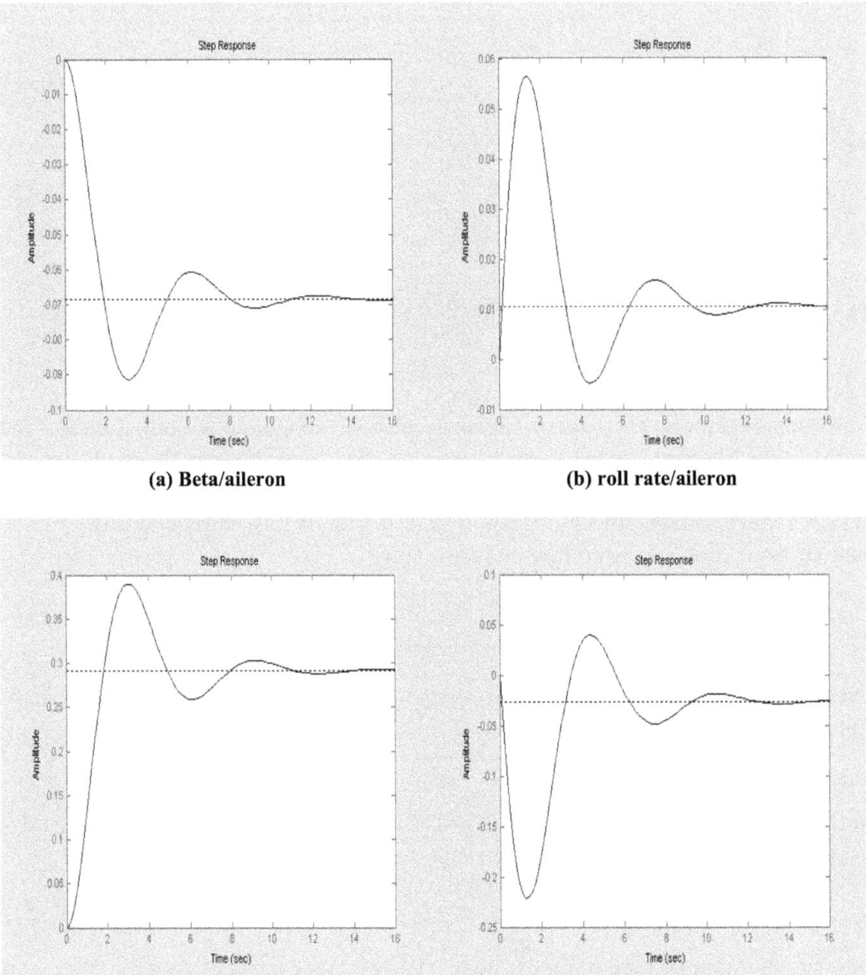

(a) Beta/aileron (b) roll rate/aileron

(c) Beta/rudder (d) roll rate/rudder

FIGURE 2.8 DR mode unit step responses of a fighter aircraft.

We see the following modes from the characteristic equation: (i) s → the spiral mode, (ii) (s-L_p) → roll subsidence mode, and (iii) quadratic DR mode that is the same as in 2DOF. The TFs obtained are $\frac{\beta(s)}{\delta(s)}, \frac{r(s)}{\delta(s)}, and \frac{\phi(s)}{\delta(s)}$.

2.8.4.1.6 3DOF Spiral and Roll Subsidence Modes

For both the modes, the sideslip motions are relatively small and the term (s-Y_v)β for the spiral mode is relatively negligible; hence, we obtain along with its characteristics equation

$$-\frac{g}{U_0}\phi + r = Y_\delta^* \delta; \quad Y_\delta^* = Y_\delta / U_0$$

$$-L_\beta \beta + s(s - L_p)\phi - (\frac{I_{xz}}{I_x} s + L_r) r = L_\delta \delta$$

$$-N_\beta \beta - (\frac{I_{xz}}{I_x} s + N_p) s\phi + (s - N_r) r = N_\delta \delta$$

$$s^2 + [-L_p + \frac{L_\beta}{N_\beta}\{(\frac{I_{xz}}{I_x} s + N_p) - g / U_0\}]s$$

$$+ g / U_0 (\frac{L_\beta}{N_\beta} N_r - L_r)$$

The possible TFs are $\frac{\beta(s)}{\delta(s)}$, $\frac{r(s)}{\delta(s)}$, and $\frac{\phi(s)}{\delta(s)}$ for aileron and rudder pedal inputs.

2.8.4.1.7 *Spiral Mode*

One of the real roots, having a small value (relatively long time period), indicates the spiral mode. The root can have a negative or positive value, making the mode convergent or divergent. This mode is dominated by rolling and yawing motions; sideslip is almost nonexistent; and this mode could be unstable, yet it is nearly coordinated. The characteristic root λ for spiral mode is given by

$$\lambda = \frac{L_\beta N_r - L_r N_\beta}{L_\beta} \tag{2.150}$$

Increasing L_β (dihedral effect) or N_r (yaw damping) will make the spiral mode more stable.

2.8.4.1.8 *Roll Mode*

In this mode, the dominant motion is roll. It is a highly damped mode with a relatively short time period. The characteristic root λ for the roll mode is given by

$$\lambda = L_p \tag{2.151}$$

Here, L_p is the roll damping derivative. The roll subsidence mode is a first-order convergent mode of relatively short time constant. In the roll mode, there is a combination of the roll damping moment and the roll moment of inertia so that the mode is conventionally considered to be essentially of single DOF.

In summary, the DR is the dominant characteristics of any aircraft. However, its dominance in the roll angle TF is very small. This is because a quadratic numerator in the roll – aileron TF – almost cancels the DR denominator. Thus, the DR excited by the aileron is more predominant in roll rate and AOSS. The rudder contribution to the roll rate is small. The DR damping is usually +ve. So, the 2DOF DR mode would be a damped oscillation or worst-case divergence/subsidence. The 3DOF DR approximation would incorporate roll angle as an additional state variable. This will then have three roots: the spiral mode ('s'), the roll subsidence mode, and DR. The 3DOF spiral and roll subsidence approximation can also be made. The roll subsidence mode is dominant in the rolling motion. In the spiral mode, the rolling and yawing motions predominate, mostly unstable; the mode has a very large time constant. It is a coordinated rolling and yawing motion mode.

2.8.4.2 Missile Aerodynamic TFs

It has already been emphasized that linearization of the EOMs can be carried out only if the small perturbation assumption is not violated. The same is true in the case of a missile; linearity constraint can be invoked if the changes in the incidence angles and angular body rates are small. To design a control system for a missile, the procedure is to define an operating point (in terms of Mach and altitude) and consider the aerodynamic derivatives pertaining to the operating point to remain unchanged within a small region of the selected flight condition. The control system is then designed to meet the specific requirements of system phase lag, damping, and bandwidth. In this manner, several test conditions are investigated. Detailed analysis is carried out by also considering the configuration changes due to change in mass and shift in C.G. as the fuel gets consumed. The force equation along the x-axis is generally omitted because it neither affects the roll, pitch, or yaw motion. We neglect the gravity terms and consider only the Y-force, Z-force, and the equations for angular accelerations, and then control equations for the missile can be expressed as [5]

$$\begin{aligned}
&\tilde{Y} = \dot{v} + rU = \tilde{Y}_v v + \tilde{Y}_r r + \tilde{Y}_{\delta_r}\delta_r && \Rightarrow \{ma_y = m(\dot{v} + ru)\} \\
&\tilde{Z} = \dot{w} - qU = \tilde{Z}_w w + \tilde{Z}_q q + \tilde{Z}_{\delta_e}\delta_e && \Rightarrow \{ma_z = m(\dot{w} - qw)\} \\
&\dot{p} = \tilde{L}_p p + \tilde{L}_{\delta_a}\delta_a && \Rightarrow \{I_x\dot{p} = L = (L_p p + L_{\delta_a}\delta_a)\} \Rightarrow \{\dot{p} = l_p p + l_{\delta_a}\delta_a\} \\
&\dot{q} = \tilde{M}_w w + \tilde{M}_q q + \tilde{M}_{\delta_e}\delta_e && \Rightarrow \{\dot{q} = m_w w + m_q q + m_{\delta_e}\delta_e\} \\
&\dot{r} = \tilde{N}_v v + \tilde{N}_r r + \tilde{N}_{\delta_r}\delta_r && \Rightarrow \{\dot{r} = n_v v + n_r r + n_{\delta_r}\delta_r\}
\end{aligned} \tag{2.152}$$

Here, $\tilde{Y}$ and $\tilde{Z}$ denote the specific forces (accelerations) given by

$$\begin{aligned}
\tilde{Y} &= Y/m \\
\tilde{Z} &= Z/m
\end{aligned} \tag{2.153}$$

and $\tilde{Y}_v, \tilde{Y}_r, \tilde{Y}_{\delta_r}, \tilde{Z}_w, \ldots, \tilde{N}_{\delta_r}$ are specific derivatives (also, $I_y = I_z$). Full values of these derivatives can be computed given the information on mass and inertia characteristics of the missile. The aileron deflection in a missile is given by one of the following formulae:

$$\frac{1}{4}(\delta_1 + \delta_2 + \delta_3 + \delta_4); \frac{1}{2}(\delta_1 + \delta_3); \frac{1}{2}(\delta_2 + \delta_4) \tag{2.154}$$

If only two surfaces act in a differential mode, then we have elevator deflection: $\frac{1}{2}(\delta_1 - \delta_2)$; rudder deflection: $\frac{1}{2}(\delta_2 - \delta_4)$. Some of the aerodynamic TFs obtained by taking Laplace of eqn. (2.152) are given as follows[5]:

2.8.4.2.1 Roll Rate/Aileron $\left(\frac{p}{\delta_a}\right)$

Consider the $\dot{p}$ equation, taking Laplace, we have

$$sp - \tilde{L}_p p = \tilde{L}_{\delta_a}\delta_a \tag{2.155}$$

Or

$$\frac{p(s)}{\delta_a(s)} = \frac{\tilde{L}_{\delta_a}}{s - \tilde{L}_p} = \frac{-\tilde{L}_{\delta_a} / \tilde{L}_p}{1 + \tau s} = \frac{\left(-l_{\delta_a}\right) / l_p}{1 + s\left(-1 / l_p\right)} \tag{2.156}$$

Then, the steady-state gain equals to $-\tilde{L}_{\delta_a} / \tilde{L}_p$ and time constant τ equals to $-1 / \tilde{L}_p$.

2.8.4.2.2 Lateral Acceleration/Rudder ($\tilde{Y} / \delta_r$)

Simplifying $\tilde{Y}$ and $\dot{r}$ equations in eqn. (5.114) by eliminating r and v and neglecting the $\tilde{Y}_r$ derivative, which is generally very small, we obtain the following expression for TF for $\tilde{Y} / \delta_r$:

$$\frac{\tilde{Y}(s)}{\delta_r(s)} = \frac{s^2\tilde{Y}_{\delta_r} - s\tilde{N}_r\tilde{Y}_{\delta_r} + U\left(\tilde{N}_v\tilde{Y}_{\delta_r} - \tilde{N}_{\delta_r}\tilde{Y}_v\right)}{s^2 - s\left(\tilde{N}_r + \tilde{Y}_v\right) + \left(U\tilde{N}_v + \tilde{N}_r\tilde{Y}_v\right)} \Rightarrow \left\{\frac{a_y(s)}{\delta_r(s)} = \frac{s^2 y_{\delta_r} - s n_r y_{\delta_r} + U\left(n_v y_{\delta_r} - n_{\delta_r} y_v\right)}{s^2 - s\left(n_r + y_v\right) + \left(U n_v + n_r y_v\right)}\right\} \tag{2.157}$$

If ω_n denotes the undamped natural frequency and ς denotes the damping ratio, then we have

$$\omega_n^2 = U\tilde{N}_v + \tilde{N}_r\tilde{Y}_v \equiv \left(U n_v + n_r y_v\right); 2\varsigma\omega_n = \tilde{N}_r + \tilde{Y}_v \tag{2.158}$$

In the expression for ω_n, the term $U\tilde{N}_v$ is generally much larger than $\tilde{N}_r\tilde{Y}_v$; hence, we have the weathercock frequency:

$$\omega_n^2 = \frac{\text{the restoring moment /unit angualr deflection} (= N_\beta)}{\text{moment of intertia about C.G.} (= I_z)} = \frac{UN_v}{I_z} = Un_v$$

The steady-state gain is given by $\dfrac{U\left(\tilde{N}_v\tilde{Y}_{\delta_r} - \tilde{N}_{\delta_r}\tilde{Y}_v\right)}{\omega_n^2}$.

2.8.4.2.3 Yaw Rate/Rudder (r/δ_r)

The TF (also called the body rate TF) r/δ_r can be obtained by eliminating v from $\tilde{Y}$ and $\dot{r}$ equations in eqn. (2.152):

$$\frac{r(s)}{\delta_r(s)} = \frac{s\tilde{N}_{\delta_r} + \left(\tilde{N}_v\tilde{Y}_{\delta_r} - \tilde{N}_{\delta_r}\tilde{Y}_v\right)}{s^2 - s\left(\tilde{N}_r + \tilde{Y}_v\right) + \left(U\tilde{N}_v + \tilde{N}_r\tilde{Y}_v\right)} \Rightarrow \frac{s n_{\delta_r} + \left(n_v y_{\delta_r} - n_{\delta_r} y_v\right)}{s^2 - s\left(n_r + n_v\right) + \left(U n_v + n_r y_v\right)}$$

$$\frac{r}{\delta_r} = \frac{-(n_{\delta_r} y_v - n_v y_{\delta_r})(T_i s + 1)}{\text{denominator}}; \; T_i = \frac{n_{\delta_r}}{-n_{\delta_r} y_v + n_v y_{\delta_r}} \approx -\frac{1}{y_v} \tag{2.159}$$

Comparing eqns. (2.157) and (2.159), one observes that ω_n and ς are the same for both the TFs. We see that the body rate is the phase advanced flight path rate, and the body rate (with 'U') is the phase advanced lateral acceleration (latex); and hence, a feedback from a rate gyro in an autopilot tends to damp the weathercock mode.

2.8.4.2.4 Sideslip/Rudder (β / δ_r)

Eliminating r from $\tilde{Y}$ and $\dot{r}$ equations in eqn. (2.152) and, for small incidence angles, assuming $\beta = v / U$, the TF for β / δ_r can be expressed as

$$\frac{\beta}{\delta_r} = \frac{s\dfrac{\tilde{Y}_{\delta_r}}{U} - \left(\dfrac{\tilde{N}_r \tilde{Y}_{\delta_r}}{U} + \tilde{N}_{\delta_r}\right)}{s^2 - s\left(\tilde{N}_r + \tilde{Y}_v\right) + \left(U\tilde{N}_v + \tilde{N}_r \tilde{Y}_v\right)} \tag{2.160}$$

Once again, the ω_n and ς for the above TF is same as that for the TFs expressed in eqns. (2.157) and (2.159).

Example 2.10

For a surface-to-air missile (SAM) with rear controls, the yaw aerodynamic derivatives for a certain flight condition (Mach =1.4, H=1.5 Km, and U=467 m/s) [5] are given as

$$Y_v = -2.74;\ N_v = 0.309;\ Y_\delta = 197;\ N_\delta = -534;\ N_r = -2.89.$$

The other details are length = 2 m, mass = 53 kg, moment of inertia, I_z = 13.8 kg m².

The missile latex (lateral acceleration) rudder control TF is given as

$$\frac{a_y(s)}{\delta(s)} = \frac{Y_\delta s^2 - Y_\delta N_r s - U(N_\delta Y_v - N_v Y_\delta)}{s^2 - (Y_v + N_r)s + Y_v N_r + U N_v} \tag{2.161}$$

With the values given, obtain the missile TF, its characteristic modes, frequency responses, and step input responses.

Solution 2.10

The following TF is obtained:

$$\frac{a_y(s)}{\delta(s)} = \frac{197s^2 + 569.33s - 467(1463.16 - 60.87)}{s^2 + 5.63s + (7.92 + 144.3)} \tag{2.162}$$

The static margin= $\dfrac{N_v}{Y_v} = \dfrac{n_v I_z}{y_v m}$ = 0.029 or 1.5% of the length of the missile; a large static margin would result in (i) a small steady-state gain, (ii) a high weathercock frequency, and (iii) a very low damping ratio. The latex dynamic mode is second-order oscillatory: −2.815 ±− 12.0123i; natural frequency = 12.4 rad/s; the damping ratio = 0.228 (this will decrease at higher altitudes); the steady-state gain is = −4300; this means that if the aerodynamics were assumed to be linear, then

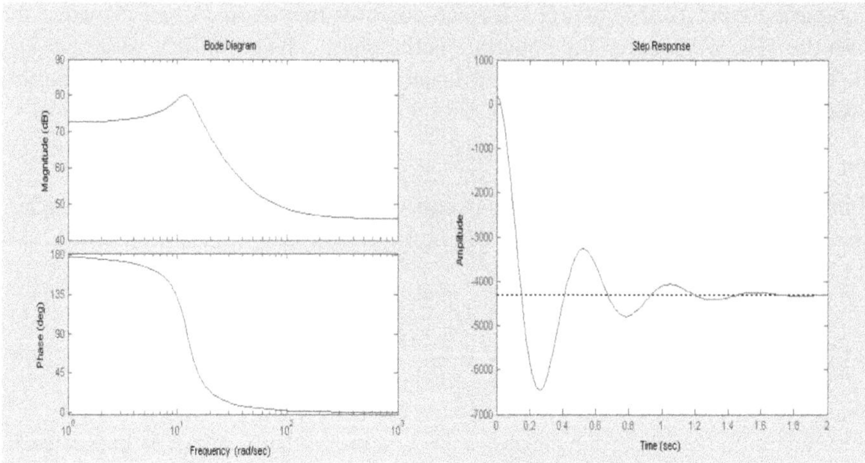

FIGURE 2.9 Latex frequency response and unit step input response for a SAM.

0.1 rad of rudder deflection would produce a lateral acceleration in the $-y$ direction of 430 m/s^2 or 4.3 g. The numerator factors are $-655{,}000(1+s/59.1)(1-s/56.2)$; one can deduce from these factors that the phase contribution of the first will be canceled by the second one; however, these zeros will become be recognizable in the range of frequencies of 60 rad/s and would be troublesome in the design of an autopilot, since these would prevent the continuous attenuation with frequency (as is usually observed in any second-order oscillatory dynamic system); all the time the rudder is moving, there will be a lateral acceleration whatever the frequency beyond 60 rad/s. The frequency response and the unit step rudder response for the latex are shown in Figure 2.9. The T_i=0.365 sec is the time constant of the zero of the yaw rate/rudder TF.

2.8.4.3 Rotorcraft Linear Modelling

A general state-space form of the model to describe rotorcraft dynamics may be written as

$$\begin{aligned} \dot{x} &= f[x(t), u(t), \Theta] + g_1 w(t); \qquad y = h[x(k), u(k), \Theta] \\ z(k) &= h[x(k), u(k), \Theta] + g_2 \upsilon(k); \qquad x(0) = x_0 \end{aligned} \tag{2.163}$$

Equation (2.163) is the mixed form of continuous-discrete state and observation equations, wherein Θ is the vector of unknown stability and control derivatives, g_1 is the state noise vector/matrix, and g_2 is the measurement noise vector/matrix.; the linear form can be expressed as

$$\begin{aligned} \dot{x} &= Ax(t) + Bu(t) + G_1 w(t) \\ z(k) &= Hx(k) + Du(k) + G_2 \upsilon(k); \quad x(0) = x_0 \end{aligned} \tag{2.164}$$

A and B are the matrices containing the stability and control derivatives, and C and D are the matrices that relate the measured quantities to the rotorcraft states and control variables. The 6DOF EOMs for modelling rotorcraft dynamics were discussed in Chapter 3; from that and from the state-space model described in eqn. (5.126), it is evident that the state, measurement, and control vectors will include the following motion variables:

$$
\begin{aligned}
x &= [u, v, w, p, q, r, \phi, \theta, h] \\
y &= [u_m, v_m, w_m, p_m, q_m, r_m, \phi_m, \theta_m, h_m, a_{xm}, a_{ym}, a_{zm}, \dot{p}_m, \dot{q}_m, \dot{r}_m,] \\
u &= [\delta_{lon}, \delta_{lat}, \delta_{col}, \delta_{ped}]
\end{aligned}
\tag{2.165}
$$

Assuming small variations in u, v, w, ϕ and θ, the linearized form in the state-space formulation can be expressed as [15,16]:

$$
\begin{bmatrix} \dot{u} \\ \dot{v} \\ \dot{w} \end{bmatrix} = \begin{bmatrix} \tilde{X} \\ \tilde{Y} \\ \tilde{Z} \end{bmatrix} + g \begin{bmatrix} -\sin\theta_0 - \Delta\theta\cos\theta_0 \\ \Delta\phi\cos\theta_0 \\ \cos\theta_0 - \Delta\theta\sin\theta_0 \end{bmatrix} + \begin{bmatrix} -w_0 q + v_0 r \\ -u_0 r + w_0 p \\ -v_0 p + u_0 q \end{bmatrix} \tag{2.166}
$$

Here, the subscript '0' represents the quantity at the reference condition. We have already studied that for a fixed-wing aircraft, decoupled longitudinal and lateral model equations can be used to model the aircraft dynamics without much loss of accuracy. In the case of helicopters, the degree of coupling between the longitudinal and LD motion is generally stronger, and therefore, a 6DOF model is preferred [6]. The dynamics of the main rotor in helicopter, however, introduces additional complexities, and it becomes necessary to go for the higher order models.

2.8.4.3.1 Rotor Plus Body Models

As mentioned above, rotorcraft applications require at least a 6DOF model to represent the rigid body motion. Such models are generally adequate to describe helicopter dynamics associated with low frequency (such as phugoid). But the prediction of flight dynamic behavior from such models in the high-frequency range is poor. This is due to the absence of the rotor DOFs in the model which affect the helicopter motion. Some improvements can be obtained by approximating rotor dynamics via equivalent time delay effects [21]. However, this approach cannot adequately represent the rotor influences [22]. The other alternative is to augment the 6DOF models with additional DOFs that explicitly model the rotor dynamic effects, e.g. longitudinal and lateral flapping and coning. An extended model of 8DOF can be obtained from eqn. (2.165) by including longitudinal and lateral flapping in the state and observation equations:

$$
\begin{aligned}
x^T &= (u, v, w, p, q, r, \phi, \theta, \psi, h, a_{1s}, b_{1s}) \\
y^T &= (a_{xm}, a_{ym}, a_{zm}, p_m, q_m, r_m, \phi_m, \theta_m, u_m, v_m, w_m, \dot{p}_m, \dot{q}_m, \dot{r}_m, \psi_m, h_m, a_{1s}, b_{1s}) \\
u^T &= (\delta_{lon}, \delta_{lat}, \delta_{col}, \delta_{ped})
\end{aligned}
\tag{2.167}
$$

In comparison with the 6DOF rigid body model, this extended model structure can provide an in-depth insight into the helicopter dynamics. However, the augmented model will have larger number of unknowns to be estimated which can lead to serious convergence problems and make identification difficult. The rotorcraft MM should be such that it gives a realistic representation of the helicopter behavior, and at the same time, it is simple and mathematically tractable. To this end, it is essential to determine the lowest order model that would best fit the flight test data. Linear models of different order, which may include body and rotor DOFs, can be tried out in identification. Any one of the following model structures, consistent with the frequency range of interest, can be selected for characterizing rotorcraft dynamics:

i. Body longitudinal dynamics alone – fourth-order model (3DOF).
ii. Body lateral dynamics alone – fourth-order model (3DOF).
iii. Body coupled dynamics – eighth-order model (6DOF).
iv. Body dynamics with first-order flapping dynamics (longitudinal and lateral tip path plane tilts) – 11th-order model (9DOF).
v. Body dynamics with rotor flapping dynamics – 15th-order model (10DOF).
vi. Body dynamics with rotor flapping and lead-lag dynamics – 21st-order model (13DOF).
vii. Body dynamics with rotor flapping and lead-lag dynamics and with inflow dynamics – 24th-order model (16DOF).

2.8.4.3.2 *Stability Derivative Models*

The linearized equations for the rigid body and rotor dynamics, in the state-space form, can be expressed as

$$\begin{aligned}\dot{x}_b &= A_b x_b + A_{rb} x_r + D_b u \\ \dot{x}_r &= A_r x_r + A_{rb} x_b + D_r u\end{aligned} \tag{2.168}$$

In (2.168), x_b represents the basic 6DOF rigid body state vector, x_r represents the higher order dynamics that may include flapping and lead-lag dynamics, and u represents the control input vector, with the following descriptions:

$$\begin{aligned}x_b &= (u, v, w, p, q, r, \phi, \theta, \psi, h) \\ x_r &= (u, v, w, p, q, r, \phi, \theta, \psi, h, a_{1s}, b_{1s}, \ldots) \\ u &= (\delta_{\text{lon}}, \delta_{\text{lat}}, \delta_{\text{col}}, \delta_{\text{ped}})\end{aligned} \tag{2.169}$$

The matrices A_b and D_b denote the rigid body dynamics in the absence of rotor dynamics, while the matrices A_r and D_r represent the rotor flapping dynamics. The matrices A_{rb} and A_{br} represent the rotor/body and body/rotor coupling. Taking Laplace transform of eqn. (5.130) and solving for x_r, we get

$$\begin{aligned}sx_r &= A_{rb} x_b + A_r x_r + D_r u \\ (sI - A_r) x_r &= A_{rb} x_b + D_r u\end{aligned} \tag{2.170}$$

$$x_r = (sI - A_r)^{-1} A_{rb} x_b + (sI - A_r)^{-1} D_r u$$

Expanding $(sI - A_r)^{-1}$ in a series form, we get

$$(sI - A_r)^{-1} = -A_r^{-1} [I + A_r^{-1} s + A_r^{-2} s^2 + A_r^{-3} s^3 + \cdots] \quad (2.171)$$

Substituting the series expansion for $(sI - A_r)^{-1}$ in eqn. (5.132) and taking inverse Laplace yields, the following solution for x_r

$$x_r = -A_r^{-1} A_{rb} x_b - A_r^{-2} A_{rb} \dot{x}_b - A_r^{-3} A_{rb} \ddot{x}_b - \ldots - A_r^{-1} D_r u - A_r^{-2} D_r \dot{u} - A_r^{-3} D_r \ddot{u} - \ldots \quad (2.172)$$

Equation (2.172) contains the control input and body states along with their higher order derivatives. At equilibrium, the higher order derivatives vanish and the rotor state x_r can be expressed as

$$x_r = -A_r^{-1} A_{rb} x_b - A_r^{-1} D_r u \quad (2.173)$$

Substituting x_r from eqn. (2.173) into eqn. (2.168) yields the quasi-static model for x_b:

$$\dot{x}_b = A_b x_b + A_{br}[-A_r^{-1} A_{rb} x_b - A_r^{-1} D_r u] + D_b u \quad (2.174)$$

Rearranging, we get

$$\dot{x}_b = [A_b - A_{br} A_r^{-1} A_{rb}] x_b + [D_b - A_{br} A_r^{-1} D_r] u \quad (2.175)$$

$$\dot{x}_b = A_{QS} x_b + B_{QS}\, u \quad (2.176)$$

In (2.176), A_{QS} and B_{QS} represent the quasi-static state and control matrices that comprise the rotor and rigid body stability and control derivatives.

2.8.4.3.3 Rotor Response Decomposition Models

These models are formulated by decomposing the response of higher order states into two components: (i) self-induced response and (ii) response from the rigid body motion. For example, the rotor state can be decomposed as

$$x_r = x_{r_{MR}} + x_{r_B} \quad (2.177)$$

Here, $x_{r_{MR}}$ indicates the response due to main rotor dynamics, and x_{r_B} is the response due to body dynamics; further, we get

$$\dot{x}_{r_{MR}} + \dot{x}_{r_B} = A_r (x_{r_{MR}} + x_{r_B}) + A_{rb} x_b + D_r u \quad (2.178)$$

In the decomposed form, we have

$$\begin{aligned} \dot{x}_{r_{MR}} &= A_r x_{r_{MR}} + D_r u \\ \dot{x}_{r_B} &= A_r x_{r_B} + A_{rb} x_b \end{aligned} \quad (2.179)$$

The approach used for obtaining the expression for x_r in eqn. (5.134) can be applied to find a solution to eqn. (5.141); the contribution to the rotor state from the body motion is then given by

$$x_{r_B} = -A_r^{-1}A_{rb}x_b - A_r^{-2}A_{rb}\dot{x}_b - A_r^{-3}A_{rb}\ddot{x}_b - \ldots \quad (2.180)$$

Assuming that the body dynamics are low-frequency dynamics compared to the rotor dynamics, the higher order derivatives in eqn. (2.180) can be neglected to obtain the instantaneous rotor response to body motion; then, we have

$$x_r = x_{r_{\mathrm{MR}}} - A_r^{-1}A_{rb}x_b \quad (2.181)$$

Substituting x_r from eqn. (2.181) into the expression for $\dot{x}_b$ in eqn. (5.130), we obtain

$$\dot{x}_b = A_b x_b + A_{br}[x_{r_{\mathrm{MR}}} - A_r^{-1}A_{rb}x_b] + D_b u; \text{ or } \dot{x}_b = [A_b - A_r^{-1}A_{rb}]x_b + A_{br}x_{r_{\mathrm{MR}}} + D_b u \quad (2.182)$$

The rotor response decomposition model can now be expressed as

$$\begin{aligned} \dot{x}_b &= A_{QS'}x_b + A_{br}x_{r_{\mathrm{MR}}} + D_b u \\ \dot{x}_{r_{\mathrm{MR}}} &= A_r x_{r_{\mathrm{MR}}} + D_r u \end{aligned} \quad (2.183)$$

The elements of the matrices in eqn. (2.183) can be identified either from WT testing or from flight testing.

2.9 UAV DYNAMICS

The unmanned/uninhibited aerial vehicles (UAVs) could perform the tasks that are more difficult to do by the manned aircraft, e.g. chemical and/or biological warfare missions. UAVs can also be used to test and verify technologies for UAV and for general aviation aircraft. Utilization and operation of UAV entails less risk to humans in combat. UAVs can also be freed from the limitation of human operators. This would improve the performance of UAVs. UAVs utilize various advanced technologies: (i) data/signal processing, (ii) off-board/on-board sensors, (iii) communications links, and (iv) integrated avionics and flight control. UAVs could be remotely piloted or autonomous with an autopilot. They can operate at altitudes above 70,000 feet, have higher maneuverability and longer endurance, and can carry optical sensors and radars. They, however, need large communications bandwidths. The UAVs could also be equipped with aircraft like controls: elevators, ailerons, rudders, flaps, extensions of wings, and undercarriage.

In an autonomous UAV, human has no direct control over the vehicle. One form of autonomy is the radio-controlled model airplane. The most complicated UAV would be that utilizes a rule-based fuzzy logic to detect, identify, and attack a moving target. Flight dynamic modelling of UAVs presents some unique challenges. UAVs are lighter compared to manned aircraft, and hence, UAVs have higher natural frequencies. To simulate the dynamic motion, the nonlinear fully coupled EOMs of an aircraft can be used. The simplified EOM discussed in this chapter can also be utilized to carry out preliminary studies relating to UAV performance evaluation. Any one of the following model forms can be selected for UAV dynamic analysis:

a. nonlinear fully coupled equations as described in eqns. (2.22), (2.23), and (2.27)
b. linear coupled, eqn. (2.108)
c. nonlinear decoupled, eqns. (2.109) and (2.138)
d. linear decoupled, eqns. (2.121), (2.131), (2.142)

Working with the fully coupled EOMs in option (a) can be time-consuming and complex. Therefore, for control system design, the designers usually make use of linearized and/or decoupled models given in option (b), (c), and (d). Of course, one has to keep in mind the limitations of these models. The linearized EOMs are valid in the small range around the trim point when excursions from the reference flight condition are small. Decoupling is also valid when there is negligible interaction between the longitudinal and LD motion. In using the said set of EOM for UAVs, the following assumptions are made:

(a) It is a rigid body, and flexibility effects are not considered; (b) it has a conventional configuration (i.e. aft tail); (c) symmetry about the *XZ* plane is assumed; and (d) the effect of thrust on the LD motion is neglected. Ref. [17] gives more details about the dynamic modelling of UAVs.

2.10 MAV DYNAMICS

Biologically inspired micro air vehicles (MAVs) are small autonomous flying vehicles. Their dimension could be smaller than 15 cm, or sometimes smaller than 300 cm, latter toward mini air vehicles (MiAVs). The MAVs are used for reconnaissance over lands, in/over buildings, for collecting evidence of industrial faults, and survey over hazardous places or nuclear sites. For design and development of these vehicles, one needs to understand aerodynamics, structural aspects, and propulsion physics at a very small scale. Due to their small size, there would be limited or no space to carry avionics that would otherwise carry out various navigational tasks [18].

The operational challenges of MAVs are related to the low Reynolds number ($\sim 10 \times 10^4$), presence of gust and obstacles at low altitudes, micro size and weight, need of compact power pack, micro sensors and actuators, and aero-elastic coupling. MAVs (and MiAVs) could be fixed wing, rotary wing, or flapping wing. Flapping (wing) flight is exhibited by millions and millions of species (of bats, birds, and insects). A flapping wing MAV can be viewed as a single nonlinear dynamical system with integrated fluid, structure, and control system components. The research in the areas of MAVs can offer new and computationally efficient solutions to problems of design, propulsion/energy efficiency, structural aspects, machine vision, and navigation. Currently, the research is focused in three main areas to study: (i) the aerodynamic characteristics of low aspect ratio wings at low Reynolds number; (ii) WT experiments for flow visualization – aerodynamic coefficients in the form of look-up tables

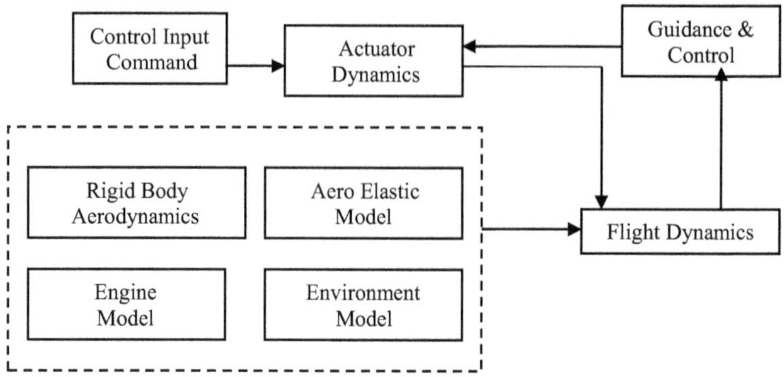

FIGURE 2.10 Schematic for MAV simulation.

of stability, dynamic, and control derivatives are also obtained from such experiments; and (iii) numerical simulation based on CFD and WT experiments.

Today, autonomous aerodynamic control of MAVs is of much interest. The issues of flight mechanics and flight control for MAVs are of particular concern. In general, MAV dynamics can be modeled using the well-established set of rigid body EOMs for aircraft. Since MAVs are extremely light and susceptible to gusts, a possible way to avoid the degradation in their performance is to build MAVs with higher flexibility [19]. The flexible MAVs will have the ability to absorb the wind energy and reduce the effect of gust thereby improving the MAV's overall performance. Models for simulation of MAV dynamics would therefore also require appropriate representation of the flexibility effects in the aerodynamic models. Figure 2.10 shows a simplistic layout of the simulation model for MAV.

2.11 LIGHTER-THAN-AIR VEHICLE/BLIMP DYNAMICS

Lighter-than-air vehicles (LAVs) have several advantages and find increasing applications in various tasks such as surveillance and advertising. Modern aerostats use helium for lift and consist of a power supply system. A two-way fiber optic datalink is provided for sensor data and control. The low-pressure differential allows them to sustain for a long time. These are also difficult to detect with radar. Due to these features, the demand for the use of aerostats for surveillance is continually growing, and being less expensive, they offer a viable and inexpensive alternative to satellites.

In this section, we briefly discuss a MM of an airship. As for the fixed-wing aircraft, the basic derivation of the EOMs of an airship is based on rigid body dynamics [20]. One noticeable difference in developing the model equations for an aircraft and an airship is the consideration of buoyancy in the latter case. To account for the apparent mass effect due to the large volume of air mass displaced by an airship, the EOMs are usually written about the center of buoyancy. In ref. [21], effects of the change in the position of the C.G. as a result of venting or intake of the ballast air are also modeled. Pashilkar [22] suggested simplification in the model equations by determining the change in the position of the C.G. and translating all the forces, moments, and inertias to the C.G. to compute the airship dynamics. This results in EOMs that are like those discussed for the fixed-wing aircraft. In the state-space form, these can be expressed as

$$\dot{x}(t) = f[x(t), u(t), \Theta] \tag{2.184}$$

Here, for a 12th-order system, the state vector x consists of

$$x = [u, v, w, p, q, r, \varphi, \theta, \psi, X, Y, H]$$

The vector Θ is a collection of the stability and control derivatives, and u is the control input vector. Computing the center of mass location and translating the moment of inertias to this center allow to decouple the accelerations and write the EOMs in the conventional form. Since the variables X, Y, and ψ do not couple with the rest of the equations, mostly a ninth-order linear model comprising of the longitudinal states u, w, q, θ, H and the lateral states comprising of v, p, r, φ can be used to analyze the modal characteristics of an airship.

Due to the buoyancy and apparent mass of the airship, its climb performance is different from that of the conventional fixed-wing aircraft [22]. For an aircraft, we have

$$\begin{aligned} T - D - W\sin\gamma &= 0 \\ L - W\cos\gamma &= 0 \end{aligned} \tag{2.185}$$

With the buoyancy force B in the case of an airship, the force balance equations in climb are

$$\begin{aligned} T - D + (B - W)\sin\gamma &= 0 \\ L - W\cos\gamma &= 0 \end{aligned} \tag{2.186}$$

The first of the above equation describes the force balance in the direction of flight path, while the second describes the force balance perpendicular to the flight path. The net buoyant force $(B - W)$ for an airship in climb is positive. The velocity of the airship will undergo a change if the first equation is not balanced, and the airship flight path angle will undergo a change if the second equation is imbalanced.

Linearized EOMs provide the modal characteristics of the airship. The longitudinal modes of an airship generally consist of an oscillatory mode with low damping and a heave mode. Unlike a conventional aircraft, the airship has no mode equivalent to the phugoid mode as the coupling between the forward speed and heave mode is missing in the case of an airship. The LD modes comprise the DR mode which is generally an oscillatory stable mode with high damping, and a roll mode which is an oscillatory stable mode with very low damping. Airship can also show sideslip subsidence and yaw subsidence modes. Typical aerodynamic derivatives for an aerostat/blimp configuration are given in ref. [4].

APPENDIX 2A: AIRCRAFT GEOMETRY AND COORDINATE SYSTEMS

The universally accepted aircraft nomenclature representing the translational and rotational motions, the flow angles, and the aerodynamic forces and moments acting on the aircraft are mentioned here. The forward edge of the wing is called the leading edge, and the rear edge is called the trailing edge.

2A.1 AIRCRAFT AXIS AND NOTATIONS

The line joining the leading edge to the trailing edge is called the chord. If b is the wing span and S represents the area of the wing, then the aspect ratio of the wing is given by $AR = b^2 / S$. A reference point of considerable importance is the aerodynamic center. This is the point about which there is no change in the pitching moment as the AOA is increased. Figure 2A.1 shows the aircraft primary control surfaces and the normally accepted sign conventions. All surface positions are angular in deflection. The elevator deflection causes the aircraft to pitch about the y-axis, aileron deflection causes the aircraft to roll about the x-axis, and the rudder deflection causes it to yaw about the z-axis. A canard is another control surface found in fighters, and it is ahead of the wing and is positioned to produce a positive lift. However, it interferes with the smooth flow over the wing and is therefore not found in every fighter aircraft. Other control surfaces are the strakes and trailing edge flaps (TEF). Strakes are small fins placed in the horizontal plane on the aircraft forebody projecting into the airflow. This is a passive device, which increases the vortex breakdown AOA. Flaps found on the trailing edge of the wing are called TEF. The downward deflection of these flaps causes a significant increase in lift coefficient and L/D ratio, i.e. the load factor. The most effective range of downward deflection is from 0° to 20°. The notations for the aircraft motion variables in the body axis system are put together in Table 2A.1 for better understanding.

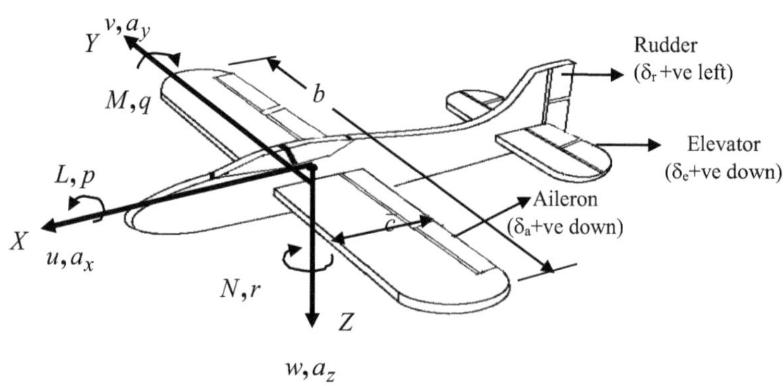

FIGURE 2A.1 Aircraft variables in body axis/control surfaces.

TABLE 2A.1
Aircraft Axis and Notations

	Longitudinal x-axis	Lateral y-axis	Vertical z-axis
Velocity components	u	v	w
Angular rates	Roll rate p	Pitch rate q	Yaw rate r
Euler angles	Roll angle φ	Pitch angle θ	Heading angle ψ
Accelerations	Longitudinal acceleration a_x	Lateral acceleration a_y	Vertical acceleration a_z
Aerodynamic forces	X	Y	Z
Aerodynamic moments	Rolling moment L	Pitching moment M	Yawing moment N
Control deflections	Elevator δ_e	Aileron δ_a	Rudder δ_r
Moment of inertia	I_{xx}	I_{yy}	I_{zz}

2A.2 AXIS/COORDINATE TRANSFORMATIONS AND QUATERNIONS

The Euler angles ψ,θ,ϕ are the basic angles of the direction cosine matrix T required to transform a vector from the earth to body axes. The relation between the Euler angles and angular rates is given by

$$\begin{bmatrix} \dot{\psi} \\ \dot{\theta} \\ \dot{\phi} \end{bmatrix} = \frac{1}{\cos\theta} \begin{bmatrix} 0 & \sin\phi & \cos\phi \\ 0 & \cos\theta\cos\phi & -\cos\theta\sin\phi \\ \cos\theta & \sin\theta\sin\phi & \sin\theta\cos\phi \end{bmatrix} \begin{bmatrix} p \\ q \\ r \end{bmatrix} \tag{2A.1}$$

The Euler angles ψ,θ,ϕ can be obtained by integrating the above equations. The matrix T of direction cosines to transform from the earth (e) to the body (b) axis is given by

$$T = \begin{bmatrix} T_{11} & T_{12} & T_{13} \\ T_{21} & T_{22} & T_{23} \\ T_{31} & T_{32} & T_{33} \end{bmatrix}; \quad T_{be} = T \tag{2A.2}$$

Here,

$$\begin{aligned} &T_{11} = \cos\theta\cos\psi && T_{12} = \cos\theta\sin\psi && T_{13} = -\sin\theta \\ &T_{21} = \sin\theta\sin\phi\cos\psi - \cos\phi\sin\psi && T_{22} = \sin\theta\sin\phi\sin\psi + \cos\phi\cos\psi && T_{23} = \cos\theta\sin\phi \\ &T_{31} = \sin\theta\cos\phi\cos\psi + \sin\phi\sin\psi && T_{32} = \sin\theta\cos\phi\sin\psi - \sin\phi\cos\psi && T_{33} = \cos\theta\cos\phi \end{aligned} \tag{2A.3}$$

The singularity encountered as pitch angle θ approaches 90° can be avoided with the use of quaternions. The four components of quaternions are expressed in terms of Euler angles and to the body axis angular rates as follows:

$$\begin{aligned}
a_1 &= \cos(\psi/2)\cos(\theta/2)\cos(\phi/2)+\sin(\psi/2)\sin(\theta/2)\sin(\phi/2)\\
a_2 &= \cos(\psi/2)\cos(\theta/2)\sin(\phi/2)-\sin(\psi/2)\sin(\theta/2)\cos(\phi/2)\\
a_3 &= \sin(\psi/2)\cos(\theta/2)\sin(\phi/2)+\cos(\psi/2)\sin(\theta/2)\cos(\phi/2)\\
a_4 &= \sin(\psi/2)\cos(\theta/2)\cos(\phi/2)-\cos(\psi/2)\sin(\theta/2)\sin(\phi/2)
\end{aligned} \tag{2A.4}$$

$$\begin{bmatrix}\dot{a}_1\\ \dot{a}_2\\ \dot{a}_3\\ \dot{a}_4\end{bmatrix} = \begin{bmatrix}0 & p & q & r\\ -p & 0 & -r & q\\ -q & r & 0 & -p\\ -r & -q & p & 0\end{bmatrix}\begin{bmatrix}a_1\\ a_2\\ a_3\\ a_4\end{bmatrix} \tag{2A.5}$$

The four components a_1, a_2, a_3, a_4 satisfy the constraint

$$a_1^2 + a_2^2 + a_3^2 + a_4^2 = 1 \tag{2A.6}$$

The elements of the transformation matrix T, in terms of the four components of quaternions, are

$$\begin{aligned}
&T_{11} = 2\left(a_1^2 + a_2^2\right) - 1 \qquad && T_{12} = 2\left(a_2a_3 + a_1a_4\right) \qquad && T_{13} = 2\left(a_2a_4 - a_1a_3\right)\\
&T_{21} = 2\left(a_2a_3 - a_1a_4\right) && T_{22} = 2\left(a_1^2 + a_3^2\right) - 1 && T_{23} = 2\left(a_3a_4 + a_1a_2\right)\\
&T_{31} = 2\left(a_1a_3 + a_2a_4\right) && T_{32} = -2\left(a_1a_2 + a_3a_4\right) && T_{33} = 2\left(a_1^2 + a_4^2\right) - 1
\end{aligned} \tag{2A.7}$$

The Euler angles can be obtained from the quaternions as follows:

$$\psi = \tan^{-1}\left[\frac{T_{12}}{T_{11}}\right];\ \theta = \tan^{-1}\left[-\frac{T_{13}}{\sqrt{T_{11}^2 + T_{12}^2}}\right];\ \phi = \tan^{-1}\left[\frac{T_{23}}{T_{33}}\right] \qquad \text{(2A.8) (A.33)}$$

The sequence of rotation is important for accurate axis transformation. The first rotation is through angle ψ about the z-axis, then through θ about the y-axis, and finally through angle ϕ about the x-axis.

2A.3 TRANSFORMATION FROM THE BODY TO THE EARTH AXIS

To transform from the body axis to the earth axis, the transformation matrix T_{eb} is given by

$$T_{eb} = \begin{bmatrix} T_{11} & T_{12} & T_{13} \\ T_{21} & T_{22} & T_{23} \\ T_{31} & T_{32} & T_{33} \end{bmatrix}; \quad T_{eb} = T_{be}^{'}; \tag{2A.9}$$

Since the matrices are orthogonal, the inverse is done by the transpose of the matrix:

$$\begin{aligned}
&T_{11} = \cos\theta\cos\psi \quad && T_{12} = \sin\theta\sin\phi\cos\psi - \cos\phi\sin\psi \quad && T_{13} = \sin\theta\cos\phi\cos\psi + \sin\phi\sin\psi \\
&T_{21} = \cos\theta\sin\psi && T_{22} = \sin\theta\sin\phi\sin\psi + \cos\phi\cos\psi && T_{23} = \sin\theta\cos\phi\sin\psi - \sin\phi\cos\psi \\
&T_{31} = -\sin\theta && T_{31} = \cos\theta\sin\phi && T_{33} = \cos\theta\cos\phi
\end{aligned} \tag{2A.10}$$

2A.4 TRANSFORMATION FROM THE STABILITY AXIS TO THE BODY AXIS

The origin of the body (b) and stability axes (s) is the same, and the only rotation is the inclination α of that of the XZ plane of the stability axis with the body axes. By putting $\psi = \phi = 0$ and $\theta = \alpha$ in the elements of matrix T in eqn. (2A.10), one obtains the matrix T_{sb} :

$$T_{sb} = \begin{bmatrix} \cos\alpha & 0 & \sin\alpha \\ 0 & 1 & 0 \\ -\sin\alpha & 0 & \cos\alpha \end{bmatrix} \tag{2A.11}$$

The transpose of this matrix can be used to transform from the body axis to the stability axis.

2A.5 TRANSFORMATION FROM THE STABILITY AXIS TO THE WIND AXIS (W)

This transformation is given by the following matrix:

$$T_{ws} \tag{2A.12}$$

2A.6 TRANSFORMATION FROM THE BODY AXIS TO THE WIND AXIS

The origin of the wind and body axis is the same, and the two-axis systems are related through the flow angles α and β. Putting $\psi = \beta$, $\theta = \alpha$, and $\phi = 0$ in the elements of matrix T in eqn. (2A.10), one obtains the matrix T_{wb} to transform from the body axis to the wind axis.

$$T_{\text{wb}} = \begin{bmatrix} \cos\alpha\cos\beta & \sin\beta & \sin\alpha\cos\beta \\ -\cos\alpha\sin\beta & \cos\beta & -\sin\alpha\sin\beta \\ -\sin\alpha & 0 & \cos\alpha \end{bmatrix} \tag{2A.13}$$

Thus, using the transformation matrix T defined in eqn. (2A.10) and knowing angles ψ,θ,ϕ, one can obtain a transformation from one axis to another.

APPENDIX 2B: TRANSFORMATIONS OF AERODYNAMIC DERIVATIVES

Derivatives from the stability to the body axis are given as follows:

$$C_{N_\alpha} = C_{L_\alpha}\cos\alpha + C_{D_\alpha}\sin\alpha + C_c$$

$$C_{c_\alpha} = C_{D_\alpha}\cos\alpha - C_{L_\alpha}\sin\alpha - C_N$$

$$C_{m_\alpha} = \left(C_{m_\alpha}\right)_s$$

$$C_{l_\beta} = \left(C_{l_\beta}\right)_s \cos\alpha - \left(C_{n_\beta}\right)_s \sin\alpha$$

$$C_{l_p} = \left(C_{l_p}\right)_s \cos^2\alpha + \left(C_{n_r}\right)_s \sin^2\alpha - \left(C_{n_p} + C_{l_r}\right)_s \sin\alpha\cos\alpha$$

$$C_{l_{\dot\beta}} = \left(C_{l_{\dot\beta}}\right)_s \cos\alpha - \left(C_{n_{\dot\beta}}\right)_s \sin\alpha$$

$$C_{l_r} = \left(C_{l_r}\right)_s \cos^2\alpha - \left(C_{n_p}\right)_s \sin^2\alpha - \left(C_{n_r} - C_{l_p}\right)_s \sin\alpha\cos\alpha$$

$$C_{l_\delta} = \left(C_{l_\delta}\right)_s \cos\alpha - \left(C_{n_\delta}\right)_s \sin\alpha$$

$$C_{n_\beta} = \left(C_{n_\beta}\right)_s \cos\alpha + \left(C_{l_\beta}\right)_s \sin\alpha$$

$$C_{n_r} = \left(C_{n_r}\right)_s \cos^2\alpha + \left(C_{l_p}\right)_s \sin^2\alpha + \left(C_{n_p} + C_{l_r}\right)_s \sin\alpha\cos\alpha$$

$$C_{n_{\dot\beta}} = \left(C_{n_{\dot\beta}}\right)_s \cos\alpha + \left(C_{l_{\dot\beta}}\right)_s \sin\alpha$$

$$C_{n_p} = \left(C_{n_p}\right)_s \cos^2\alpha - \left(C_{l_r}\right)_s \sin^2\alpha - \left(C_{n_r} - C_{l_p}\right)_s \sin\alpha\cos\alpha$$

$$C_{n_\delta} = \left(C_{n_\delta}\right)_s \cos\alpha + \left(C_{l_\delta}\right)_s \sin\alpha$$

Derivatives from body to stability axes are given as

$$C_{L\alpha} = C_{N\alpha}\cos\alpha - C_{c\alpha}\sin\alpha - C_D$$

$$C_{D\alpha} = C_{c\alpha}\cos\alpha + C_{N\alpha}\sin\alpha + C_L$$

$$\left(C_{m\alpha}\right)_s = C_{m\alpha}$$

$$\left(C_{l\beta}\right)_s = C_{l\beta}\cos\alpha + C_{n\beta}\sin\alpha$$

$$\left(C_{l_r}\right)_s = C_{l_r}\cos^2\alpha - C_{n_p}\sin^2\alpha + \left(C_{n_r} - C_{l_p}\right)\sin\alpha\cos\alpha$$

$$\left(C_{l\dot{\beta}}\right)_s = C_{l\dot{\beta}}\cos\alpha + C_{n\dot{\beta}}\sin\alpha$$

$$\left(C_{l_p}\right)_s = C_{l_p}\cos^2\alpha + C_{n_r}\sin^2\alpha + \left(C_{l_r} + C_{n_p}\right)\sin\alpha\cos\alpha$$

$$\left(C_{l\delta}\right)_s = C_{l\delta}\cos\alpha + C_{n\delta}\sin\alpha$$

$$\left(C_{n\beta}\right)_s = C_{n\beta}\cos\alpha - C_{l\beta}\sin\alpha$$

$$\left(C_{n_r}\right)_s = C_{n_r}\cos^2\alpha + C_{l_p}\sin^2\alpha - \left(C_{l_r} - C_{n_p}\right)\sin\alpha\cos\alpha$$

$$\left(C_{n\dot{\beta}}\right)_s = C_{n\dot{\beta}}\cos\alpha - C_{l\dot{\beta}}\sin\alpha$$

$$\left(C_{n_p}\right)_s = C_{n_p}\cos^2\alpha - C_{l_r}\sin^2\alpha + \left(C_{n_r} - C_{l_p}\right)\sin\alpha\cos\alpha$$

$$\left(C_{n\delta}\right)_s = C_{n\delta}\cos\alpha - C_{l\delta}\sin\alpha$$

EXERCISES

2.1 If the total momentum is given as $H = \sum r\,\delta m\, x\, V + \sum [r\, x\, (\omega\, x\, r)]\delta m$, where r is the distance from the center, m is mass of the body, and x defines the vector cross-product, the first term would be zero. Explain why?

2.2 If $\omega = pi + qj + rk$ and $r = xi + yj + zk$, then obtain the cross-product $(\omega \times r)$.

2.3 What are merits and demerits of small disturbance theory?

2.4 Is $(r \times \omega) = (\omega \times r)$?

2.5 By rotating the stability axis through positive AOA, which axis system is obtained? About which axis is this rotation?

2.6 How many and what consecutive rotations are required to transform a vector from the earth reference axis system to the body axis system?

2.7 What are the normal limits of the Euler angles?

2.8 Use the equation F=mass x acceleration and mass flow concept for the momentum theory of the helicopter/rotor and compare sizes of the rotor (disk area) and the induced velocities. The mass is the mass flow of the air being pumped through the rotor every second.

2.9 Can you guess what would be (helicopter) disk loading, and its relation with the dynamic pressure? Explain by formulae.

2.10 There would be aircraft with variable wing area. Define wing loading so that the comparison between such aircraft would be easy.

2.11 Obtain the formula for the landing speed of an aircraft.

2.12 What is the significance of this landing speed?

2.13 What is the relation between the coefficients of normal force and Z-force? Under what condition, $C_L = C_N$?

2.14 What is the relation between the coefficients of the axial force and X force?

2.15 What is the result of the cross-product of the angular (rate) and linear velocity:

$$kR \otimes (iU + jV)$$

2.16 The plane XZ is regarded as a plane of symmetry. What is its significance from the moment of inertias in this plane wrt to the y-axis?

2.17 Show that $\tan\beta = \tan\alpha_f \cos\alpha$.

2.18 What are the effects of the gravity forces on a missile under the following conditions? (i) only pitch attitude is zero; (ii) only bank angle is zero; and (iii) both the pitch and bank angle are zero.

2.19 Determine an approximate relation for the rate of change of sideslip angle $\dot{\beta}$ under the following conditions: (a) AOA is zero and sideslip is very small; and (b) bank angle is also zero or small.

2.20 In the polar coordinate velocity form of the EOM, if α and β are both negligibly small, obtain an approximate expression for the total velocity.

2.21 If gravity and thrust effects are assumed small, how does the roll rate contribute to sideslip?

2.22 Which is the most important merit of the non-dimensional form of the aerodynamic derivatives?

2.23 Verify that the unit for M_w is 1/mt–s by doing dimensional analysis.

2.24 Based on the proper disposition of lift force L and drag force D, obtain the relationships (i) $X = L\sin\alpha - D\cos\alpha$ and $C_x = C_L \sin\alpha - C_D \cos\alpha$; and (ii) $Z = -(L\cos\alpha + D\sin\alpha)$ and $C_z = -(C_L\cos\alpha + C_D\sin\alpha)$.

2.25 Let the yawing moment derivative wrt yaw rate be defined as $N_r = \dfrac{1}{I_z}\dfrac{\partial N}{\partial \text{r}}$, define $C_{n_r} = \dfrac{\partial C_n}{\partial\,(\text{br/2U})}$, then obtain N_r in terms of C_{n_r}.

2.26 Establish relation between $\dfrac{\Delta\ddot{\psi}}{\Delta\beta}$ and N_β.

2.27 Establish relation between $\dfrac{\Delta a_z}{\Delta \alpha}$ and $Z_{\dot{w}}$.

2.28 Given $Z_M = -\dfrac{\rho S}{2m}\left(\mathrm{U}^2 \dfrac{\partial \mathrm{C_L}}{\partial M} + 2U^2\mathrm{C_L} / M\right)$, obtain the relation $Z_u a$.

2.29 From polar coordinate velocity form of the EOM, obtain the expression for total velocity in terms of the aerodynamic derivatives in the wind axis when AOA and angle of sideslip are very small. Ignore thrust and gravity effects.

2.30 Determine an approximate relation between the roll rate damping derivative C_{l_p} and the cross-derivative C_{n_p}. Assume cross-coupling inertia is negligible and roll rate and yaw rate are small, under steady-state roll and yaw motion.

2.31 How would you get the flight path rate from the acceleration and subsequently the body rate?

2.32 Will large or small changes would occur in the airspeed in the SP transient mode?

2.33 What is the main distinction between the SP and phugoid mode characteristic from the motion of the aircraft and the interplay of the forces and moments?

2.34 If you want to add another state variable in the phugoid model, which one would you add?

2.35 In the SP mode/model, which aerodynamic derivative would have the largest influence on the natural frequency?

2.36 In the SP mode, which aerodynamic derivatives would have influence on the damping ratio?

2.37 What is inertial and aerodynamic 'damping' in the SP mode?

2.38 In the phugoid mode, which is the most influential aerodynamic derivative and in what way?

2.39 Which aerodynamic derivatives have most significant influence on the DR frequency?

2.40 Which are most effective aerodynamic derivatives in spiral and roll subsidence modes?

2.41 Which is most effective DR damping aerodynamic derivative?

2.42 Why the name spiral divergence?

REFERENCES

1. Mcruer, D. T., Ashkenas, I., and Graham, D. *Aircraft Dynamics and Automatic Control.* Princeton University Press, Princeton, NJ, 1973.
2. Nelson, R. C. *Flight Stability and Automatic Control.* WCB/McGraw Hill Book Company, New York, 1998.
3. Cook, M. V. *Flight Dynamics Principles.* John Wiley & Sons, Inc., New York, 1997.
4. Raol, J. R., and Singh, J. *Flight Mechanics Modeling and Analysis.* Revised edition. CRC Press, Taylor and Francis Gr., Boca Raton, FL, 2023.
5. Garnell, P. *Guided Weapon Control Systems.* 2nd edition. Pergamon Press, New York, 1980.

6. Seddon, J. *Basic Helicopter Dynamics*. 2nd edition. AIAA Education Series, AIAA, Reston, VA, 2001.
7. Venkatesan, C. (Ed.). *Short Course on Helicopter Technology – Lecture Notes*. Department of Aerospace Engineering, IIT Kanpur, India, 15–18 Oct., 1997.
8. Prouty, R. W. *Practical Helicopter Aerodynamics*. PJS Publications, Peoria, IL, p. 86, 1982.
9. Hamel, P. G. (Ed.). *Rotorcraft System Identification*. AGARD LS-178, NATO, France, 1991.
10. Kaletka, J., Tischler, M. B., von Grünhagen, W., and Fletcher, J. Time and frequency identification and verification of BO 105 dynamic models. *J. Am. Helicopter Soc.*, 36(4), 25–38, 1991.
11. Tischler, M. B., and Cauffman, M. G. Frequency response method for rotorcraft system identification: Flight applications to the BO 105 coupled rotor/fuselage dynamics. *J. Am. Helicopter Soc.*, 37(3), 3–17, 1992.
12. Yechout, T. R., Morris, S. L., Bossert, D. E., and Hallgren, W. F. *Introduction to Aircraft Flight Mechanics – Performance, Static Stability, Dynamic Stability, and Classical Feedback Control*. AIAA Education Series, AIAA, Reston, VA, 2003.
13. Raol, J. R., Girija, G., and Singh, J. *Modelling and Parameter Estimation of Dynamic Systems. IEE Control Book Series*, Vol. 65, IEEE, London, UK, 2004.
14. Maine, R. E. and Iliff, K. W. Application of parameter estimation to aircraft stability and control – The output error approach, NASA RP-1168, 1986.
15. De Leeuw, J. H. and Hui, K. The application of linear maximum-likelihood estimation of aerodynamic derivatives for the Bell-205 and Bell-206, NAE-AN-48, NRC No. 28442, Ottawa, Canada, 1987.
16. Anon. Rotorcraft System Identification, AGARD LS-178, 1991.
17. Sadraey, M., and Colgren, R. UAV flight simulation: Credibility of linear decoupled vs. nonlinear coupled equations of motion, *AIAA Modeling and Simulation Technologies Conference and Exhibit*, 15–18 August 2005, San Francisco, CA, AIAA 2005-6425.
18. Anon. Indo-US Workshop on Micro Air Vehicles, National Aerospace Laboratories, Bangalore, India, Aug. 1–2, 2005.
19. Stewart, K., Abate, G., and Evers, J. Flight mechanics and control issues for micro air vehicles, *AIAA Atmospheric Flight Mechanics Conference and Exhibit*, 21–24 August 2006, Keystone, CO, AIAA 2006-6638.
20. Khoury, G. A., and Gillet, J. D. (Eds.). *Airship Technology, Cambridge Aerospace Series: 10*, Cambridge University Press, United Kingdom, 1999.
21. An, J., et al. Aircraft dynamic response to variable wing sweep geometry. *J. Aircr.*, 25(1), 216–221, 1988.
22. Pashilkar, A. A. Modeling, analysis and autopilot design for an airship. *J. Aero. Sci. Technol*, 55(3), 186–192, 2003.

3 Models Used in Target Tracking and Data Fusion

3.1 INTRODUCTION

The problem of tracking of moving objects (including targets, mobile robots, and other vehicles) using measurements from sensors, mounted on these vehicles, is of considerable interest in many military and civil applications, which use radar, sonar systems, and EOTs for tracking from flight testing of aircraft (and missiles, unmanned aerial vehicles (UAVs), micro/mini air vehicles (MAVs), and rotorcrafts). It also has good use in nonmilitary applications such as robotics, air traffic control and management (ATC/M), air surveillance, and ground vehicle tracking (GVT). In practice, scenarios for target tracking could include maneuvering, crossing, and splitting (meeting and separating) targets. Various algorithms are available to achieve target tracking for such scenarios, and the selection of the algorithms is in general application-dependent and is also based on the merits of the algorithm, complexity of the problem (data corrupted by ground clutter/noise processes, etc.), and computational burden. Target tracking comprises estimation of the current state of a target usually based on noisy measurements. The problem is complex even for single target tracking because of uncertainties in target's mathematical model, especially for maneuvering targets (which needs more than one model and transition model, etc.), and process (state)/measurement noises. The complexity of the tracking problem increases for multiple targets using measurements from multiple sensors.

In the process of the target tracking, the target track is updated by correlating measurements with the existing tracks or initiating new tracks by using measurements coming from different sensors. The methods/process of gating and data association enable proper and accurate tracking in the multi-sensor multi-target (MSMT) scenario [1–3]. Gating helps in deciding if an observation (which could include clutter, false alarms, and electronic countermeasures) is a probable candidate for track maintenance or track update. Data association is a step to associate the measurements to the targets with certainty when several targets are in the same neighborhood. In practice, measurements arriving from the sensors may not be true due to the effect of clutter, false alarms, and interference from other targets, limited resolution capability (spatial coverage limitation of the sensor), and several targets in neighborhood.

Gating is used to screen out spurious signal such as clutter, whereas association algorithms are used for automatic track initiation, measurement-to-track correlation, and track-to-track correlation. In measurement-to-track association, the sensor data are associated with an existing number of tracks to determine which sensor data or observation belongs to which target. Once a determination has been made that there are more than one observation for a particular target, then these observations are

DOI: 10.1201/9781003429869-3

combined at raw level using measurement fusion (in a typical sequential estimation technique such as Kalman filter (KF)). The steps needed to implement measurement fusion algorithm are described below. It is a technique for eliminating unlikely observation-to-track pairings. The gating process determines if an observation belongs to a previously established target or to a new target. Gates (rectangular, circular, or ellipsoidal in shape) are defined for one or more existing tracks. If an observation satisfies the gate, which means it falls in the gate, it becomes a candidate for association with that track. The region enclosed by the gate is called the validation/confirmation region.

Evaluation of performance of a data fusion system, algorithms, and software is very important for establishing confidence in the DF system and its results. Filter tracking performance is evaluated by computing several metrics [4]. The performance of various tracking filters and related algorithms is evaluated when data analysis is supported with image data and or acoustic data.

3.2 INTERACTING MULTIPLE MODEL (IMM) KALMAN FILTER ALGORITHM

Target tracking is more difficult if the target in motion is also maneuvering. KF has some ability to adapt to maneuvers by tuning the process noise covariance (i.e. Q, and perhaps to some extent measurement noise covariance, R) matrices, if the target is mildly maneuvering. The filter tuning could be either manual or automatic; manual tuning means that the KF has available values of Q and R, and it is evaluated off-line from post-flight data. However, for any new scenario this might not be the case. In such situations, the trial-and-error method or automatic tuning is to be employed. Thus, in a multi-target tracking system, KF can be used for relatively benignly maneuvering maneuvers. In most situations, an interacting multiple model KF (IMMKF) has been found to give better performance than KF [5,6]. The IMMKF uses a few target motion models (constant velocity (CV), constant acceleration (CA), coordinate turn model, etc.). For example, IMMKF may use one model for a straight and level flight and different models for maneuvers or turns. The IMMKF always maintains the repository of all the models and blends their outputs with certain weighting coefficients that are computed probabilistically as part of the algorithm. In addition to the state estimates for each motion model, the IMMKF maintains an estimate of the probability with which the target is moving in accordance with each model.

3.2.1 IMMKF Algorithm

Thus, the IMMKF uses i) several possible models for the target's motion, and ii) a probabilistic switching mechanism that does an appropriate switching between these models. It is implemented with multiple parallel (not in the real sense of parallelism/or on parallel computers; however, this is feasible and should be implemented in this way for saving computational times) filters, where each of the filters would correspond to one of the multiple models. Due to the switching between different models, there is an exchange of some information between the filters. During each sampling period, it is likely that all the filters of the IMMKF are in operation. The overall state estimate is a combination of the state estimates from the individual filters. Let us consider $M_1,\ldots,M_r$ as the r models of IMMKF and let $M_j(k)$ mean that the model M_j is in effect during the sampling period ending at frame k. During the event $M_j(k+1)$, the state of the target evolves according to

$$X(k+1) = F_j X(k) + w_j(k) \tag{3.1}$$

The measurement equation is given by

$$z(k+1) = H_j X(k+1) + v_j(k+1) \tag{3.2}$$

The variables have the usual meanings from the state-space literature. Figure 3.1 depicts the information flow of a cycle of IMMKF iteration [7]. For simplicity, two-model IMMKF is considered.

3.2.1.1 Interaction and Mixing

For the event $M_j(k+1)$, the mixed estimate $X_{0j}(k|k)$ and the covariance matrix $P_{0j}(k|k)$ are computed as

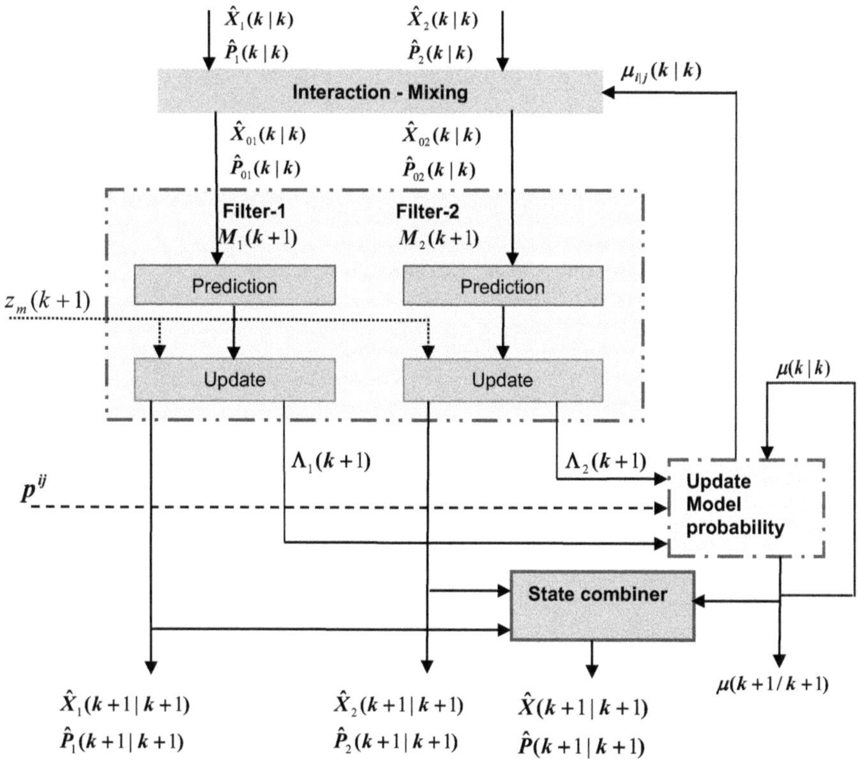

FIGURE 3.1 IMMKF – a cycle of operation.

$$\hat{X}_{0j}(k|k) = \sum_{i=1}^{r} \mu_{i|j}(k|k)\hat{X}_i(k|k)$$

$$\hat{P}_{0j}(k|k) = \sum_{i=1}^{r} \mu_{i|j}(k|k)\left\{\hat{P}_i(k|k) + \left[\hat{X}_i(k|k) - \hat{X}_{0j}(k|k)\right]\left[\hat{X}_i(k|k) - \hat{X}_{0j}(k|k)\right]^T\right\} \tag{3.3}$$

The mixing probabilities $\mu_{i|j}(k|k)$ are given by

$$\mu_{i|j}(k|k) = \frac{1}{\mu_j(k+1|k)}\, p_{ij}\, \mu_i(k|k) \tag{3.4}$$

Here, the predicted mode probability $\mu_j(k+1|k)$ is computed by

$$\mu_j(k+1|k) = \sum_{i=1}^{r} p_{ij}\, \mu_i(k|k) \tag{3.5}$$

The mode switching process is done by the Markov process and is specified by the following mode transition probabilities:

$$p_{ij} = \Pr\left\{M_j(k+1)|M_i(k)\right\} \tag{3.6}$$

Here, $\Pr\{.\}$ denotes the probability of an event, which means p_{ij} is the probability that M_i model at k^{th} instant is switching over to M_j model at $(k+1)^{\text{th}}$ instant. This is used to calculate the model probabilities for the final output. The usual KF equations are used with appropriate target motion models to update the mixed state estimates with current measurement. The innovation covariance is given by

$$S_j = H_j\tilde{P}_j(k+1|k)H_j^T + R_j \tag{3.7}$$

The innovations sequence is given by

$$\upsilon_j = z(k+1) - \tilde{z}_j(k+1|k) \tag{3.8}$$

The likelihood function for matched filter j is a Gaussian pdf of innovations $\upsilon_j(k+1)$ with zero mean and covariance S_j. It is used for updating the probabilities of the various models:

$$\Lambda_j = \frac{1}{(2\Pi)^{0.5n}\sqrt{|S_j|}}\exp\left\{-0.5\upsilon_j^T S_j^{-1}\upsilon_j\right\} \tag{3.9}$$

Here, n denotes the dimension of the innovation vector υ.

3.2.1.1.1 Mode Probability Update

Once each model has been updated with measurement $z(k+1)$, the mode probability $\mu_j(k+1 \mid k+1)$ is updated using mode likelihood Λ_j and the predicted mode probabilities $\mu_j(k+1 \mid k)$ for $M_j(k+1)$

$$\mu_j(k+1|k+1) = \frac{1}{c}\mu_j(k+1|k)\Lambda_j \tag{3.10}$$

Here, the normalization factor

$$c = \sum_{i=1}^{r} \mu_i(k+1|k)\Lambda_i \tag{3.11}$$

3.2.1.1.2 State Estimate and Covariance Combiner

The estimated states $\hat{X}_j(k+1|k+1)$ and covariance $\hat{P}_j(k+1|k+1)$ from each filter are combined using the updated mode probability $\mu_j(k+1|k+1)$ to produce overall state estimate $\hat{X}(k+1|k+1)$ and the associated covariance $\hat{P}(k+1|k+1)$ as given below:

$$\hat{X}(k+1|k+1) = \sum_{j=1}^{r} \mu_j(k+1|k+1)\hat{X}_j(k+1|k+1) \tag{3.12}$$

$$\hat{P}(k+1|k+1) = \sum_{j=1}^{r} \mu_j(k+1|k+1) \left\{ \begin{aligned} &\hat{P}_j(k+1|k+1) + \\ &\left[\hat{X}_j(k+1|k+1) - \hat{X}(k+1|k+1)\right]\left[\hat{X}_j(k+1) - \hat{X}(k+1|k+1)\right]^T \end{aligned} \right\} \tag{3.13}$$

3.2.1.2 Target Motion Models

The most common forms of target motion models are (i) 2 degree of freedom (DOF) kinematic model (constant velocity model [CVM]), and (ii) 3DOF kinematic model (constant acceleration model [CAM]) [8–10].

3.2.1.2.1 Constant Velocity Model

The 2DOF model with position and velocity components in each of the three Cartesian coordinates x, y, and z has the following transition and process noise gain matrices:

$$F_{CV} = \begin{bmatrix} \Phi_{CV} & 0 & 0 \\ 0 & \Phi_{CV} & 0 \\ 0 & 0 & \Phi_{CV} \end{bmatrix} G_{CV} = \begin{bmatrix} \varsigma_{CV} & 0 & 0 \\ 0 & \varsigma_{CV} & 0 \\ 0 & 0 & \varsigma_{CV} \end{bmatrix} \tag{3.14}$$

Here, $\Phi_{CV} = \begin{bmatrix} 1 & T & 0 \\ 0 & 1 & 0 \\ 0 & 0 & 0 \end{bmatrix} \varsigma_{CV} = \begin{bmatrix} T^2/2 \\ T \\ 0 \end{bmatrix}$

The variations in velocity are modeled as zero-mean white noise accelerations. Low noise variance Q_{cv} is used with the model to represent the constant course and speed of the target in a non-maneuvering mode. The process noise intensity in each coordinate is generally assumed to be small and equal $\left(\sigma_x^2 = \sigma_y^2 = \sigma_Z^2\right)$, which accounts for air turbulence, slow turns, and small linear acceleration. Although the 2DOF model is primarily used to track the non-maneuvering mode of a target, the use of higher level of process noise variance will allow the model to track maneuvering targets as well, of course to a limited extent.

3.2.1.2.2 Constant Acceleration Model

The 3DOF model, with position, velocity, and acceleration components in each of the three Cartesian coordinates x, y, and z, has the following transition and process noise gain matrices:

$$F_{CA} = \begin{bmatrix} \Phi_{CA} & 0 & 0 \\ 0 & \Phi_{CA} & 0 \\ 0 & 0 & \Phi_{CA} \end{bmatrix} G_{CA} = \begin{bmatrix} \varsigma_{CA} & 0 & 0 \\ 0 & \varsigma_{CA} & 0 \\ 0 & 0 & \varsigma_{CA} \end{bmatrix} \tag{3.15}$$

Here, $\Phi_{CA} = \begin{bmatrix} 1 & T & T^2/2 \\ 0 & 1 & T \\ 0 & 0 & 1 \end{bmatrix}; \varsigma_{CA} = \begin{bmatrix} T^3/6 \\ T^2/2 \\ T \end{bmatrix}$

The acceleration increments over a sampling period are a discrete time zero-mean white noise. A low value of process noise variance Q_2 (but relatively higher than Q_1) will yield nearly a CA motion. The noise variance in each coordinate is assumed to be equal $\left(\sigma_x^2 = \sigma_y^2 = \sigma_z^2\right)$.

3.2.1.3 Implementation and Validation

The algorithm implemented in MATLAB consists of both IMMKF and conventional KF. Two-model IMMKF (one for non-maneuver mode and another for maneuver mode, i.e. $r = 2$) is considered to describe the principles and steps of the IMMKF algorithm. The observation matrix used is given as

$$H = \begin{bmatrix} 1 & 0 & 0 & 0 & 0 & 0 & 0 & 0 & 0 \\ 0 & 0 & 0 & 1 & 0 & 0 & 0 & 0 & 0 \\ 0 & 0 & 0 & 0 & 0 & 0 & 1 & 0 & 0 \end{bmatrix}$$

The initial mode probabilities $\mu = \left[\mu_1 \;\; \mu_2\right]^T$ corresponding to the non-maneuver and maneuver mode are taken as 0.9 and 0.1, respectively. p_{ij} is the Markov chain transition matrix, which takes care of the switching from mode i to mode j. This is a design parameter and can be chosen by the user. The switching probabilities are generally known to depend upon sojourn time. For example, consider the following Markov chain transition matrix between the two modes of the IMMKF [5]: $p_{ij} = \begin{bmatrix} 0.9 & 0.1 \\ 0.33 & 0.67 \end{bmatrix}$. The basis for selecting $p_{12} = 0.1$ is that initially the target is likely to be in the non-maneuvering mode and the probability to switch over to the maneuvering mode will be then relatively low, whereas p_{22} is selected based on the number of sampling periods during which the target is expected to maneuver (target's sojourn time). If the target maneuver lasts for three sampling periods ($\tau = 3$), the probability p_{22} is

$$p_{22} = 1 - \frac{1}{\tau} = 0.67 \tag{3.16}$$

The performance of the IMMKF with respect to conventional KFcv (KF with only CVM) and KFca (KF with only CAM) is evaluated by computing certain performance metrics, in addition to the root sum variance in position (RSvarP):

$$\text{RSvarP} = \sqrt{P_x + P_y + P_z} \tag{3.17}$$

Here, $P_x, P_y, \& P_z$ are the diagonal elements in P corresponding to positions x, y & z; similarly, it can be used for other elements. Similar computations are carried out for root sum variance in velocity (RSvarV) and root sum variance in acceleration (RSvarA).

The data for maneuvering target are simulated using the 2DOF and 3DOF models with sampling interval of 1 s. for 500 seconds. The simulation proceeds with the following parameters: (i) initial state $(x, \dot{x}, \ddot{x}, y, \dot{y}, \ddot{y}, z, \dot{z}, \ddot{z})$ of the target = 11097.6, −6.2, 0, 3425, −299.9, 0, 40, 0, 0); (ii) a low-level process noise variance of $Q_1 = 0.09$ is considered for model 1; (iii) a higher level process noise variance of $Q_2 = 36$ is considered for model 2; (iv) noise variance for both models in each coordinate is assumed to be equal, i.e. $Q_{xx} = Q_{yy} = Q_{zz} = Q$; and (v) measurement noise variance of $R = 100$ is considered, and it is also assumed to be equal $R_{xx} = R_{yy} = R_{zz} = R$.

At scan 100, additional acceleration $\ddot{x} = 27.4$ m/s^2 is given in the x-direction to simulate the first maneuver, and at scan 350, $\ddot{y} = -99.4$ m/s^2 is given in the y-direction to simulate the second maneuver. The acceleration profiles are shown in Figure 3.2. The target spends most of the time in the non-maneuvering sate. The first maneuver starts at scan 100 and ends at scan 150, and the second maneuver starts at scan 350 and ends at scan 400. KFcv uses a single CVM with $Q_1 = 0.09$ and $R = 100$, whereas KFca uses a single CAM with $Q_2 = 36$ and $R = 100$, and IMMKF with two-mode (CVM with $Q_1 = 0.09$ and $R = 100$ for the non-maneuvering mode and CAM with $Q_2 = 36$ and $R = 100$ for the maneuvering mode) filters to track the target through both the modes. The Markov chain transition matrix is

$$p_{ij} = \begin{bmatrix} 0.9 & 0.1 \\ 0.05 & 0.95 \end{bmatrix}$$

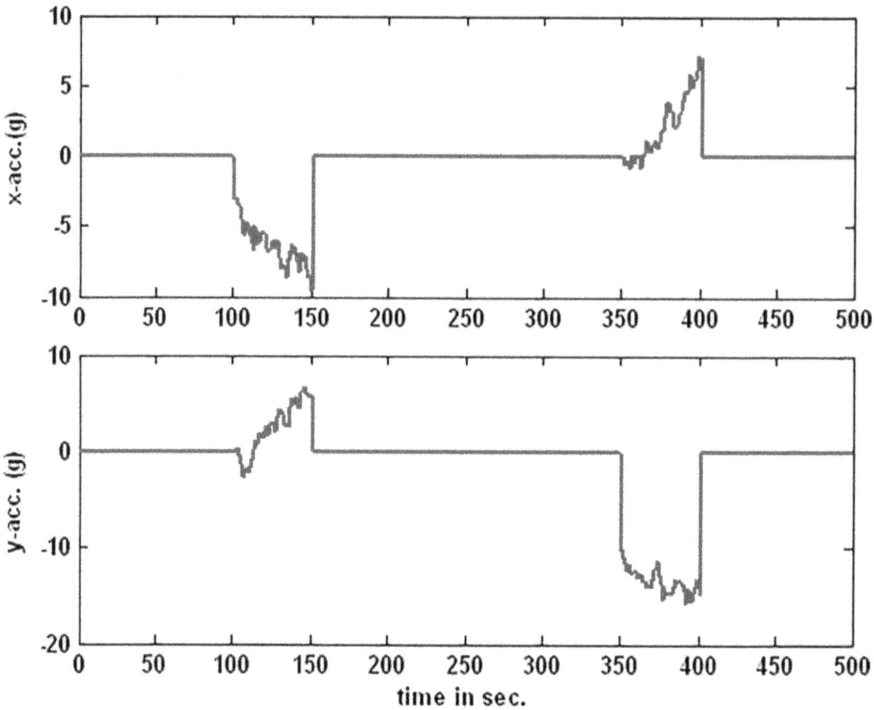

FIGURE 3.2 Acceleration profiles in X- and Y-axes.

This choice is made keeping in view the sojourn time $(\tau = 50)$ of the target. Equation (3.16) is used to arrive at the p_{22} value in the transition matrices. The enlarged display of some portion of the trajectories using KFcv, KFca, and IMMKF is shown in Figure 3.3. It is observed that the track tracked by KFcv is away from the true track during the maneuver. The tracks tracked by both KFca and IMMKF are very close to the true track. The mode probabilities are shown in Figure 3.4; it is seen that there are two maneuvers in the trajectory with start and end time of the maneuvers, i.e. which model or the weighted combination of models is active at a given time. The root sum variance (RSvar) from KFcv, KFca, and IMMKF is shown in Figure 3.5. The RSvar in KFcv and KFca is constant throughout the trajectory. Due to high process noise covariance, the RSvar is high in KFca than in KFcv. But it is high during maneuver and low during non-maneuver in case of IMMKF. This information could be useful in track fusion, and it is one of the features of IMMKF. This shows that maneuvers can detect using the IMMKF very well. The RSS position errors (RSSPE) from KFcv, KFca, and IMMKF are shown in Figure 3.6. It is very high during the maneuver in case of KFcv, and low throughout the trajectory in case of KFca. The performance metrics are shown in Table 3.1. These results are obtained from 50 Monte Carlo simulations. From these results, it is observed that the tracking performance is worst in case of KFcv during maneuver. The tracking performance of both KFca and IMMKF is almost similar during maneuver. In overall, the tracking performance of IMMKF is acceptable.

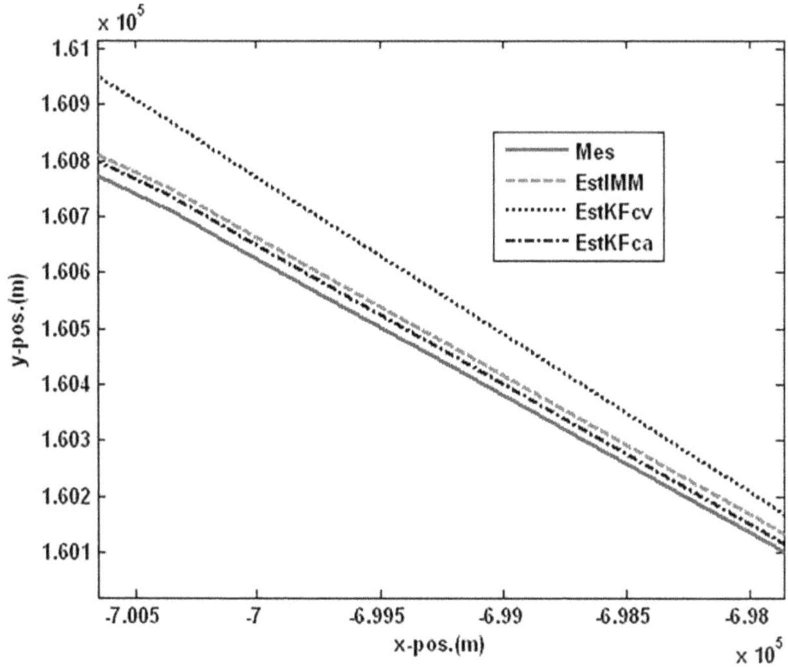

FIGURE 3.3 Enlarged display of the trajectories.

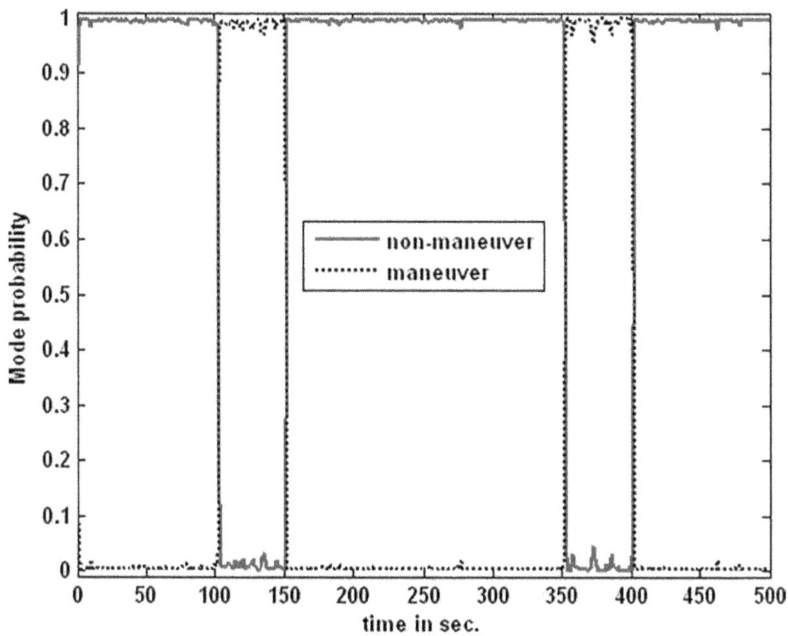

FIGURE 3.4 Mode probabilities for a maneuver.

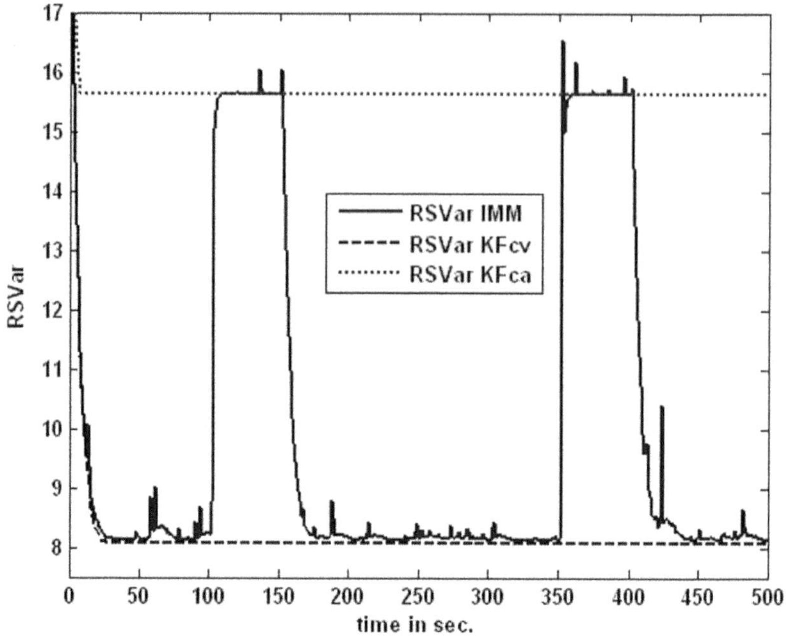

FIGURE 3.5 RSvar for KFcv, KFca, and IMM filters.

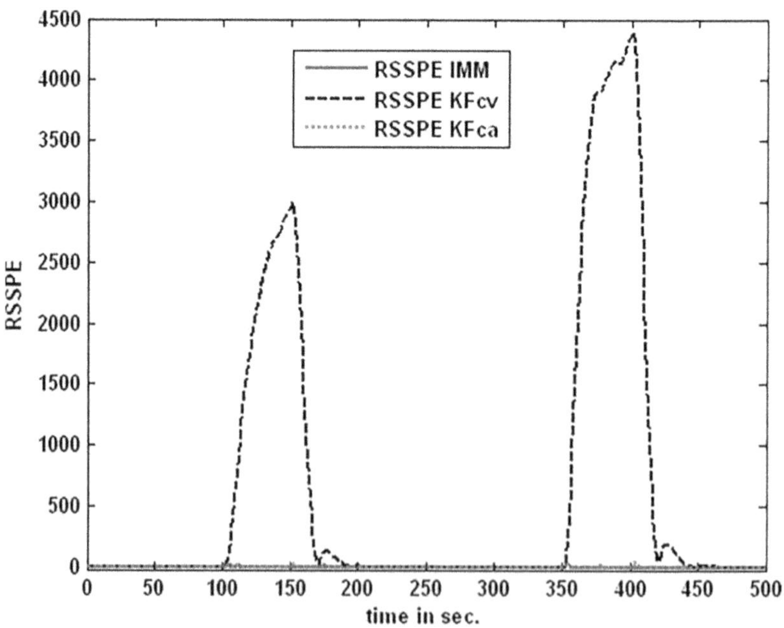

FIGURE 3.6 RSSPE for KFcv, KFca, and IMM filters.

TABLE 3.1
Performance metrics for KFcv, KFca, and IMMKF

Filter	PFE_x	PFE_y	PFE_z	RMSPE	RMSVE	RMSAE
KFcv	0.11	0.84	0.11	824.4	220.78	29.49
KFca	~0	~0	~0	2.75	9.15	7.72
IMMKF	~0	~0	~0	2.78	6.88	5.83

3.3 A MISSILE SEEKER ESTIMATOR

In interceptors, active radar seeker is used to measure (i) relative range, (ii) relative range rate, (iii) line-of-sight (LOS) angles, and (iv) rates between the interceptor and evader. The measurements are contaminated by high degree of noise due to glint, radar cross section (RCS) fluctuations, and thermal noise. The true LOS rates are to be estimated for proportional navigation (PN) guidance of the interceptor. An advanced proportional navigation (APN) guidance control system would use target acceleration also. An estimator or seeker filter is required that processes the seeker measurements recursively to obtain the signals required for guidance of the interceptor toward the evader or target. The design of the estimator is complex because the cumulative effects of the noise are non-Gaussian and time-correlated. Also, there would be a periodic loss of seeker measurements due to target eclipsing effects. Since the seeker measurements are available in the inner Gimbal frame, one would need an extended KF (EKF) for the seeker estimator. In order to handle non-Gaussian noise, an augmented EKF (AEKF) is used. In interceptor-evading target engagement scenario or game, the evader would execute (perhaps unpredictable) maneuvers to avoid the interceptor. To track such a maneuvering target, an IMM is generally used. The IMM is an adaptive estimator based on the assumption of the use of a finite number of models [11]. In this section, an IMM based on the soft switching between a set of predefined (evader target) models such as CV, CA, and constant jerk (CJ) [12] is described. At each sampling time, the mode probability is computed for each model using an innovation vector and a covariance matrix. Each of the mode-matched filter is based on the AEKF. The performance evaluation of an IMM-AEKF in a closed loop using MATLAB-simulated data of a typical interceptor-evader engagement scenario is discussed.

3.3.1 IMM-Augmented EKF

The state vector of the respective mode-matched filter is augmented with additional states for handling glint noise and RCS fluctuations.

3.3.1.1 State Model

The state vector consists of 18 states (with six states related to glint noise and RCS fluctuations):

$$\begin{bmatrix} \Delta x & \Delta V_x & a_{tx} & j_{tx} & \Delta y & \Delta V_y & a_{ty} & j_{ty} & \Delta z & \Delta V_z & a_{tz} & j_{tz} & \text{RCS \& GLINT states} \end{bmatrix}$$

The mathematical models for glint noise and RCS are given next [13–15]:

State Model 1: Constant Velocity (CV) Model

$$\begin{aligned} &\Delta\dot{x} = \Delta V_x;\ \Delta\dot{V}_x = 0;\ \dot{a}_{tx} = 0;\ \dot{j}_{tx} = 0; \\ &\Delta\dot{y} = \Delta V_y;\ \Delta\dot{V}_y = 0;\ \dot{a}_{ty} = 0;\ \dot{j}_{ty} = 0; \\ &\Delta\dot{z} = \Delta V_z;\ \Delta\dot{V}_z = 0;\ \dot{a}_{tz} = 0;\ \dot{j}_{jz} = 0; \end{aligned} \tag{3.18}$$

State Model 2: Constant Acceleration (CA) Model

$$\begin{aligned} &\Delta\dot{x} = \Delta V_x;\Delta\dot{V}_x = a_{tx} - a_{mx};\dot{a}_{tx} = -\left(\frac{a_{tx}}{\tau_x}\right);\dot{j}_{tx} = 0;\Delta\dot{y} = \Delta V_y; \\ &\Delta\dot{V}_y = a_{ty} - a_{my};\dot{a}_{ty} = -\left(\frac{a_{ty}}{\tau_y}\right);\dot{j}_{ty} = 0; \\ &\Delta\dot{z} = \Delta V_z;\Delta\dot{V}_z = a_{tz} - a_{mz};\dot{a}_{tz} = -\left(\frac{a_{tz}}{\tau_z}\right);\dot{j}_{jz} = 0; \end{aligned} \tag{3.19}$$

State Model 3: Constant Jerk (CJ) Model

$$\begin{aligned} &\Delta\dot{x} = \Delta V_x;\ \Delta\dot{V}_x = a_{tx} - a_{mx};\ \dot{a}_{tx} = j_{tx};\ \dot{j}_{tx} = -\left(\frac{j_{tx}}{\tau_x}\right); \\ &\Delta\dot{y} = \Delta V_y;\ \Delta\dot{V}_y = a_{ty} - a_{my};\ \dot{a}_{ty} = j_{ty};\ \dot{j}_{ty} = -\left(\frac{j_{ty}}{\tau_y}\right); \\ &\Delta\dot{z} = \Delta V_z;\ \Delta\dot{V}_z = a_{tz} - a_{mz};\ \dot{a}_{tz} = j_{tz};\ \dot{j}_{jz} = -\left(\frac{j_{tz}}{\tau_z}\right); \end{aligned} \tag{3.20}$$

In these equations, $\Delta x, \Delta y, \Delta z$ are the relative positions, $\Delta V_x, \Delta V_y, \Delta V_z$ are the relative velocities of target with respect to missile, a_{tx}, a_{ty}, a_{tz} are the target accelerations, j_{tx}, j_{ty}, j_{tz} are the target jerks, a_{mx}, a_{my}, a_{mz} are the missile accelerations, and τ_x, τ_y, τ_z are the correlation time constants.

3.3.1.2 Measurement Model

Measurement vector is given as $\begin{bmatrix} \rho & \dot{\rho} & \varphi_y & \varphi_z & \dot{\varphi}_y & \dot{\varphi}_z \end{bmatrix}$ during the non-eclipsing period and $\begin{bmatrix} \rho & \dot{\rho} & \varphi_y & \varphi_z \end{bmatrix}$ during the eclipsing period. Here, ρ is range-to-go, $\dot{\rho}$ is range rate, φ_y and φ_z are Gimbal angles in yaw and pitch planes, respectively, and $\dot{\varphi}_y$, and $\dot{\varphi}_z$ are the respective LOS rates in inner Gimbal frame. The relative position and velocity states of the target with respect to missile in the inertial frame are transformed to the LOS frame using the following formulae:

$$\rho = \sqrt{\Delta x^2 + \Delta y^2 + \Delta z^2}; \quad \dot{\rho} = \frac{\Delta x \Delta \dot{x} + \Delta y \Delta \dot{y} + \Delta z \Delta \dot{z}}{\rho}$$

$$\lambda_e = \tan^{-1}\left(\frac{\Delta z}{\sqrt{\Delta x^2 + \Delta y^2}}\right); \quad \dot{\lambda}_e = \frac{\Delta \dot{z}\left(\Delta x^2 + \Delta y^2\right) - \Delta z\left(\Delta x \Delta \dot{x} + \Delta y \Delta \dot{y}\right)}{\rho^2 \sqrt{\Delta x^2 + \Delta y^2}} \tag{3.21}$$

$$\lambda_a = \tan^{-1}\left(\frac{\Delta y}{\Delta x}\right); \quad \dot{\lambda}_a = \frac{\left(\Delta x \Delta \dot{y} - \Delta y \Delta \dot{x}\right)}{\sqrt{\Delta x^2 + \Delta y^2}}$$

The measurement model during the non-eclipsing period is given as

$$\begin{bmatrix} \rho \\ \dot{\rho} \end{bmatrix}_m = \begin{bmatrix} \rho \\ \dot{\rho} \end{bmatrix}; \begin{bmatrix} \varphi_y \\ \varphi_z \end{bmatrix}_m = \begin{bmatrix} \varphi_y \\ \varphi_z \end{bmatrix}; \begin{bmatrix} 0 \\ \dot{\varphi}_y \\ \dot{\varphi}_z \end{bmatrix}_m = C_f^g \; C_b^f \; C_i^b \; C_l^i \begin{bmatrix} -\dot{\lambda}_a \sin \lambda_e \\ \dot{\lambda}_e \\ \dot{\lambda}_a \cos \lambda_e \end{bmatrix} \tag{3.22}$$

The measurement model during the eclipsing period is given as

$$\begin{bmatrix} \rho \\ \dot{\rho} \end{bmatrix}_m = \begin{bmatrix} \rho \\ \dot{\rho} \end{bmatrix}; \begin{bmatrix} \varphi_y \\ \varphi_z \end{bmatrix}_m = \begin{bmatrix} \varphi_y \\ \varphi_z \end{bmatrix} \tag{3.23}$$

Here,

$$\varphi_y = \tan^{-1}\left(\frac{m}{l}\right)$$

$$\varphi_z = \tan^{-1}\left(\frac{n}{\sqrt{l^2 + m^2}}\right) \text{and} \begin{bmatrix} l \\ m \\ n \end{bmatrix} = C_b^f C_i^b C_l^i \begin{bmatrix} 1 \\ 0 \\ 0 \end{bmatrix} \tag{3.24}$$

The direction cosine matrix (DCM) for LOS-to-inertial frame transformation is given as

$$C_l^i = \begin{bmatrix} \cos \lambda_e \cos \lambda_a & -\sin \lambda_a & -\sin \lambda_e \cos \lambda_a \\ \cos \lambda_e \sin \lambda_a & \cos \lambda_a & -\sin \lambda_e \sin \lambda_a \\ \sin \lambda_e & 0 & \cos \lambda_e \end{bmatrix} \tag{3.25}$$

The DCM for inertial-to-body frame transformation is given as

$$C_i^b = \begin{bmatrix} q_4^2 + q_1^2 - q_2^2 - q_3^2 & 2(q_1q_2 + q_3q_4) & 2(q_1q_3 - q_2q_4) \\ 2(q_1q_2 - q_3q_4) & q_4^2 - q_1^2 + q_2^2 - q_3^2 & 2(q_2q_3 + q_1q_4) \\ 2(q_1q_3 + q_2q_4) & 2(q_2q_3 - q_1q_4) & q_4^2 - q_1^2 - q_2^2 + q_3^2 \end{bmatrix} \tag{3.26}$$

Here, q_1, q_2, q_3, q_4 are the attitude quaternion of the missile coordinate system. The DCM for body-to-fin frame transformation is given as

$$C_b^f = \begin{bmatrix} 1 & 0 & 0 \\ 0 & \frac{1}{\sqrt{2}} & \frac{1}{\sqrt{2}} \\ 0 & -\frac{1}{\sqrt{2}} & \frac{1}{\sqrt{2}} \end{bmatrix} \tag{3.27}$$

The DCM for fin-to-inner Gimbal frame transformation is given as

$$C_f^g = \begin{bmatrix} \cos\varphi_z \cos\varphi_y & \cos\varphi_z \sin\varphi_y & \sin\varphi_z \\ -\sin\varphi_y & \cos\varphi_z & 0 \\ -\sin\varphi_z \cos\varphi_y & -\sin\varphi_z \sin\varphi_y & \cos\varphi_z \end{bmatrix} \tag{3.28}$$

3.3.2 Interceptor-Evader Engagement Simulation

The close-loop simulation was carried for a typical interceptor-evader terminal phase scenario [15]. Typical 6DOF missile dynamics were simulated, and relative measurements containing target information were generated in inner Gimbal frame in terms of range, range rate, two Gimbal angles, and two LOS rates in pitch and yaw planes. For the terminal guidance measurement, noise is added to true $(r, \dot{r}, \varphi_g, \gamma_g, \dot{\varphi}_g, \dot{\gamma}_g)$ to obtain the noisy measurements. The Gaussian noise is added to the range and range rate measurements, random colored noise with bore sight error is added to Gimbal angle measurements, and thermal Gaussian, correlated glint, and RCS fluctuation noises are added to the LOS rate measurements. Data loss in LOS rates due to pulse repetition frequency (PRF) and closing velocity is also simulated. The data are sampled at 0.01 second. The LOS rates also have the component of body rates affecting it. The measured data are then processed by the IMM-AEKF algorithm. The guidance commands are generated and fed back to missile autopilot, which in turn provides sufficient fin deflections to steer the missile toward the evading target.

3.3.2.1 Evader Data Simulation

To simulate realistic missile-target engagement, the aspects considered are as follows: (i) in the presence of adversary, the target generally executes a turn and accelerates at short range-to-go (≈ 2 km), (ii) target velocity is to be maintained at 300–400 km/h, (iii) target turns (velocity vector) at the rate of 20–25 deg/s, and (iv) target undergoes maximum roll rate (≈ 270 deg/s) to generate glint effect. The evader performing evasive maneuver is simulated within the permissible "g" limit of a typical fighter aircraft at different altitudes. The target speeds are chosen to achieve the desired turn rates in the range of 20–25 deg/s. Different data sets have been generated by allowing the target to maneuver continuously; however, only a particular result is presented here.

3.3.3 Performance Evaluation

The performance of the algorithm is evaluated in a closed-loop simulation of missile-target engagement. The filter initialization parameters are given as [15]

i. initial state vector:

$$\Delta\hat{x} = \Delta x + 100; \Delta\hat{\dot{x}} = \Delta\dot{x} + 20; \hat{a}_{t_x} = 0; \hat{j}_{t_x} = 0;$$

$$\Delta\hat{y} = \Delta y - 50; \Delta\hat{\dot{y}} = \Delta\dot{y} - 5; \hat{a}_{t_y} = 0; \hat{j}_{t_y} = 0;$$

$$\Delta\hat{z} = \Delta z + 50; \Delta\hat{\dot{z}} = \Delta\dot{z} - 5; \hat{a}_{t_z} = 0; \hat{j}_{t_z} = 0;$$

ii. initial state error covariance: $\hat{P}(0/0) = 10000 * \text{eye}(\text{ns}, \text{ns})$

iii. initial process noise covariance:

For CV model:

Q.v = diag[0.0 0.005555 0.05 0.005555 0.0 0.005555 0.05 0.005555 0.0 0.005555 0.05 0.005555 0.001 0.001 0.001 0.001 0.001 0.001]

For CA model: Q.a = 100*Q.v

For CJ model: Q.j = 10*Q.a

iv. Measurement noise covariance:

$$R = \text{diag}\left[\; 2.5\times10^{3} \quad 1\times10^{2} \quad 7.6\times10^{-5} \quad 7.6\times10^{-5} \quad 2.467\times10^{-2} \quad 2.467\times10^{-2} \;\right]$$

v. Initial mode probability: $\mu = \left[0.3 \quad 0.3 \quad 0.4\right]$

vi. Mode transition probability:

$$p = \begin{bmatrix} 0.800 & 0.100 & 0.100 \\ 0.0009 & 0.999 & 0.0001 \\ 0.0009 & 0.0001 & 0.999 \end{bmatrix}$$

A 3DOF model-based AEKF is integrated with 6DOF missile simulation model in a closed loop [15]. Innovation smoothing using a low-pass (LP) filter with a cutoff frequency of 3 Hz is used in the filter. For a particular case where the maneuver is initiated at $R_{\text{to-go}} = 10$ km at an altitude of 0.5 km (other parameters being target speed = 135 m/s, maneuver g's = 6 in pitch and −6 in yaw, and turn rate = 24.6 deg/s).

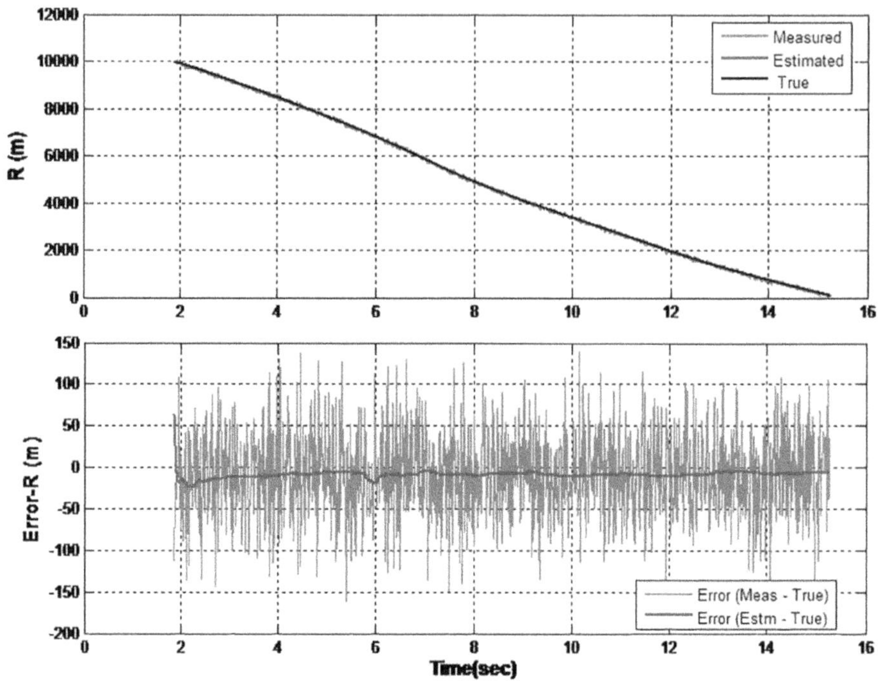

FIGURE 3.7 Measurements: estimation errors for a particular case.

We have the following closed-loop performance result: the miss distance is 5.44 m, and the time to intercept is 15.41 s for the IMM-AEKF (with glint states). Figure 3.7 shows the comparison of estimated, true, and noisy measurement signals along with the estimation error, and Figure 3.8 shows the comparison of typical estimated state with true values. The mode probabilities for the three models are shown in Figure 3.9 for engagement at an altitude of 0.5 km and with target maneuvers initiated at $R_{\text{to-go}} = 10\,\text{km}$, and comparison of noise attenuation factors (NAF) in the measurements is shown in Figure 3.10, and it is observed that the NAF computed using a sliding window of length 10 is well within the stipulated limit of 0.1. Some points out of these limits especially in the rate measurements are perhaps due to eclipsing effects in the measurements.

3.4 TRACKING OF MANEUVERING TARGETS WITH PROBABILITY DATA ASSOCIATION FILTER

The performance of IMM probability data association filter (IMMPDAF) for estimation of multiple maneuvering target trajectories in clutter is investigated using simulated data – the targets being tracked by single sensor. The scheme can be extended to more than one sensor. Sensitivity of the filter to the choice of the models and process noise covariance matrix values is evaluated. The use of the IMMPDAF algorithm for

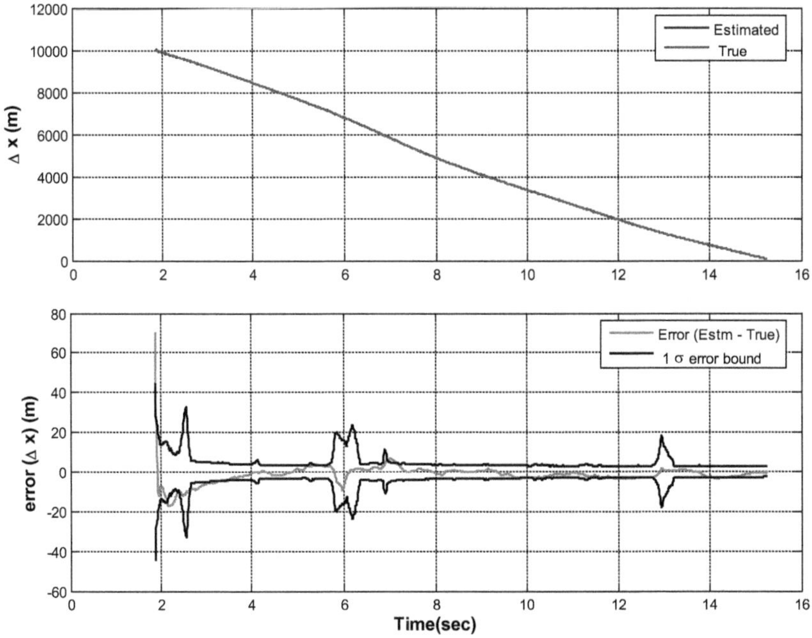

FIGURE 3.8 Estimated states and state error with bounds.

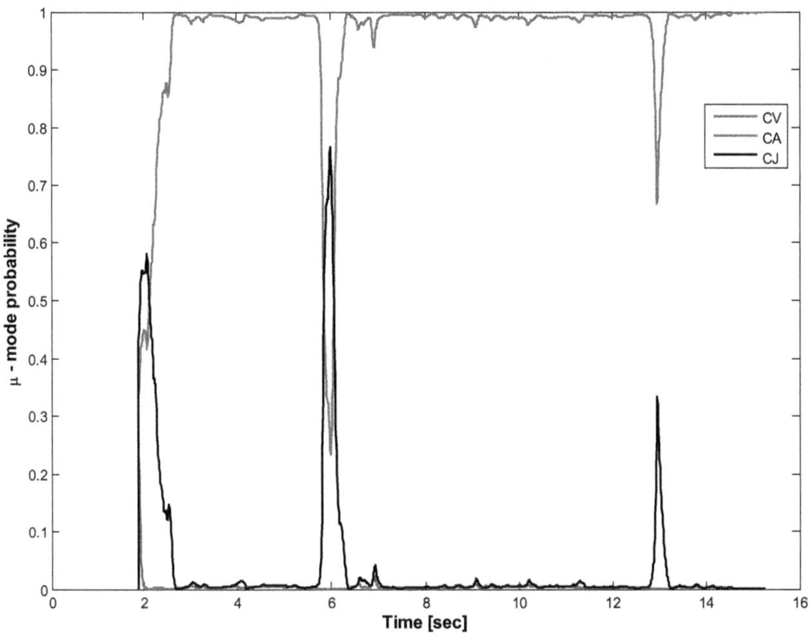

FIGURE 3.9 Mode probabilities for three models.

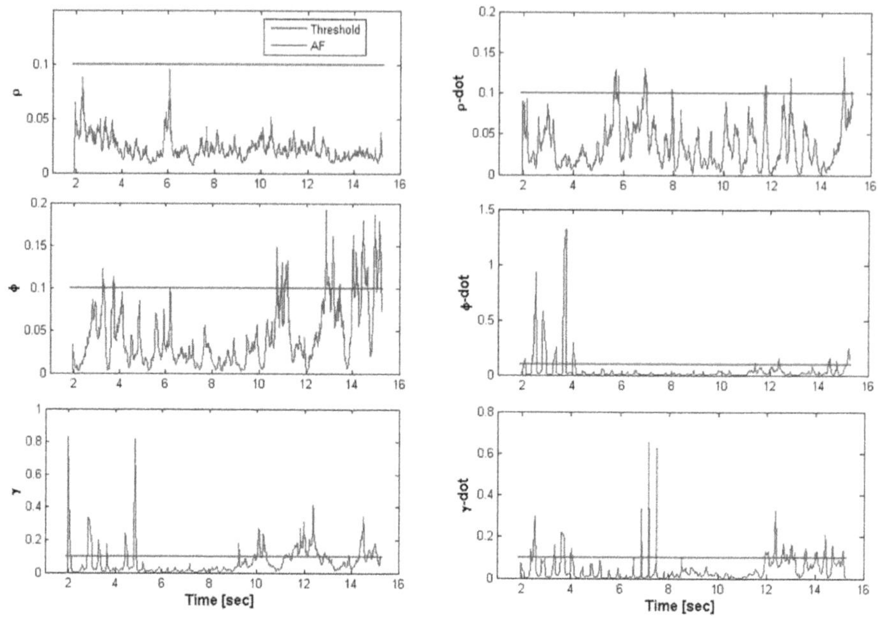

FIGURE 3.10 Noise attenuation factors (NAF).

tracking and fusion of simulated data of four targets whose positions are measured by two sensors located at different spatial positions is also discussed. Thus, the algorithm is evaluated for the MSMT scenario using the fusion method. It was observed that IMMPDAF gives a realistic confidence in the estimates during maneuvers and low RSSPE during the non-maneuvering phase of the targets.

3.4.1 IMM Algorithm

IMM and related concepts are versatile methods for adaptive state estimation in systems whose behavior changes with time and the system obeys one of a finite number of models. Two or more models are run in concurrence (of course, this is mainly true in case of parallel computer-based implementations) to achieve better performance of maneuvering targets. The features of IMM can be combined with PDAF to obtain the IMMPDAF algorithm. Figure 3.11 depicts a cycle of a two-model IMMPDAF algorithm, which is applicable to single-sensor multi-target scenario, which can also be extended to multiple sensors. Figure 3.12 depicts the MSMT fusion process.

3.4.1.1 Automatic Track Formation

The objective of track initiation is to compute initial state estimates of all possible tracks along with the computations of the associated state covariance matrices. There is every possibility of false tracks being generated due to the presence of spurious measurements and/or multiple targets. A technique to initiate track based on a logic

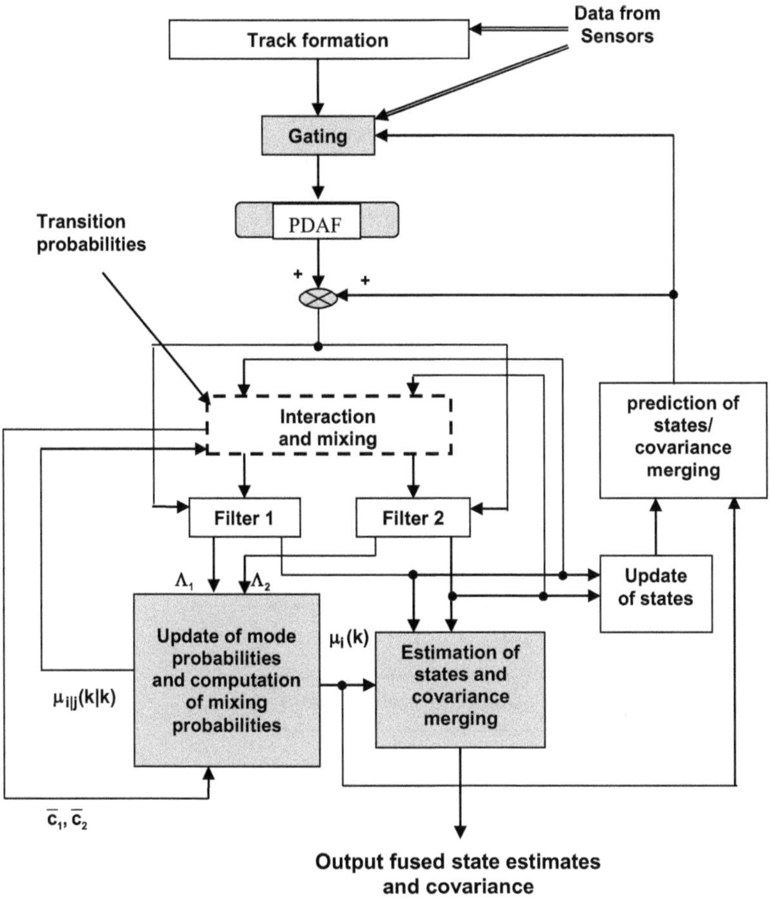

FIGURE 3.11 IMMPDAF for two models.

that requires M detections out of N scans in the gate can be used. A tentative track is formed on all two-measurement pairs with expected motions of the targets.

3.4.1.2 Gating and Data Association

After the formation of the track, the measurement screening is carried out to get potential candidates for corresponding tracks. The combined innovation sequence v is computed using the steps, as in the PDAF, which has two additional blocks (compared to nearest neighbor KF (NNKF)/KF) for computation of association probabilities $\beta_i(k)$ and combined innovation sequences $v(k)$. If $v_i(k)$ corresponds to innovation on measurement i, then the merged innovation is given by

$$v(k) = \sum_{i=1}^{m(k)} \beta_i(k) v_i(k) \tag{3.29}$$

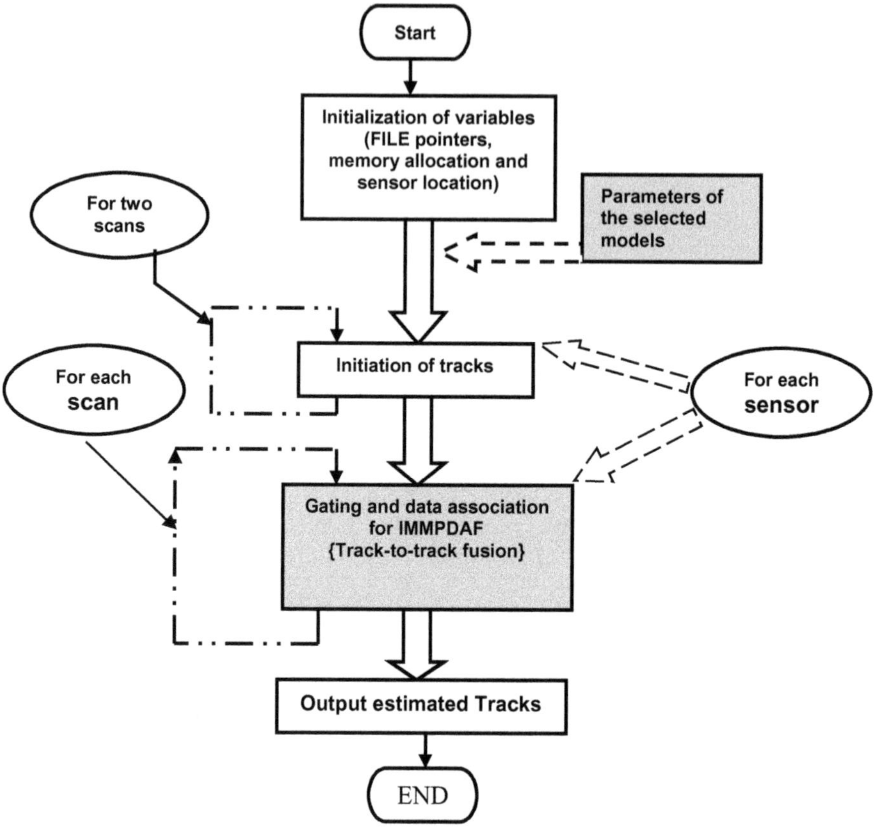

FIGURE 3.12 MSMT fusion process – flow structure.

Here, $m(k)$ denotes the number of detections in kth scan, and the probability $\beta_i(k)$ that the ith validated measurement is correct one is given by

$$\beta_i(k) = \begin{cases} \dfrac{e_i}{b + \sum_{j=1}^{m(k)} e_j} & i = 1,2,\ldots,m(k) \\[2ex] \dfrac{b}{b + \sum_{j=1}^{m(k)} e_j} & i = 0 \end{cases} \tag{3.30}$$

Here, $\beta_0(k)$ is the probability that none of the measurements are correct, and in eqn. (3.30)

$$e_i \cong e^{-\frac{1}{2}v_i(k)'S(k)^{-1}v_i(k)} \tag{3.31}$$

$$b \cong \left(\frac{2\pi}{\gamma}\right)^{\frac{n_z}{2}} m(k)c_{n_z}{}^{-1}\frac{(1-P_DP_G)}{P_D} \tag{3.32}$$

Here, n_z is the dimension of the measurement, c_{n_z} is the volume of the unit hyper-sphere of dimensions ($c_1=2$, $c_2=\pi$, $c_3=4\pi/3$, etc.), and P_G and P_D represent the gate probability and probability of target detection. The computed $\beta_i(k)$ and $v(k)$ are used to estimate the states and covariance matrices of the updating cycle. The resultant measurement that is next used in the mode-conditioned filtering is computed with the following equations:

$$Z(k) = H\tilde{X}(k / k-1) + v(k) \tag{3.33}$$

Here, $\tilde{X}(k / k-1)$ is predicted target states at the kth scan.

3.4.1.3 Interaction and Mixing in IMMPDAF

In this step using the mixing probabilities $\mu_{i|j}(k-1|k-1)$ as weighting coefficients, the estimates of $\hat{X}_i(k-1|k-1)$ and $\hat{P}_i(k-1|k-1)$ from the previous cycle are employed to obtain the initial conditions $\hat{X}_{0j}(k-1|k-1)$ and $\hat{P}_{0j}(k-1|k-1)$ for the mode-matched filters M_1 and M_2 of the present cycle. For all $i, j \in M$, the initial conditions for the filters are computed from the following equations:

$$\hat{X}_{0j}(k-1|k-1) = \sum_{i=1}^{r} \hat{X}_i(k-1|k-1)\mu_{i|j}(k-1|k-1) \tag{3.34}$$

$$P_{0j}(k-1|k-1) = \sum_{i=1}^{r} \begin{bmatrix} P_i(k-1|k-1) + \\ [\{\hat{X}_i(k-1|k-1) - \hat{X}_{0j}(k-1|k-1)\}x \\ \{\hat{X}_i(k-1|k-1) - \hat{X}_{0j}(k-1|k-1)\}^T] \end{bmatrix} \mu_{i|j}(k-1|k-1) \tag{3.35}$$

Here, mode-matched filters run as $j = 1,...,r$; the models run as $i = 1,...,r$; and $r = 2$ for the 2-model IMM method.

3.4.1.4 Mode-Conditioned Filtering

With r parallel (though used in a sequential mode only!) mode-matched KFs, the states and covariance matrices are estimated using the normal prediction and update steps.

$$\hat{X}_j(k|k-1) = F_j(k-1)\hat{X}_{0j}(k-1|k-1) + G_j(k-1)w_j(k-1)$$

$$P_j(k|k-1) = F_j(k-1)P_{0j}(k-1|k-1)F_j(k-1)^T + G_j(k-1)Q_j(k-1)G_j(k-1)^T$$

$$\hat{X}_j(k|k) = \hat{X}_j(k|k-1) + K_j(k)v_j(k)$$

$$P_j(k|k) = P_j(k|k-1) - K_j(k)S_j(k)K_j(k)^T \tag{3.36}$$

The measurement prediction is computed by

$$\hat{Z}_j(k|k-1) = H_j(k)\hat{X}_j(k|k-1) \tag{3.37}$$

The residual $v_j(k)$, residual covariance $S_j(k)$, and the filter gain are computed by

$$\begin{aligned} v_j(k) &= Z(k) - \hat{Z}_j(k|k-1) \\ S_j(k) &= H_j(k)P_j(k|k-1)H_j(k)^T + R_j(k) \\ K_j(k) &= P_j(k|k-1)H_j(k)^T S_j(k)^{-1} \end{aligned} \tag{3.38}$$

The process and measurement noise covariance matrices are Q and R, and these might differ from mode to mode. The likelihood function for mode-matched filter j is a Gaussian pdf of residual v with zero mean and covariance matrix S, and is computed as

$$\Lambda_j(k) = \frac{1}{\sqrt{|S_j(k)|}\,(2\pi)^{n/2}} e^{-0.5[v_j(k)^T S_j(k)^{-1} v_j(k)]} \tag{3.39}$$

Here, n is the dimension of the measurement vector $\mathbf{Z}$.

3.4.1.5 Probability Computations

The mixing probabilities are computed as follows:

$$\mu_{i|j}(k-1|k-1) = \frac{1}{\bar{c}_j} p_{ij}\mu_i(k-1) \tag{3.40}$$

$$\bar{c}_j = \sum_{i=1}^{r} p_{ij}\mu_i(k-1) \tag{3.41}$$

Here, $\mu_i(k)$ is the mode probability at time k, $\bar{c}_j$ is a normalization factor, and p_{ij} is the Markov transition probability to take care of switching from one mode to another mode. It is a design parameter chosen by the user. The switching probabilities are known to depend upon *sojourn* time; consider the Markov chain transition matrix between the two modes of the IMM: $p_{ij} = \begin{bmatrix} 0.9 & 0.1 \\ 0.33 & 0.67 \end{bmatrix}$, then the basis for selecting $p_{12} = 0.1$ is that the target is likely to be in the non-maneuvering mode in initial stages and the probability to switch over to the maneuvering mode will be relatively low. The p_{22} is selected based on the number of sampling periods for which the target is expected to be in a maneuver (*sojourn* time). This, if the target maneuver lasts for 3 sample periods ($\tau = 3$), the probability p_{22} is given by

$$p_{22} = 1 - \frac{1}{\tau} = 0.67 \tag{3.42}$$

The initial mode probabilities $\mu_i(k)$ corresponding to the non-maneuver and maneuver mode are taken as 0.9 and 0.1, respectively, to compute $\mu_{i|j}(k|k)$ and $\bar{c}_j$ in the first cycle of estimation algorithm. This assumes that the target is more likely to be in the non-maneuver mode than in the maneuver mode during the initial stages of its motion. Subsequently, the mode probabilities are updated using the following equation:

$$\mu_j(k) = \frac{1}{c}\Lambda_j(k)\bar{c}_j \quad j = 1,\ldots,r \tag{3.43}$$

Here, $\Lambda_j(k)$ represents the likelihood function corresponding to filter j, and the normalizing factor c is given as

$$c = \sum_{j=1}^{r} \Lambda_j(k)\bar{c}_j \tag{3.44}$$

3.4.1.6 Combined State and Covariance Prediction/Estimation

This has two sages as described next. In the prediction stage, the average mode probabilities obtained in eqn. (3.43) are used as weighting factors to combine the predicted state and covariance matrices for all filters ($j = 1,\ldots,r$), to obtain overall state estimate and covariance prediction:

$$\begin{aligned} \tilde{X}_j(k+1|k) &= F_j(k)\hat{X}_j(k|k) \\ \tilde{P}_j(k+1|k) &= F_j(k)\hat{P}_j(k|k)F_j(k)^T + G_j(k)Q_j(k)G_j(k)^T; j = 1,\ldots,r \end{aligned} \tag{3.45}$$

$$\tilde{X}(k+1|k) = \sum_{j=1}^{r} \tilde{X}_j(k+1|k)\mu_j(k)$$

$$\tilde{P}(k+1|k) = \sum_{j=1}^{r} \left[\tilde{P}_j(k+1|k) + \{\tilde{X}_j(k+1|k) - \tilde{X}(k+1|k)\}\{\tilde{X}_j(k+1|k) - \tilde{X}(k+1|k)\}^T \right] \mu_j(k) \tag{3.46}$$

In the estimation stage, the average mode probabilities are also used as weighting factors to combine the updated state and covariance matrices for all filters ($j = 1,\ldots,r$) to obtain overall state estimate and covariance as follows:

$$\hat{X}(k|k)=\sum_{j=1}^{r}\hat{X}_j(k|k)\mu_j(k)$$

$$P(k|k)=\sum_{j=1}^{r}\left[\,P_j(k|k)+\{\hat{X}_j(k|k)-\hat{X}(k|k)\}\{\hat{X}_j(k|k)-\hat{X}(k|k)\}^T\,\right]\mu_j(k) \quad (3.47)$$

3.4.2 Simulation Validation

The simulation is carried out in PC-MATLAB [16]. The target motion is simulated using a 2DOF CVM during the non-maneuvering phase and a 3DOF CAM during the maneuvering phase of the target. The models are described in the Cartesian coordinate system by

$$\mathbf{X}(k+1)=F\mathbf{X}(k)+Gw(k) \quad (3.48)$$

$$Z(k) = H\,\mathbf{X}(k) + v(k) \quad (3.49)$$

Here, the Cartesian state vector $\mathbf{X}$ consists of the position and velocity of the target: $\mathbf{X}=[\ x \quad \dot{x} \quad y \quad \dot{y}\ ]$ when the target is in the non-maneuvering phase and $\mathbf{X}=[\ x \quad \dot{x} \quad \ddot{x} \quad y \quad \dot{y} \quad \ddot{y}\ ]$ when it is maneuvering. The process noise w and measurement noise v are white and zero-mean Gaussian noises with covariance Q and R, respectively.

3.4.2.1 Constant Velocity Model (CVM)

The 2DOF model with position and velocity components in each of the two Cartesian coordinates x and y is given as

$$F=\begin{bmatrix} 1 & T & 0 & 0 & 0 & 0 \\ 0 & 1 & 0 & 0 & 0 & 0 \\ 0 & 0 & 0 & 0 & 0 & 0 \\ 0 & 0 & 0 & 1 & T & 0 \\ 0 & 0 & 0 & 0 & 1 & 0 \\ 0 & 0 & 0 & 0 & 0 & 0 \end{bmatrix} \quad G=\begin{bmatrix} T^2/2 & 0 \\ T & 0 \\ 0 & 0 \\ 0 & T^2/2 \\ 0 & T \\ 0 & 0 \end{bmatrix} \quad (3.50)$$

The acceleration component in the above model (though identically equal to zero) is retained for compatibility with the 3DOF model, the variations in velocity are modeled as zero-mean white noise accelerations, and low noise variance Q_1 is used with the model to represent the constant course/speed of the target in a non-maneuvering mode. The process noise intensity is assumed to be equal

$$Q_1=\sigma_{\dot{x}}^2=\sigma_{\dot{y}}^2 \quad (3.51)$$

3.4.2.2 Constant Acceleration Model (CAM)

The 3DOF model with position, velocity, and acceleration components in each of the two Cartesian coordinates x and y is given as

$$F = \begin{bmatrix} 1 & T & T^2/2 & 0 & 0 & 0 \\ 0 & 1 & T & 0 & 0 & 0 \\ 0 & 0 & 1 & 0 & 0 & 0 \\ 0 & 0 & 0 & 1 & T & T^2/2 \\ 0 & 0 & 0 & 0 & 1 & T \\ 0 & 0 & 0 & 0 & 0 & 1 \end{bmatrix};$$

$$G = \begin{bmatrix} T^2/2 & 0 \\ T & 0 \\ 1 & 0 \\ 0 & T^2/2 \\ 0 & T \\ 0 & 1 \end{bmatrix} \tag{3.52}$$

The acceleration increments are taken as a discrete time zero-mean white noise, and a low value of process noise variance Q_2 ($> Q_1$) will yield nearly a CA motion. The noise variance in each coordinate is

$$Q_2 = \sigma_{\ddot{x}}^2 = \sigma_{\ddot{y}}^2 \tag{3.53}$$

The following are the components of the target data simulation scenario: we have for the data simulation for a single sensor the following: (i) number of targets = 2; (ii) target initial states:

$$\text{Target1:}\begin{bmatrix} x & \dot{x} & \ddot{x} & y & \dot{y} & \ddot{y} \end{bmatrix} = \begin{bmatrix} 0 & 5 & 0 & 100 & 5 & 0 \end{bmatrix}$$

$$\text{Target2:}\begin{bmatrix} x & \dot{x} & \ddot{x} & y & \dot{y} & \ddot{y} \end{bmatrix} = \begin{bmatrix} 1200 & 5 & 0 & -1200 & 5 & 0 \end{bmatrix};$$

(iii) sampling time $T = 1$ second.; (iv) false alarm density = 1.0×10^{-5}; (v) measurement noise covariance $R = 25$; (vi) process noise covariance (for model 1) $Q_1 = 0.1$; (vii) process noise covariance (for model 2) $Q_2 = 0.1$; (viii) number of data points $N = 150$; and (ix) maneuvering injection time and magnitude of injection for two targets (Table 3.2).

TABLE 3.2
Maneuvering Times and Magnitudes

Target number	1	2	3
Maneuvering time in seconds/ (start, end)	(30,50)	(70,100)	(50,80)
Maneuvering magnitude (in m/s^2)/(x, y, z)	(1g,−1g,0)	(1g,−1g,0)	(−1g,1g,0)

For data simulation for multiple sensors, we have the following: (i) number of sensors = 2; (ii) sensor locations – Sensor1 location: [0,0,0], Sensor2 location: [1000,1000,0]; (iii) number of targets = 3 for sensor 1 and 3 for sensor 2; (iv) target initial states:

$$\text{Target1:}\begin{bmatrix} x & \dot{x} & \ddot{x} & y & \dot{y} & \ddot{y} \end{bmatrix}=\begin{bmatrix} 0 & 5 & 0 & 100 & 5 & 0 \end{bmatrix}$$

$$\text{Target2:}\begin{bmatrix} x & \dot{x} & \ddot{x} & y & \dot{y} & \ddot{y} \end{bmatrix}=\begin{bmatrix} 1200 & 5 & 0 & -1200 & 5 & 0 \end{bmatrix}$$

$$\text{Target3:}\begin{bmatrix} x & \dot{x} & \ddot{x} & y & \dot{y} & \ddot{y} \end{bmatrix}=\begin{bmatrix} 1200 & 5 & 0 & -1200 & 5 & 0 \end{bmatrix}$$

$$\text{Target4:}\begin{bmatrix} x & \dot{x} & \ddot{x} & y & \dot{y} & \ddot{y} \end{bmatrix}=\begin{bmatrix} 1200 & 5 & 0 & -1200 & 5 & 0 \end{bmatrix};$$

(v) sampling time $T = 1$ second; (vi) false alarm density $= 1.0 \times 10^{-7}$; (vii) measurement noise covariance $R = 5$; (viii) process noise covariance (for model 1) $Q_1 = 0.01$; (ix) process noise covariance (for model 2) $Q_2 = 0.01$; (x) number of data points $N = 100$; and (xi) target 1 maneuvers with 1g acceleration from 40 to 50s., while there is no acceleration in the other three targets.

3.4.2.3 Performance Evaluation and Discussions

Performance of the IMMPDAF for single and multiple sensors/targets is evaluated. The KF with a CVM can be used for tracking maneuvering targets if a relatively large value of Q is used during the maneuver phase. During the non-maneuver phase, higher Q might result in a degraded performance. However, if a lower value of Q is used for tracking during maneuvers, there could be filter divergence and track loss due to the filter being unable to jack its performance during the maneuver. The sensitivity of the association algorithms to Q values and models is evaluated by considering the two combinations of models and Q values as per Table 3.3. The constant parameters used are as follows: (i) probability of detection $P_D = 0.99$, (ii) gate probability $P_G = 0.99998$, (iii) gating threshold $G = 25$, (iv) sojourn time $\tau = 15$ seconds, and (v) onset mode probability $p_{12} = 0.12$.

3.4.2.4 Evaluation of IMMPDAF

Figure 3.13 shows the x-position data of the two maneuvering targets in clutter. The performance of the algorithm is evaluated for two conditions of Q in terms of: (i) estimated and true x-position in clutter, (ii) estimated standard deviations ($\sigma_{x\text{-pos}}$), and (iii) RSSPE. For case 1, process noise covariance for CVM and CAM is chosen as 0.1 and 30, respectively, whereas for case 2, process noise covariance values of 0.1

TABLE 3.3
Range of Q for Sensitivity Study – IMMPDAF

Combinations	Case 1	Case 2
Q1	Low	Low
Q2	High	Low

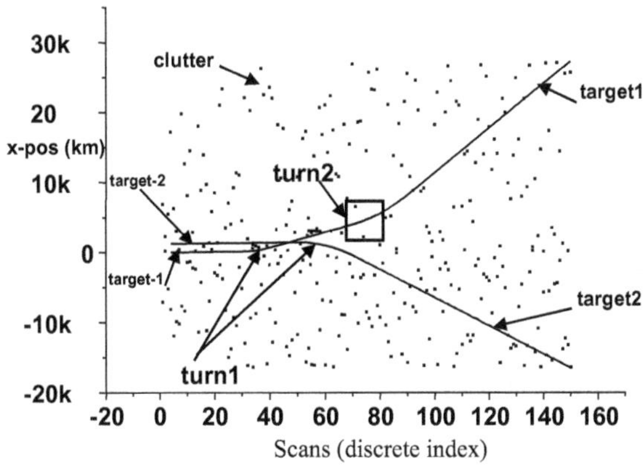

FIGURE 3.13 Simulated/true trajectories with clutter {Target-1 turns twice; target-2 turns once}.

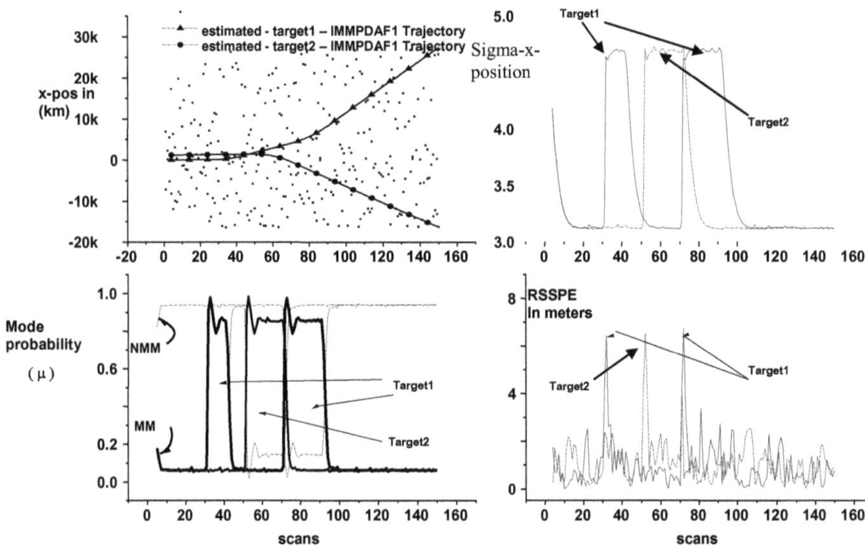

FIGURE 3.14 IMMPDAF – performance results (Q1 = 0.1 and Q2 = 30. NMM, non-maneuver mode; MM, maneuver mode).

and 2 are used. Figures 3.14 and 3.15 show the performance comparison for case 1 (IMMPDAF 1) and case 2 (IMMPDAF 2), respectively, in terms of estimated and measured tracks with clutter, mode probability of tracks, $\sigma_{x\text{-pos}}$, and RSSPE [16]. The tracking performance for both the combinations of Q is good, and the mode probability indicates the switching from the non-maneuver to the maneuver mode. The $\sigma_{x\text{-pos}}$ has a higher value during the maneuver which reflects the correct situation that the filter is adaptively tracking the target which is maneuvering. In case 2, there is delay in maneuver detection as compared to case 1.

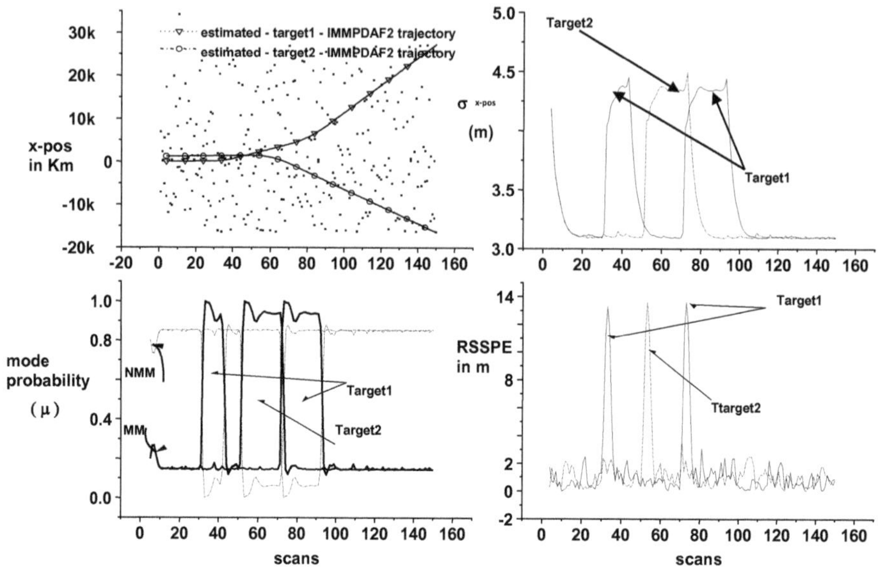

FIGURE 3.15 IMMPDAF – performance {Q1 = 0.1, and Q2 = 20}.

3.4.2.5 Multiple Sensors – Fusion of Data

The simulated data of four targets seen by two sensors are considered for fusion using the IMMPDAF (case 1 algorithm) where the data from each of the sensors are used to initiate tracks using the first two scans of measurements and the tracks are updated after measurement-to-track association with the valid measurements using IMMPDAF. It was necessary to transform the data to a common reference before fusion was carried out. Track-to-track association was used for combining similar tracks. In the scenario (among the four targets), the three targets are seen by both the sensors. Figure 3.16 depicts the measurements for sensor 1 and sensor 2, the estimated tracks, and mode probability. The four targets are seen from the estimated positions after combining similar tracks using track-to-track fusion, and it can be concluded that the event of target maneuvering is noticed in track-1 only. It is found that the IMMPDAF gives a realistic confidence in the estimates during maneuvers and lower RSSPE during the non-maneuvering phase of the targets.

3.5 TARGET TRACKING WITH IMAGING AND RADAR SENSORS

Often multiple sensors are used to enhance the target tracking capabilities. Radar can measure range with good resolution, but the angular measurements would have poor resolution. Radar provides sufficient information to track the target since it measures both angular and range to a target. The uncertainty associated with radar might be represented as a volume whose dimensions are relatively large perpendicular to the

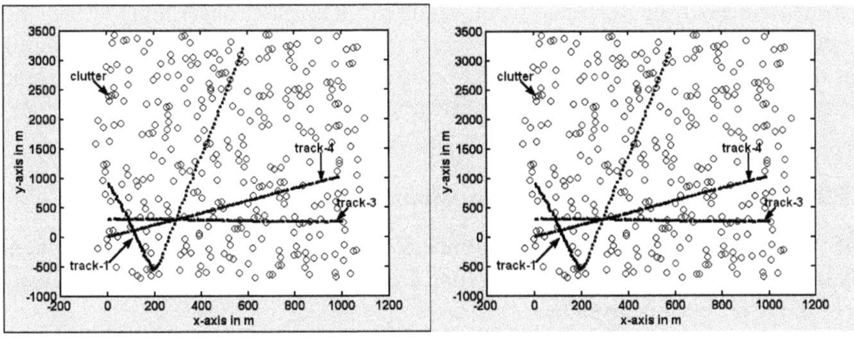

a) Observations trajectories: Sensor -1 & Sensor -2

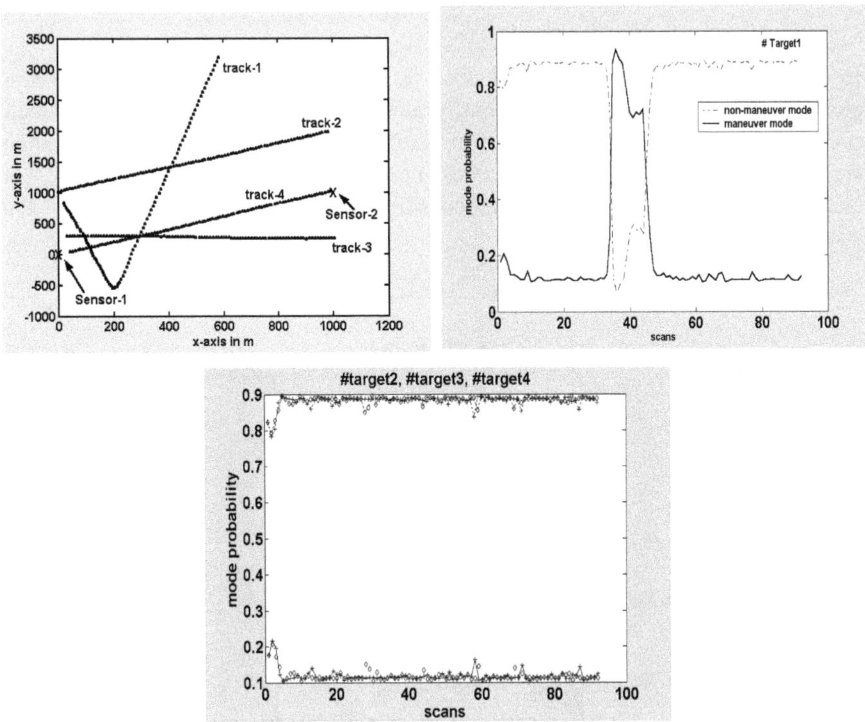

b) Estimated & fused tracks with the mode probabilities

FIGURE 3.16 Results of performance evaluation of IMMPDAF for MSDF task.

measured LOS and small along the LOS. An infrared search and track sensor (IRST) can measure azimuth and elevation of a target with good resolution. It can provide only the direction to a target but not its location because it does not provide the range. The uncertainty associated with IRST might be represented as a square whose dimensions are comparatively small perpendicular to the measured LOS. The fusion of measurements from radar and IRST would result in lesser uncertainty of the estimated position of the target. The state vector fusion of tracks obtained from imaging and radar sensors is discussed. IMMKF is used to track a maneuvering target using the measurements obtained from both imaging and radar sensors.

3.5.1 Passive Optical Sensor Mathematical Model

The imaging sensor produces a two-dimensional data array representing the intensity of the pixels [17,18]. The image at time k is $I(k)$ with size N by N. The intensity $I_{ij}(k)$ of the ij^{th} pixel is given by

$$I_{ij}(k) = \begin{cases} s_{ij}(k) + n_{ij}(k) \\ n_{ij}(k) \end{cases} \tag{3.54}$$

Here, $s_{ij}(k)$ is the target intensity of pixel (i, j) at time k, and $n_{ij}(k)$ is background intensity of pixel (i, j) at time k. Generally, an imaging sensor provides angular measurements. The centroid of the target in pixel coordinates is determined, and then, one can transform the pixel coordinates into equivalent angular coordinates. The relation between the angular and pixel coordinates is shown in Figure 3.17 [18]. The centroid Z_c can be computed as

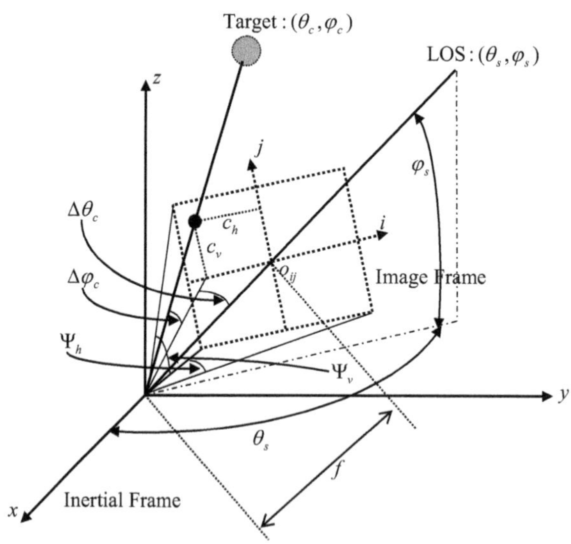

FIGURE 3.17 Angular and pixel coordinate relation.

$$Z_c(k) = \begin{bmatrix} c_h(k) \\ c_v(k) \end{bmatrix} = \frac{\sum_{i=-(N-1)/2}^{(N-1)/2} \sum_{j=-(N-1)/2}^{(N-1)/2} I_{ij}(k) \begin{bmatrix} i \\ j \end{bmatrix}}{\sum_{i=-(N-1)/2}^{(N-1)/2} \sum_{j=-(N-1)/2}^{(N-1)/2} I_{ij}(k)} \tag{3.55}$$

Here, c_h is the centroid of the target in i coordinate, and c_v is the centroid of the target in j coordinate. The distance spanned by each pixel in i and j coordinates is given as

$$u_h = f \frac{\tan\left(\frac{\psi_h}{2}\right)}{\frac{(N-1)}{2}} \text{ and } u_v = f \frac{\tan\left(\frac{\psi_v}{2}\right)}{\frac{(N-1)}{2}} \tag{3.56}$$

Here, ψ_h is the number of degrees spanned in i coordinates, ψ_v is the number of degrees spanned in j coordinates, u_h is the distance spanned in i coordinates, u_v is the distance spanned in j coordinates, and f is the focal length. The angular coordinates $(\Delta\theta_c(k), \Delta\phi_c(k))$ of the centroid $Z_c(k)$ are computes as

$$\Delta\theta_c(k) = \tan^{-1}\left(\frac{c_h(k)u_h}{f}\right) \text{ and } \Delta\phi_c(k) = \tan^{-1}\left(\frac{c_v(k)u_v}{f}\right) \tag{3.57}$$

The imaging sensor measurements in inertial frame are modeled as

$$\theta(k) = \theta_s(k) + \Delta\theta_c(k) + v_\theta(k)\,(\text{rad}) \tag{3.58}$$

$$\phi(k) = \phi_s(k) + \Delta\phi_c(k) + v_\phi(k)\,(\text{rad})$$

Here, $\theta_s(k)$ and $\phi_s(k)$ are the LOS of the sensor, and $v_\theta(k)$ and $v_\phi(k)$ are zero-mean white noise sequences with standard deviations σ_θ and σ_ϕ, respectively.

3.5.1.1 State Vector Fusion for Fusing IRST and Radar Data

Two tracking algorithms can be used to track the target independently, and then, final estimated target states can be fused using state vector fusion to get the improved target states [17,18]. The scheme is shown in Figure 3.18.

3.5.1.2 Extended Kalman Filter

A motion model for target tracking is given as

$$X(k) = \text{F}X(k-1) + \text{G}w(k-1) \tag{3.59}$$

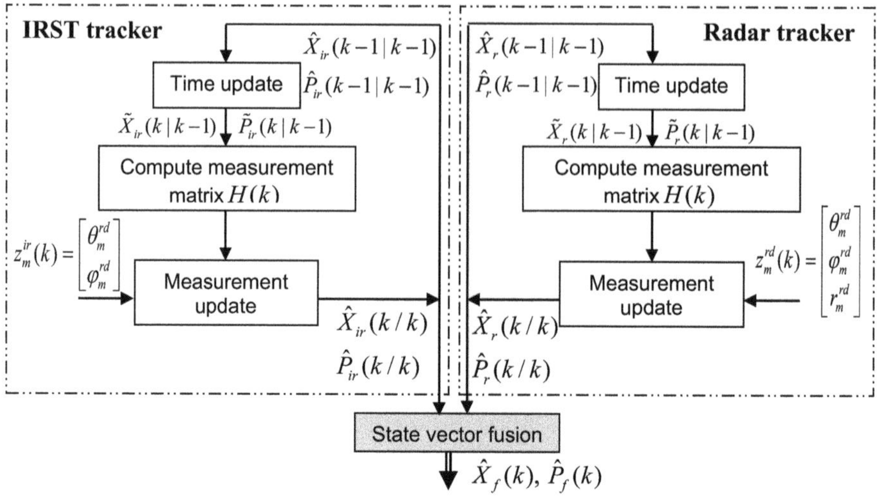

FIGURE 3.18 Track fusion: estimated from IRST and radar data.

$$z(k) = h(X(k)) + v(k) \tag{3.60}$$

Radar and IRST provide the measurements in a spherical coordinate system. In most cases, the state vector could be estimated in the Cartesian coordinate system. Consider the state vector consisting of position, velocity, and acceleration components in x-, y- and z-direction as

$$\begin{bmatrix} x & \dot{x} & \ddot{x} & y & \dot{y} & \ddot{y} & z & \dot{z} & \ddot{z} \end{bmatrix} \tag{3.61}$$

IRST sensor measurement vector consists of azimuth θ and elevation φ and is given in the vector form

$$z(k) = \begin{bmatrix} \theta & \varphi \end{bmatrix}^T \tag{3.62}$$

Radar measurement vector consists of azimuth, elevation, and range r and is given in the vector form

$$z(k) = \begin{bmatrix} \theta & \varphi & r \end{bmatrix}^T \tag{3.63}$$

The predicted state is in the form:

$$\begin{bmatrix} \tilde{x} & \dot{\tilde{x}} & \ddot{\tilde{x}} & \tilde{y} & \dot{\tilde{y}} & \ddot{\tilde{y}} & \tilde{z} & \dot{\tilde{z}} & \ddot{\tilde{z}} \end{bmatrix} = \tilde{X}(k \mid k-1) \tag{3.64}$$

The predicted measurement of IRST sensor is

$$\tilde{z}(k|k-1) = h[\tilde{X}(k|k-1)] = \begin{bmatrix} \tilde{\theta} & \tilde{\varphi} \end{bmatrix}^T \tag{3.65}$$

The predicted measurement of radar is

$$\tilde{z}(k|k-1) = h[\tilde{X}(k|k-1)] = \begin{bmatrix} \tilde{\theta} & \tilde{\varphi} & \tilde{r} \end{bmatrix}^T \tag{3.66}$$

Components in the predicted measurement are computed from the predicted state vector given in eqn. (3.64).

$$\tilde{\theta} = \tan^{-1}\left(\frac{\tilde{x}}{\tilde{y}}\right), \quad \tilde{\varphi} = \tan^{-1}\left(\frac{\tilde{z}}{\sqrt{\tilde{x}^2 + \tilde{y}^2}}\right), \quad \tilde{r} = \sqrt{\tilde{x}^2 + \tilde{y}^2 + \tilde{z}^2} \tag{3.66}$$

3.5.2 State Vector Fusion

Consider the tracks coming from IRST and radar whose state estimations and covariance matrices are at scan k as: Track from IRST sensor: $\hat{X}_1(k|k)$, $\hat{P}_1(k|k)$; Track from radar:

$$\hat{X}_2(k|k), \ \hat{P}_2(k|k) \tag{3.67}$$

These two tracks can be fused or combined using any one of the following schemes.

3.5.2.1 SVF1

Without consideration of the cross-covariance matrix, the fused state vector and covariance matrix can be given as

$$X_f(k) = \hat{X}_1(k|k) + \hat{P}_1(k|k)\left[\hat{P}_1(k|k) + \hat{P}_2(k|k)\right]^{-1}[\hat{X}_2(k|k) - \hat{X}_1(k|k)] \tag{3.68}$$

Or equivalently,

$$\begin{aligned} X_f(k) = {} & P_2(k|k)\left[\hat{P}_1(k|k) + \hat{P}_2(k|k)\right]^{-1} \hat{X}_1(k|k) \\ & + P_1(k|k)\left[\hat{P}_1(k|k) + \hat{P}_2(k|k)\right]^{-1} \hat{X}_2(k|k) \end{aligned} \tag{3.69}$$

$$\hat{P}_f(k) = \hat{P}_1(k|k) - \hat{P}_1(k|k)[\hat{P}_1(k|k) + \hat{P}_2(k|k)]^{-1} \hat{P}_1^T(k|k) \tag{3.70}$$

3.5.2.2 SVF2

With consideration of the cross-covariance matrix, the following computation is done during the time updating step. The predicted covariance matrices of individual sensors and cross-covariance matrix between two sensors are computed as

$$\tilde{P}_{11}(k|k-1) = F\hat{P}_{11}(k-1|k-1)F^T + GQG^T \tag{3.71}$$

$$\tilde{P}_{22}(k|k-1) = F\hat{P}_{22}(k-1|k-1)F^T + GQG^T \tag{3.72}$$

$$\tilde{P}_{12}(k|k-1) = F\hat{P}_{12}(k-1|k-1)F^T + GQG^T \tag{3.73}$$

$$\tilde{P}_{21}(k|k-1) = F\hat{P}_{21}(k-1|k-1)F^T + GQG^T \tag{3.74}$$

These covariance and cross-covariance matrices are updated during measurement updates. Covariance and Cross-covariance matrices are updated as

$$\hat{P}_{11}(k|k) = (I - K_1(k)H_1(k))\tilde{P}_{11}(k|k-1) \tag{3.75}$$

$$\hat{P}_{22}(k|k) = (I - K_2(k)H_2(k))\tilde{P}_{22}(k|k-1) \tag{3.76}$$

$$\hat{P}_{12}(k|k) = [I - K_1(k)H_1(k)]\tilde{P}_{12}(k|k-1)[I - H_2^T(k)K_2^T(k)] \tag{3.77}$$

$$\hat{P}_{21}(k|k) = [I - K_2(k)H_2(k)]\tilde{P}_{21}(k|k-1)[I - H_1^T(k)K_1^T(k)] \tag{3.78}$$

Here, K_1 and K_2 are the Kalman gains from IRST and radar trackers, respectively. H_1 and H_2 are the measurement matrices which are the nonlinear function of the sate vector from the IRST and radar trackers, respectively. Then, the fused state vector and covariance matrix are computed as follows:

Let

$$A_1 = \hat{P}_{11}(k|k) - \hat{P}_{12}(k|k);\ A_2 = \hat{P}_{22}(k|k) - \hat{P}_{21}(k|k) \tag{3.79}$$

The fusion covariance matrix can be computed as follows:

$$P_f(k) = P_{11}(k|k) - [P_{11}(k|k) - P_c(k|k)]P_e^{-1}(k)[P_{11}(k|k) - P_c(k|k)]^T \tag{3.80}$$

Here,

$$P_e(k) = P_{11}(k|k) + P_{22}(k|k) - P_c(k|k) - P_c(k|k)^T \tag{3.81}$$

$$\begin{aligned} P_c(k|k) = {} & (I - K_1(k)H_1(k))FP_c(k-1|k-1)F^T(I - K_2(k)H_2(k)) \\ & +(I - K_1(k)H_1(k))GQG^T(I - K_2(k)H_2(k))^T \end{aligned} \tag{3.82}$$

with $P_c(0\,|\,0) = 0$

Next, we obtain

$$P_c(k|k) = [I - K_1(k)H_1(k)]F\tilde{P}_{12}(k-1|k-1)F^T[I - H_2(k)K_2(k)] = \hat{P}_{12}(k|k) \quad (3.83)$$

$$P_c(k|k)^T = [I - K_2(k)H_2(k)]F\tilde{P}_{21}(k-1|k-1)F^T[I - H_1(k)K_1(k)] = \hat{P}_{21}(k|k) \quad (3.84)$$

$$P_e(k) = \hat{P}_{11}(k|k) + \hat{P}_{22}(k|k) - \hat{P}_{12}(k|k) - \hat{P}_{21}(k|k) = A_1 + A_2 \quad (3.85)$$

Finally, the fused covariance matrix can be written as

$$\begin{aligned} P_f(k) &= \hat{P}_{11}(k|k) - (\hat{P}_{11}(k|k) - \hat{P}_{12}(k|k))P_e^{-1}(\hat{P}_{11}(k|k) - \hat{P}_{12}(k|k))^T \\ P_f(k) &= \hat{P}_{11}(k|k) - A_1[A_1 + A_2]^{-1}A_1^T \end{aligned} \quad (3.86)$$

The fused state vector can be computed as

$$X_f(k) = \hat{X}_1(k|k) + A_1[A_1 + A_2]^{-1}[\hat{X}_2(k|k) - \hat{X}_1(k|k)] \quad (3.87)$$

Or equivalently,

$$X_f(k) = A_2[A_1 + A_2]^{-1}\hat{X}_1(k|k) + A_1[A_1 + A_2]^{-1}\hat{X}_2(k|k) \quad (3.88)$$

3.5.2.3 Numerical Simulation

The 3DOF kinematic model with position, velocity, and acceleration components in each of the three Cartesian coordinates x, y, and z has the following transition and process noise gain matrices:

$$F = \text{diag}\begin{bmatrix} \Phi & \Phi & \Phi \end{bmatrix} \quad G = \text{diag}\begin{bmatrix} \zeta & \zeta & \zeta \end{bmatrix} \quad (3.89)$$

Here, $\Phi = \begin{bmatrix} 1 & T & T^2/2 \\ 0 & 1 & T \\ 0 & 0 & 1 \end{bmatrix} \quad \zeta = \begin{bmatrix} T^3/6 \\ T^2/2 \\ T \end{bmatrix}$

and T is a sampling interval, F is a state transition matrix, and G is a process noise gain matrix. The algorithm is validated using the simulated data which utilizes the parameters: (i) sampling interval = 1 second, (ii) process noise variance = 0.00001^2, (iii) measurement noise variance as per Table 3.4, and (iv) duration of simulation = 500 seconds. The simulated IRST and radar measurements are shown in Figure 3.19. The initial state vector is

TABLE 3.4
Measurement Variance

Sensor	Azimuth (rad)	Elevation (rad)	Range (m)
IRST	10^{-6}	10^{-6}	–
Radar	10^{-2}	10^{-2}	50

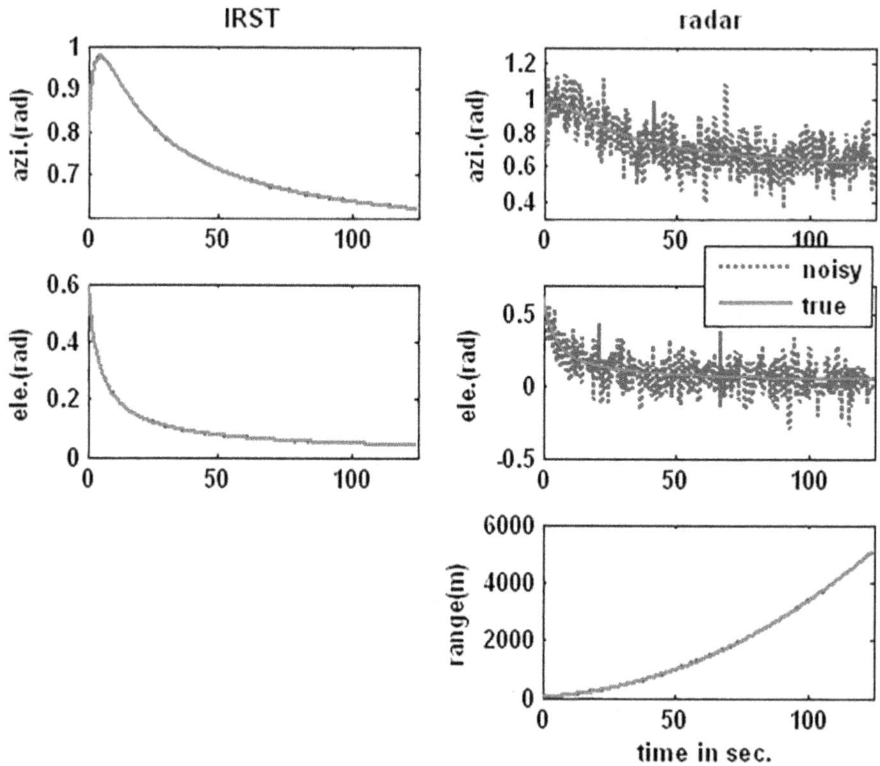

FIGURE 3.19 Measurements from IRST and radar.

$$\begin{bmatrix} x & \dot{x} & \ddot{x} & y & \dot{y} & \ddot{y} & z & \dot{z} & \ddot{z} \end{bmatrix}$$

$$= \begin{bmatrix} 100 & -0.2 & -0.05 & 100 & -2 & 0.01 & 100 & -0.5 & 0.1 \end{bmatrix}$$

The tracking algorithm is evaluated using 75 Monte Carlo runs. The PFE and MAE metrics for x-, y-, and z-positions are given in Table 3.5 and the RMS errors in position (RMSPE), RMS errors in velocity (RMSVE), and RMS errors in acceleration (RMSAE) metrics in Table 3.6. It is observed that after fusion, the PFE is comparatively very less which indicates the true and estimated positions are well matched. The RSS acceleration errors are shown in Figure 3.20, and it is observed that after

TABLE 3.5
PFE and MAE Metrics

	PFE*x*	PFE*y*	PFE*z*	MAE*x*	MAE*y*	MAE*z*
IRST	9.07	9.07	9.07	130	100	8.94
Radar	0.36	0.6	4.21	5.29	6.36	4.09
SVF1	0.05	0.08	0.33	0.67	0.76	0.26

TABLE 3.6
RMSPE, RMSVE, and RMSAE Metrics

	RMSPE	RMSVE	RMSAE
IRST	123.99	2.415	0.031
Radar	7.0	0.247	0.108
SVF1	0.853	0.108	0.009

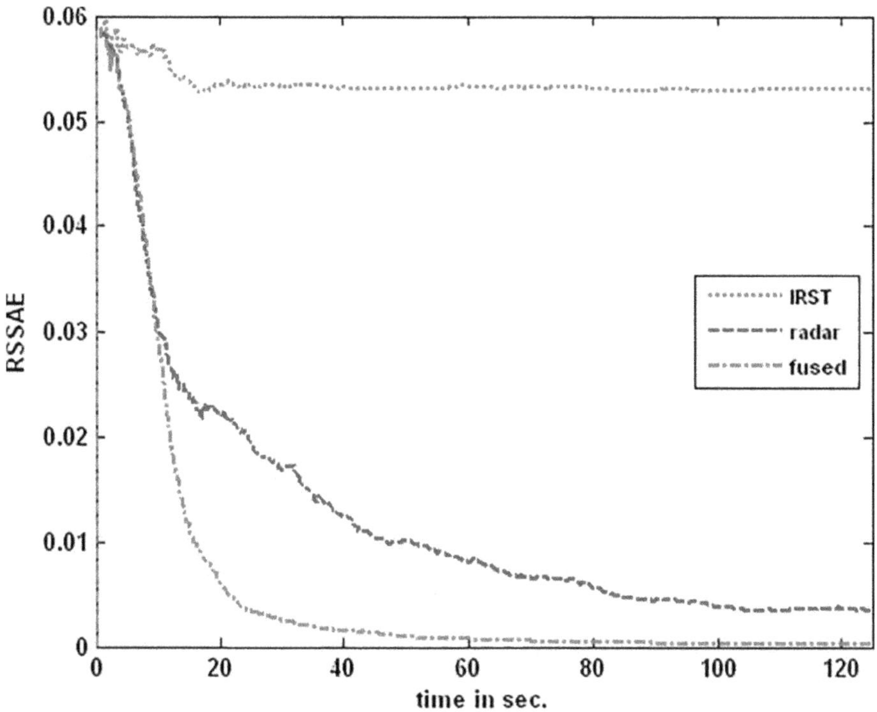

FIGURE 3.20 RSS acceleration errors.

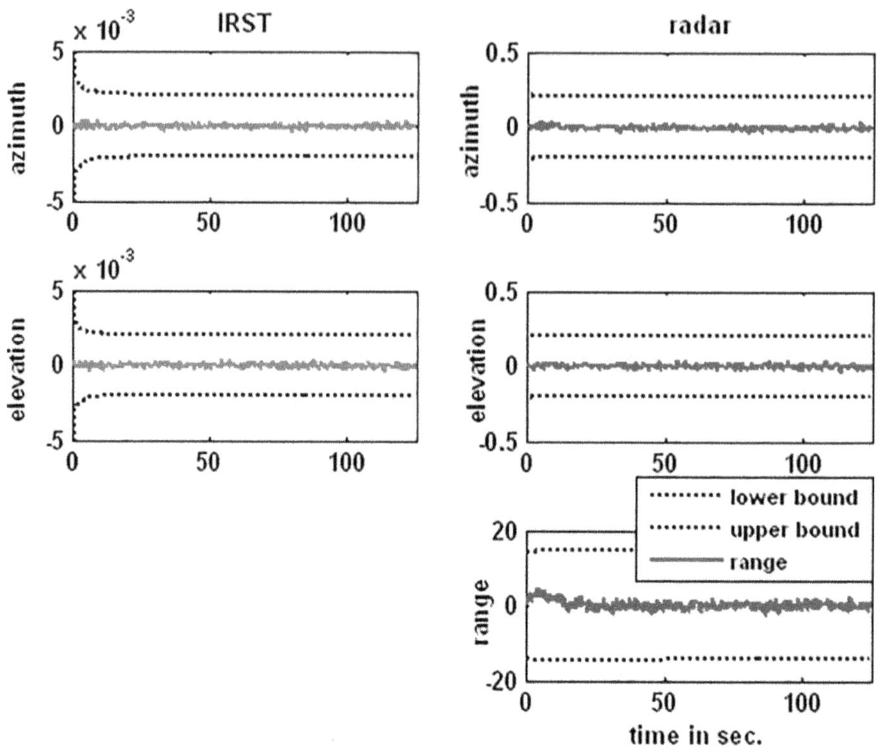

FIGURE 3.21 Innovation sequences.

fusion, these errors are comparatively very less and the estimated trajectory follows the true trajectory. After fusion, the uncertainty in estimation is also comparatively reduced. The innovation sequences with their bounds for both IRST and radar trackers are shown in Figure 3.21.

3.5.3 Measurement Fusion

In this section, two-measurement vector fusion schemes are given. In first method (MF1), the measurements from IRST and radar are merged into an augmented measurement vector, and in the second method (MF2), the measurements from IRST and radar are combined by using covariance matrices as weights. The information flow diagram is shown in Figure 3.22.

3.5.3.1 MF1

The measurement vectors $z_m^{ir}(k)$ and $z_m^{rd}(k) \in$ from the IRST and radar are merged into an augmented measurement vector as

$$z_m(k) = \begin{bmatrix} z_m^{ir}(k) \\ z_m^{rd}(k) \end{bmatrix} \tag{3.90}$$

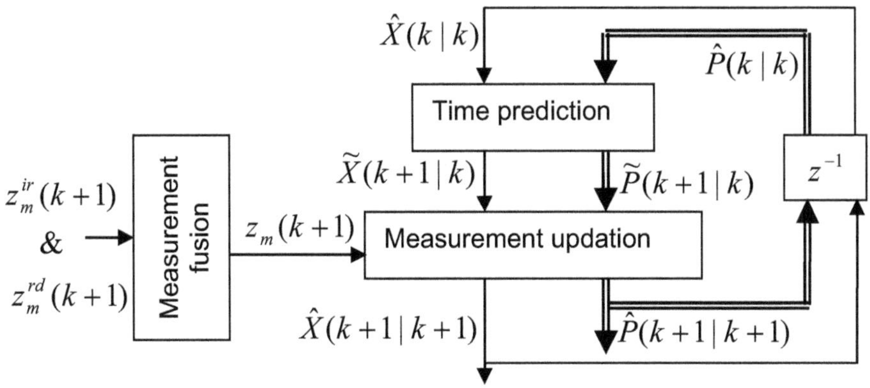

FIGURE 3.22 Information flow diagram for MF.

Here, $z_m^{ir}(k) = \begin{bmatrix} \theta_m^{ir}(k) & \phi_m^{ir}(k) \end{bmatrix}^T$, and $z_m^{rd}(k) = \begin{bmatrix} \theta_m^{rd}(k) & \phi_m^{rd}(k) & r_m^{rd}(k) \end{bmatrix}^T$.

The observation matrices of IRST and radar are merged into the observation matrix as

$$h(X(k)) = \begin{bmatrix} h^{ir}(X(k)) \\ h^{rd}(X(k)) \end{bmatrix} \tag{3.91}$$

The measurement noise covariance matrices of IRST and radar are merged as

$$R = \begin{bmatrix} R^{ir} & 0 \\ 0 & R^{rd} \end{bmatrix} \tag{3.92}$$

Here, $R^{ir} = \begin{bmatrix} (\sigma_\theta^{ir})^2 & 0 \\ 0 & (\sigma_\phi^{ir})^2 \end{bmatrix}$, and $R^{rd} = \begin{bmatrix} (\sigma_\theta^{rd})^2 & 0 & 0 \\ 0 & (\sigma_\phi^{rd})^2 & 0 \\ 0 & 0 & (\sigma_r^{rd})^2 \end{bmatrix}$.

3.5.3.2 MF2

The weighted combination of measurements based on covariance matrices is considered to fuse measurement vectors [18]:

$$z_m(k) = \frac{c_1 z_m^{ir}(k) + c_2 z_m^{rd}(k)}{c_1 + c_2} \tag{3.93}$$

Here, c_1 and c_2 are the weights. The weights in the above equation are computed from the measurement noise covariance matrix as

$$c_1 = \frac{1}{R^{ir}} \text{ and } c_2 = \frac{1}{R^{rd}} \tag{3.94}$$

$$z_m(k) = \frac{\frac{1}{R^{ir}} z_m^{ir}(k) + \frac{1}{R^{rd}} z_m^{rd}(k)}{\frac{1}{R^{ir}} + \frac{1}{R^{rd}}}$$

The MF2 is

$$z_m(k) = \frac{R^{rd} z_m^{ir}(k) + R^{ir} z_m^{rd}(k)}{R^{ir} + R^{rd}} \tag{3.95}$$

Or

$$\begin{aligned} z_m(k) &= \frac{R^{rd} z_m^{ir}(k)}{R^{ir} + R^{rd}} + \frac{R^{ir} z_m^{rd}(k)}{R^{ir} + R^{rd}} + \frac{R^{ir} z_m^{ir}(k)}{R^{ir} + R^{rd}} - \frac{R^{ir} z_m^{ir}(k)}{R^{ir} + R^{rd}} \\ z_m(k) &= z_m^{ir}(k) + R^{ir}(R^{ir} + R^{rd})^{-1}(z_m^{rd}(k) - z_m^{ir}(k)) \end{aligned} \tag{3.96}$$

The associated measurement noise covariance matrix (R) of the fused measurement vector (eqn. 3.96) is computed as follows:

$$\frac{1}{R} = \frac{1}{R^{ir}} + \frac{1}{R^{rd}} \tag{3.97}$$

$$R = \left((R^{ir})^{-1} + (R^{rd})^{-1}\right)^{-1} \tag{3.98}$$

Or

$$\begin{aligned} R &= \frac{R^{ir} R^{rd}}{R^{ir} + R^{rd}} = \frac{R^{ir} R^{rd}}{R^{ir} + R^{rd}} - R^{ir} + R^{ir} \\ R &= R^{ir} + \frac{R^{ir} R^{rd} - R^{ir}(R^{ir} + R^{rd})}{R^{ir} + R^{rd}} \\ R &= R^{ir} - R^{ir}(R^{ir} + R^{rd})^{-1}(R^{ir})^T \end{aligned} \tag{3.99}$$

3.5.4 Maneuvering Target Tracking

In tracking applications for a target moving with CV, the state model includes the first derivative of the position, and for a target moving with CA, the state model includes second derivative of position. Models with second-order derivatives are preferred for tracking maneuvering targets and referred to as accelerations models. The IMMKF is an adaptive state estimator, which is based on the assumption that a finite number of models are required to characterize the target motion at all times.

3.5.4.1 Motion Models

The two types of models in the Cartesian frame are considered for tracking a maneuvering target:

i. CV model (2DOF kinematic model with position and velocity) with the state transition and process noise gain matrices is given as

$$F_1 = F_{CV} = \begin{bmatrix} \Phi_V & 0_{3\times3} & 0_{3\times3} \\ 0_{3\times3} & \Phi_V & 0_{3\times3} \\ 0_{3\times3} & 0_{3x3} & \Phi_V \end{bmatrix} G_1 = G_{CV} = \begin{bmatrix} \varsigma_V & 0_{3\times1} & 0_{3\times1} \\ 0_{3\times1} & \varsigma_V & 0_{3\times1} \\ 0_{3\times1} & 0_{3\times1} & \varsigma_V \end{bmatrix} \tag{3.100}$$

Here,

$$\Phi_V = \begin{bmatrix} 1 & T & 0 \\ 0 & 1 & 0 \\ 0 & 0 & 0 \end{bmatrix} \varsigma_V = \begin{bmatrix} T^2/2 \\ T \\ 0 \end{bmatrix} 0_{3\times3} = \begin{bmatrix} 0 & 0 & 0 \\ 0 & 0 & 0 \\ 0 & 0 & 0 \end{bmatrix} 0_{3\times1} = \begin{bmatrix} 0 \\ 0 \\ 0 \end{bmatrix}$$

$$\varsigma_V = \begin{bmatrix} T^2/2 \\ T \\ 0 \end{bmatrix}$$

The state vector is given as $X = \begin{bmatrix} x & \dot{x} & 0 & y & \dot{y} & 0 & z & \dot{z} & 0 \end{bmatrix}^T$. The process noise intensity (air turbulence, slow turns, and small linear acceleration) in each axis is generally assumed to be small and equal in all three axes $\left(\sigma_x^2 = \sigma_y^2 = \sigma_z^2\right)$. It is noted that the acceleration components in the model, though identically equal to zero, have been retained for dimensional compatibility with the third-order acceleration model.

ii. The CA model (3DOF order kinematic model with position, velocity, and acceleration) has the following state transition and process noise gain matrices:

$$F_2 = F_{CA} = \begin{bmatrix} \Phi_A & 0_{3\times3} & 0_{3\times3} \\ 0_{3\times3} & \Phi_A & 0_{3\times3} \\ 0_{3\times3} & 0_{3\times3} & \Phi_A \end{bmatrix} G_2 = G_{CA} = \begin{bmatrix} \varsigma_A & 0_{3\times1} & 0_{3\times1} \\ 0_{3\times1} & \varsigma_A & 0_{3\times1} \\ 0_{3\times1} & 0_{3\times1} & \varsigma_A \end{bmatrix} \tag{3.101}$$

Here,

$$\Phi_A = \begin{bmatrix} 1 & T & T^2/2 \\ 0 & 1 & T \\ 0 & 0 & 1 \end{bmatrix} \varsigma_A = \begin{bmatrix} T^3/6 \\ T^2/2 \\ T \end{bmatrix} 0_{3\times3} = \begin{bmatrix} 0 & 0 & 0 \\ 0 & 0 & 0 \\ 0 & 0 & 0 \end{bmatrix} 0_{3\times1} = \begin{bmatrix} 0 \\ 0 \\ 0 \end{bmatrix}$$

The state vector in this model is $X = \begin{bmatrix} x & \dot{x} & \ddot{x} & y & \dot{y} & \ddot{y} & z & \dot{z} & \ddot{z} \end{bmatrix}^T$. A moderate value of process noise variance Q_{CA} (relatively higher than Q_{CV}) will yield nearly a CA motion. The noise variance is assumed as $\left(\sigma_x^2 = \sigma_y^2 = \sigma_z^2\right)$.

3.5.4.2 Measurement Model

The measurement vector of the IRST is azimuth (θ) and elevation (φ) as given by

$$z_m^{ir}(k) = \begin{bmatrix} \theta_m^{ir}(k) & \phi_m^{ir}(k) \end{bmatrix}^T \tag{3.102}$$

The measurement vector of the radar is range (r), azimuth, and elevation and is given by

$$z_m^{rd}(k) = \begin{bmatrix} \theta_m^{rd}(k) & \phi_m^{rd}(k) & r_m^{rd}(k) \end{bmatrix}^T \tag{3.103}$$

These are related to states in a nonlinear fashion as

$$\begin{aligned} x(k) &= r(k)\cos\theta(k)\cos\phi(k) \\ y(k) &= r(k)\sin\theta(k)\cos\phi(k) \\ z(k) &= r(k)\sin\phi(k) \end{aligned} \tag{3.104}$$

The measurement equation is given as follows:

$$z_m(k+1) = h(X(k+1) + v(k+1) \tag{3.105}$$

3.5.4.3 Numerical Simulation

The target trajectory is simulated in Cartesian coordinates with a sampling interval of 0.25 second using the CV model for the first 50 seconds followed by the CA model for 25 s and then the CV model for 50 seconds. Acceleration magnitudes of 3 and –3 m/s^2 are injected at the time of maneuver in the x- and y-axis, respectively. Initial sate vector is $X = [x \; \dot{x} \; \ddot{x} \; y \; \dot{y} \; \ddot{y} \; z \; \dot{z} \; \ddot{z}]^T = [100 \; 50 \; 0 \; 100 \; 50 \; 0 \; 10 \; 1 \; 0]^T$. The range, azimuth, and elevation data of radar are obtained from the following transformations:

$$\begin{aligned} r_m^{rd}(k) &= \sqrt{x^2(k) + y^2(k) + z^2(k)} + v_r^{rd}(k) \\ \theta_m^{rd}(k) &= \tan^{-1}\left(\frac{y(k)}{x(k)}\right) + v_\theta^{rd}(k) \\ \phi_m^{rd}(k) &= \tan^{-1}\left(\frac{z(k)}{\sqrt{x^2(k) + y^2(k)}}\right) + v_\phi^{rd}(k) \end{aligned} \tag{3.106}$$

The azimuth and elevation data of IRST are obtained from the following transformations:

$$\theta_m^{ir}(k) = \tan^{-1}\left(\frac{y(k)}{x(k)}\right) + v_\theta^{ir}(k)$$
$$\varphi_m^{ir}(k) = \tan^{-1}\left(\frac{z(k)}{\sqrt{x^2(k)+y^2(k)}}\right) + v_\phi^{ir}(k) \tag{3.107}$$

Measurement noise variance is given in Table 3.7. Duration of simulation is 125 s. The resultant noisy IRST and radar measurements are shown in Figure 3.23.

TABLE 3.7
Measurement Variance

Sensor	Azimuth (rad)	Elevation (rad)	Range (m)
IRST	1×10^{-6}	1×10^{-6}	–
Radar	1×10^{-2}	1×10^{-2}	100

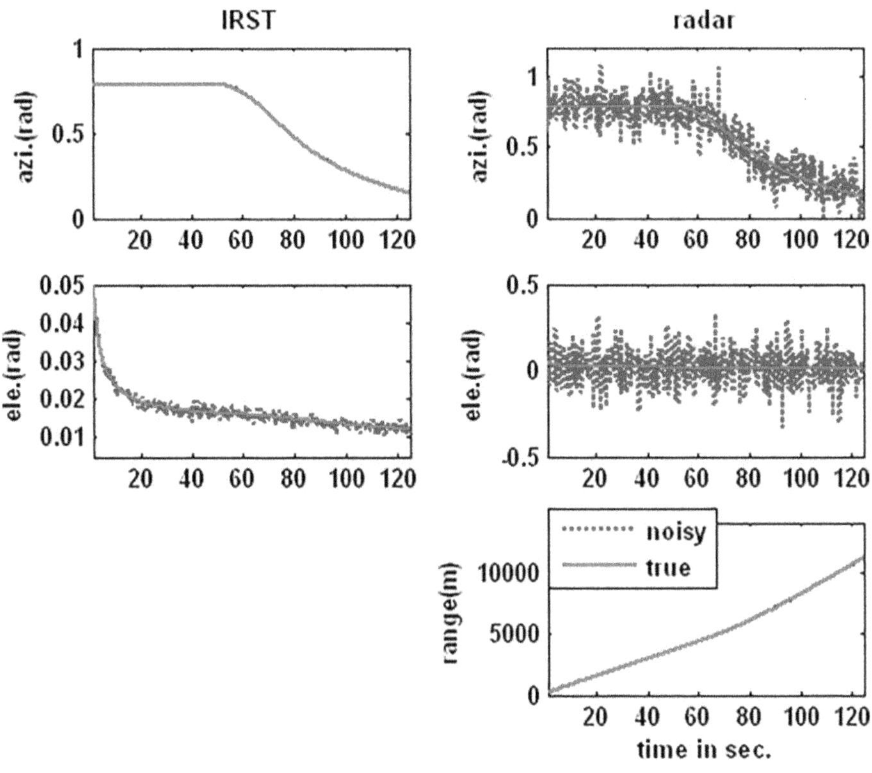

FIGURE 3.23 Simulated measurements from IRST and radar.

The process noise variance of $\sigma_x^2 = \sigma_y^2 = \sigma_z^2 = 0.00001$ is considered for the CVM, and $\sigma_x^2 = \sigma_y^2 = \sigma_z^2 = 3$ is considered for the CAM. The Markov chain transition matrix is $p = \begin{bmatrix} 0.99 & 0.01 \\ 0.01 & 0.99 \end{bmatrix}$, and the initial mode probabilities $\mu = \begin{bmatrix} \mu_1 & \mu_2 \end{bmatrix}^T$ corresponding to the non-maneuver and maneuver mode can be taken as 0.9 and 0.1, respectively.

The true and estimated velocities and accelerations in all axes are shown in Figures 3.24 and 3.25. The estimated positions, velocities, and accelerations match well with the true trajectories. There is a little delay in the acceleration estimation. It is because of only position measurements utilized in the state estimation. The velocity and acceleration estimates in the z-axis during the maneuvering portion are a little degraded. It could be due to the active CAM even though there is no maneuver in the z-axis. It could be due to the high process noise. This problem can be solved by providing less process noise to this axis. The RSS errors in position, velocity, and acceleration are shown in Figure 3.26. These errors are high during the maneuvering period. The state errors in the *x*-, *y*-, and *z*-position with theoretical bounds are shown in Figure 3.27. These errors are within the bounds. Similar observations are made for velocities and acceleration states. It is observed that the bounds are high since the CA model is active during this portion. RSvarP, RSvarV, and RSvarA are shown

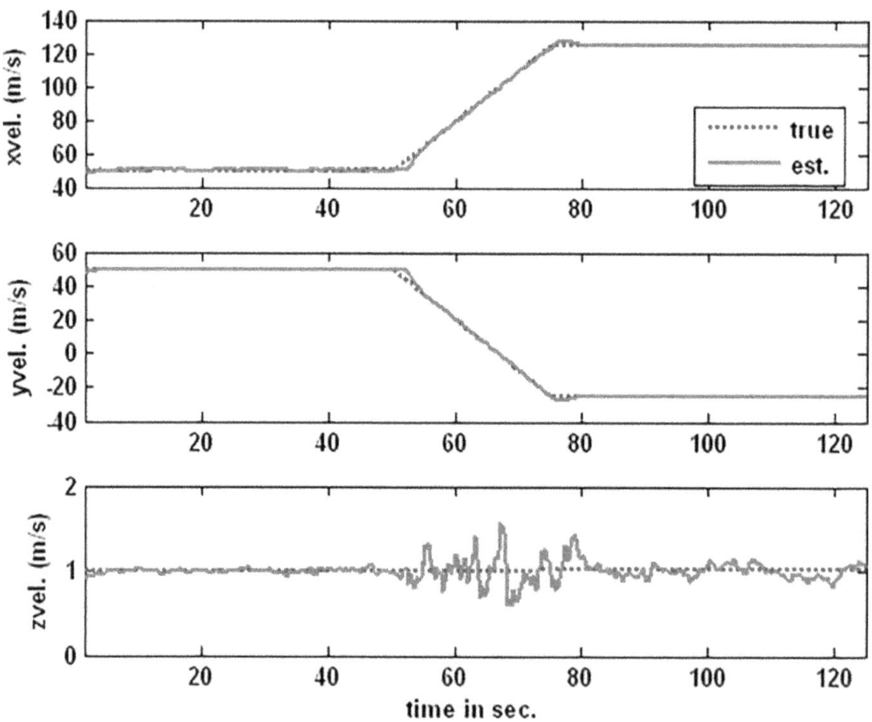

FIGURE 3.24 True and estimated velocities.

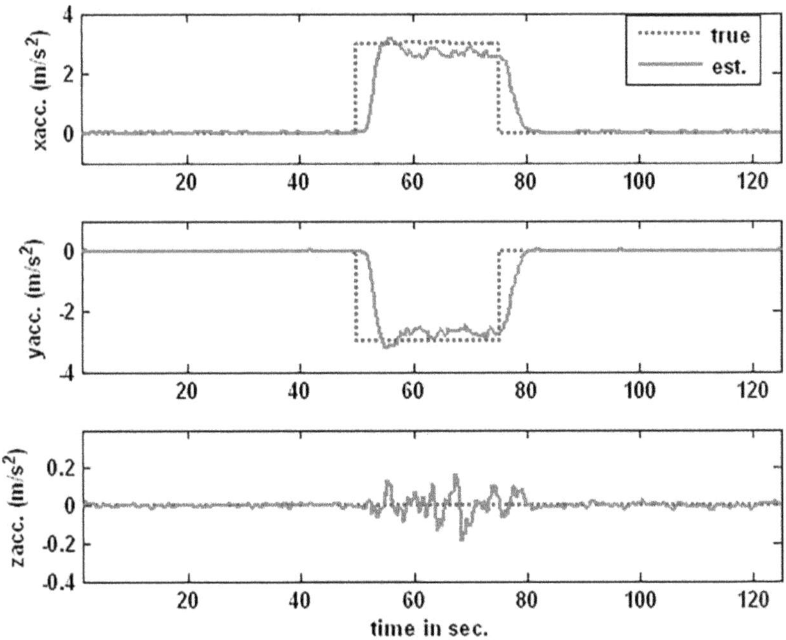

FIGURE 3.25 True and estimated accelerations.

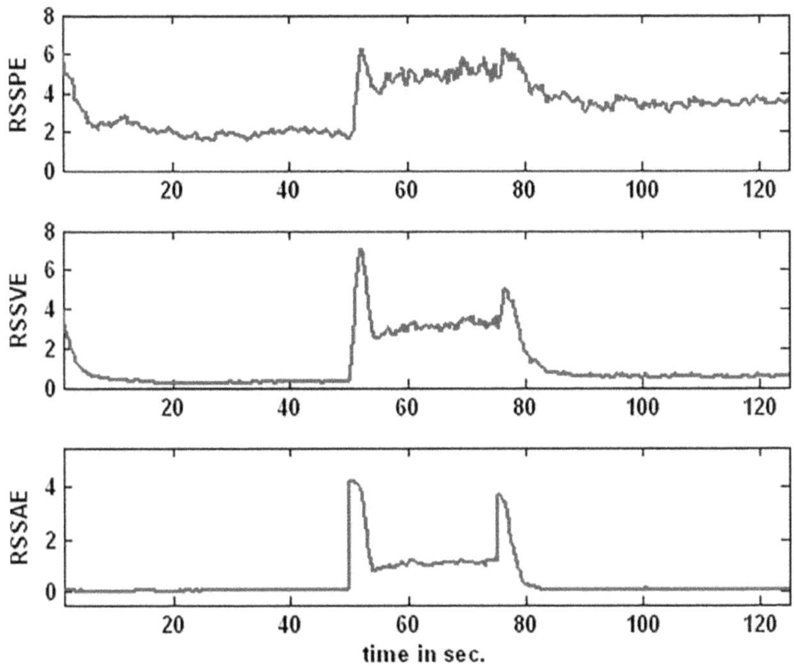

FIGURE 3.26 RSS errors in position, velocity, and acceleration.

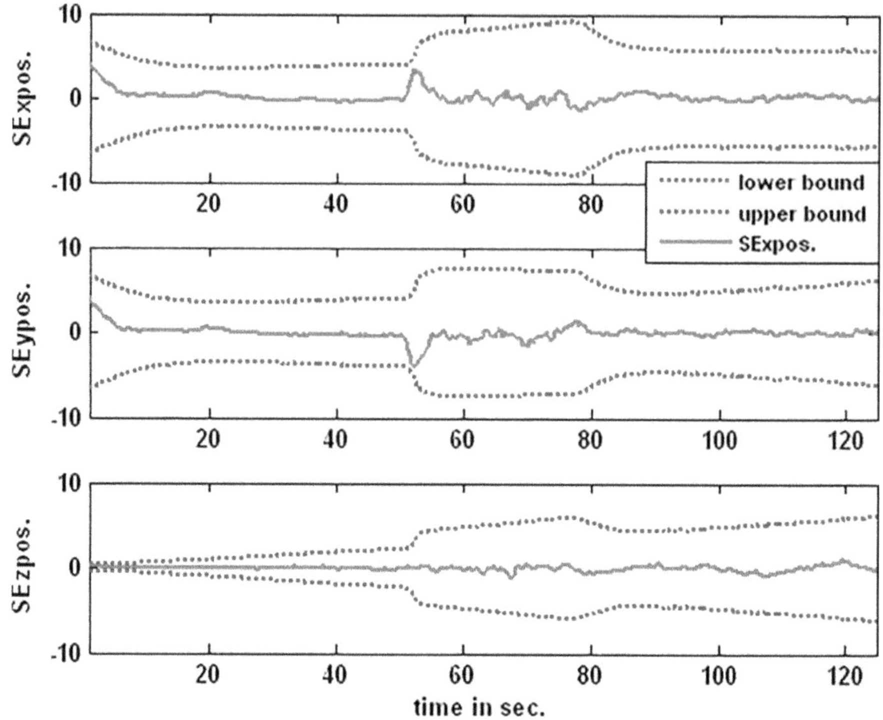

FIGURE 3.27 Sate errors in positions with bounds.

in Figure 3.28. As expected, the variance is very low during the non-maneuvering portion and high during the maneuvering portion. It could be due to the following reasons: the CV model is active during the non-maneuvering portion, and the CA model is active during the maneuvering portion. This information could be very useful in the fusion process because it is realistic representation of the maneuver. The mode probabilities are shown in Figure 3.29, showing the switching between the models. This information is useful in knowing which model is active in a given time (interval), and useful in our confidence in state and state error covariance matrix (combination). The PFE metrics for positions, velocities, and acceleration of all three axes are given in Table 3.8a. The MA errors in positions, velocities, and accelerations of all axes are shown in Table 3.8b. The RMSPE, RMSVE, and RMSAE are given in Table 3.8c.

3.6 TARGET TRACKING WITH ACOUSTIC ARRAY SENSORS AND IMAGING SENSOR DATA

An approach to joint audio-video tracking based on acoustic sensor array (ASA) modelling, uniform linear array sensor data generation, and tracking a target in Cartesian coordinates is discussed. Also, the joint audio-video tracking of a target in 3DOF Cartesian coordinates using decentralized KF is discussed.

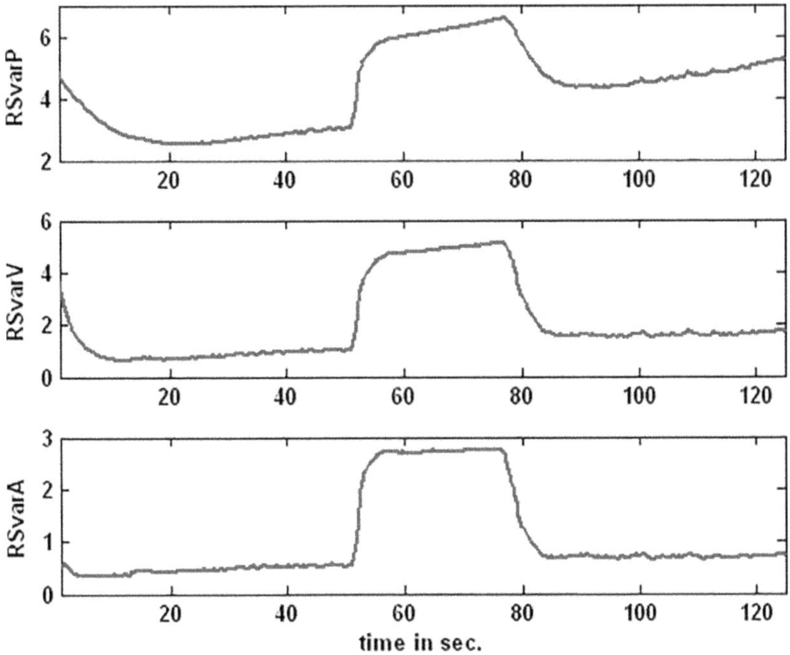

FIGURE 3.28 Position (RSvarP), velocity (RSvarV), and acceleration (RSvarA) errors.

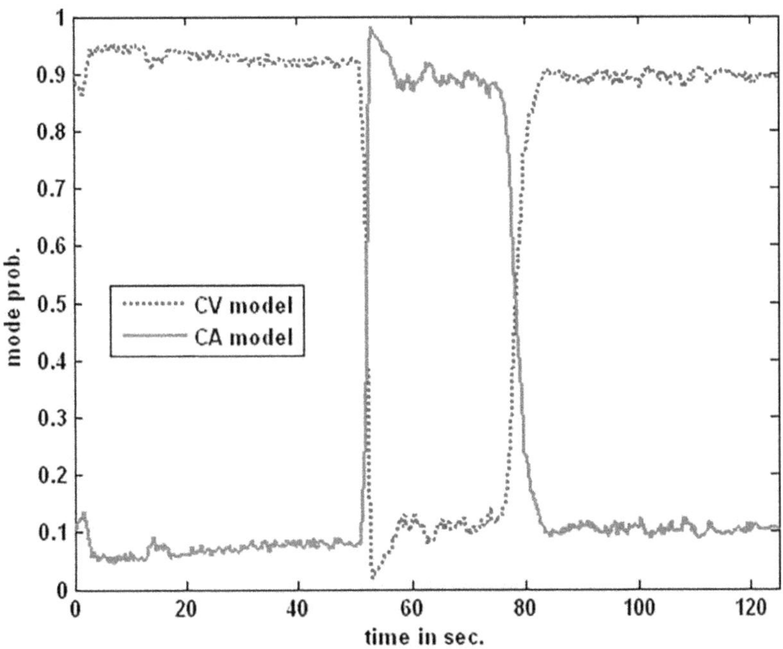

FIGURE 3.29 Mode probabilities.

TABLE 3.8A
PFE Metrics in *x*-, *y*-, and *z*-Positions, Velocities, and Accelerations

PFEx	PFEy	PFEz	PFE$\dot{x}$	PFE$\dot{y}$	PFE$\dot{z}$	PFE$\ddot{x}$	PFE$\ddot{y}$	PFE$\ddot{z}$
0.049	0.113	2.175	1.43	3.56	78.11	46.947	46.987	61,212.7

TABLE 3.8B
MA Error Metrics in *x*-, *y*-, and *z*-Positions, Velocities, and Accelerations

MAEx	MAEy	MAEz	MAE$\dot{x}$	MAE$\dot{y}$	MAE$\dot{z}$	MAE$\ddot{x}$	MAE$\ddot{y}$	MAE$\ddot{z}$
1.1995	1.864	1.23	0.734	0.714	0.428	0.245	0.245	0.11

TABLE 3.8C
RMS Errors in Position, Velocity, and Acceleration Metrics

RMSPE	RMSVE	RMSAE
2.339	1.185	0.536

3.6.1 Tracking with Multi-Acoustic Array Sensors

Target tracking using acoustic array sensor is very useful in air traffic control, air defense, mobile user location in cellular communications, military, underwater tracking, and acoustic source localization [19]. In ASA tracking, the sensor of the array would receive the acoustic signal emitted by the moving target. ASA-based target tracking is inexpensive compared to other modalities and requires less power, and the system may include magnetic sensors and imaging sensors. When the target is in the vicinity of the acoustic array, acoustic sounds could be used to determine the location, range, and angle of the target in polar coordinates, and when the target is in the far-field of the acoustic array, only bearing (angle) information could be extracted. Hence, multiple acoustic array sensors are needed in order to determine the target location. ASA output data are analyzed by well-known multiple signal classification (MUSIC) algorithm to estimate the direction of arrival (DoA) of the emitted acoustic signal [20,21]. The DoA estimation based on the batch processing such as MUSIC, minimum norm, does not reuse the information from the previous batches to help refine the estimates at the current batch. Since ASA provides only the angle (azimuth), it would be difficult to track the target in Cartesian coordinates with single angle information. Hence, multi-acoustic sensor arrays (MASA) are required to provide multiple angle measurements of the target at the same instant of time and the target location then could be computed by the triangulation method [22,23]. It is assumed that the target is in the far-field of the array. Two such arrays are used to get the angle-bearing of the target. Three tracking schemes for an acoustic target in Cartesian coordinates are studied here: (i) a digital filter with least square (LS) estimation is used for estimation of the target position in Cartesian coordinates; (ii) KF with LS estimation is used to track the target in Cartesian coordinates. It could be straightforward since the time propagation and measurement updating are in polar coordinates; and (iii) EKF is used where the state estimation is in Cartesian coordinates and measurement updating in polar coordinates.

3.6.1.1 Modelling of Acoustic Sensors

The ASA contains m sensors uniformly placed (with distance d) along a line as in Figure 3.30, with $d = \lambda/2$ that prevents spatial aliasing (λ is the carrier wavelength) [24]. It is assumed that the sound waves reaching each sensor are coming in parallel. The direction perpendicular to the ASA is the broadside direction (BD). When a target is moving, its DoA would be measured with respect to BD. The target is assumed to be moving with a CV (with minor random perturbations) and emitting narrow-band signals of wavelength λ which impinge on ASA (which are passive) at an angle of θ. The signal from the target reaches the sensors at different times because each sound wave travels a different distance to reach the different sensors. For example, the signal incident on sensor S_{r1} has to travel an extra distance of $d\sin(\theta_1)$ as compared to the signal incident on sensor S_{r2} (the delay being $-d\sin(\theta_1)/c$) (here, c is the rate of propagation of the signal). Let l ($l < m$) be the number of targets in the test scenario; then, one can express the jth sensor output as the sum of the shifted versions of the source signals represented as [18]

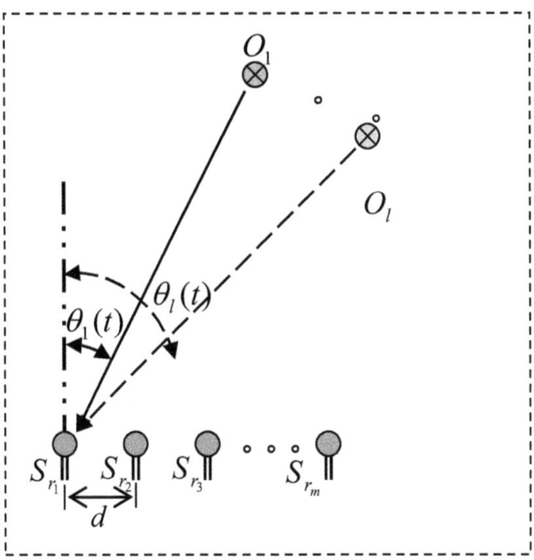

FIGURE 3.30 ASA geometry.

$$X_{a_j}(t) = \sum_{i=1}^{l} S_{a_i}(t + \tau_{ij}(t)) + e_{a_j}(t) \tag{3.108}$$

$$\tau_{ij}(t) = \frac{1}{c}\underline{z}_j \underline{k}_i(t)$$

$$\text{with } i = 1,2,...,l \quad j = 1,2,...,m$$

Here, we have the following: (i) $X_{a_j}(t)$ as the jth sensor output at time t, (ii) $S_{a_i}(t)$ as i^{th} target signal waveform, (iii) $\tau_{ij}(t)$ as relative time delay induced by the ith target signal in the jth sensor, (iv) $e_{a_j}(t)$ as additive noise at j^{th} sensor, (v) $\underline{k}_i(t) = [\sin\theta_i(t)]$ for $\theta \in = [-90^0,...,0,...,90^0]$ as i^{th} direction vector, (vi) $\underline{z}_j$ as j^{th} sensor position, (vii) c as velocity of the acoustic signal in the medium, and (viii) τ as the time delay between any two neighboring sensors in the array. Thus, the array outputs are written as

$$X_a(t) = A[\theta(t)]S_a(t) + e_a(t) = Y \tag{3.109}$$

Here, we have the following: (i) $X_a(t) = \left[X_{a_1}(t)...X_{a_m}(t)\right]^T$ as an m vector of complex envelopes of the sensor outputs, (ii) $S_a(t) = \left[S_{a_1}(t)...S_{a_l}(t)\right]^T$ as an l vector of complex envelopes of the target signals, (iii) $e_a(t) = \left[e_{a_1}(t)\cdots e_{a_m}(t)\right]^T$ as an m vector of complex envelopes of the additive measurement noises (with usual assumptions), (iv) $A[\theta(t)] = \left[a(\theta_1)\cdots a(\theta_l)\right]$ as the $m \times l$ Vandermonde matrix (steering matrix), (v) as the steering vector for the ith source/target signal with unit-magnitude elements, and (vi) $\theta(t) = \left[\theta_1(t)\cdots\theta_l(t)\right]^T$ as the unknown DoA vector (of the l targets/objects at time t) to be estimated. The $\theta(t)$ is a slowly varying function. The change in $\theta(t)$ is either zero or negligible in each interval of $[nT,(n+1)T]$, $n = 0,1,2,....$, and we have

$$\theta(t) \approx \theta(nT) \text{ for } t \in [nT,(n+1)T], \quad n = 0,1,2,...., \tag{3.110}$$

In an interval, N snapshots of sensor data are available for processing. Based on the assumptions made in eqn. (3.110), these N snapshots of sensor data are expressed as

$$X_a(k) \approx A[\theta(n)]S_a(k) + e_a(k), \qquad k = n, n+1, n+2,...,n+N-1 \tag{3.111}$$

Equation (3.111) is the discrete form of eqn. (3.109) with a sampling interval of $\frac{T}{N}$.

3.6.1.1.1 DoA Estimation

The MUSIC is one of the popular subspace methods used in spectral estimation [20]. Let us consider the ASA output data of eqn. (3.109):

$$Y = AS_a + e_a \tag{3.112}$$

If the signals from different targets/sources are uncorrelated, the correlation matrix of Y is written as

$$R_a = E\{YY^T\} = E\{AS_aS_a{}^T A^T\} + E\{e_ae_a{}^T\} \tag{3.113}$$

$$R_a = AR_sA^T + \sigma^2 I = Z + \sigma^2 I$$

Here, $R_s = E\{S_aS_a{}^T\}$ is $l \times l$ autocorrelation matrix with rank l, and I is $l \times l$ identity matrix. The σ^2 is noise variance, and Z is $l \times l$ the signal covariance matrix. The singular value decomposition (SVD) is represented as [18]

$$\text{UDU}^T = \text{svd}(R_a) \tag{3.113}$$

Here, $U = \begin{bmatrix} u_1 & u_2 & \cdots & u_m \end{bmatrix}$,

$$D = \begin{bmatrix} \lambda_1 + \sigma^2 & 0 & \cdots & 0 & 0 & \cdots & 0 \\ 0 & \vdots & \cdots & \vdots & \vdots & \cdots & 0 \\ \vdots & \vdots & \ddots & \vdots & \vdots & \cdots & \vdots \\ 0 & 0 & \cdots & \lambda_l + \sigma^2 & 0 & \cdots & 0 \\ 0 & 0 & \cdots & 0 & \sigma^2 & \cdots & 0 \\ \vdots & \vdots & \cdots & \vdots & \vdots & \ddots & \vdots \\ 0 & 0 & \cdots & 0 & 0 & \cdots & \sigma^2 \end{bmatrix},$$

The matrix U could be partitioned into a matrix U_s with l columns corresponding to l signal values and a matrix U_e with $m-l$ columns corresponding to the noise singular values: U_s as the signal subspace, and U_e as the noise subspace, both being orthogonal to each other. All noise eigenvectors are orthogonal to the signal steering vectors, and the pseudo-spectrum is computed as

$$P_M(\theta) = \frac{1}{A^T U_e U_e^T A} \tag{3.114}$$

If $\theta = \text{DoA}$, the denominator would become zero causing peaks in function $P_M(\theta)$, the accuracy being limited by the discretization (to which P is computed). The search algorithm being computationally demanding the estimation of DoA from the pseudo-spectrum is not very practical. These limitations are overcome by Root-MUSIC (RM) algorithm, a model-based parameter estimation method. The RM algorithm uses a model (steering vector) of the received signal as a function of the DoA which is the parameter in the model. DoA (or θ) is estimated based on the model and received signal. The RM algorithm is described as follows [18]:

Let $z = e^{jkd\cos\theta}$ and $A[\theta] = \begin{bmatrix} 1 & z & z^2 & . & . & . & z^{m-1} \end{bmatrix}^T$, then we have

$$u_p^T A = \sum_{q=0}^{m-1} u_{pq}^{*} z^q = u_p(z) \tag{3.115}$$

The DoA is obtained where $u_p \perp A[\theta],\ \ p = l+1, l+2, ..., m$. Also, we have

$$P_M^{-1} = A^T(\theta) U_e U_e^T A(\theta) = A^T(\theta) C A(\theta) \tag{3.116}$$

$$P_M^{-1} = \sum_{p=0}^{m-1} \sum_{q=0}^{m-1} z^q C_{pq} z^{-p} = \sum_{p=0}^{m-1} \sum_{q=0}^{m-1} z^{(q-p)} C_{pq} \tag{3.117}$$

The double summation can be avoided by setting $r = q - p$; i.e., eqn. (3.117) is written as

$$P_M^{-1}(\theta) = \sum_{r=-(m-1)}^{m+1} C_r z^r \tag{3.118}$$

With

$$C_r = \sum_{q-p=r} C_{pq} \tag{3.119}$$

Here, C_r is the sum of the elements on the qth diagonal. And since z and $\frac{1}{z^*}$ have the same phase and reciprocal magnitude, one zero is within the unit circle and the other is outside the circle. The phase carries the desired information. Without noise, the roots fall on the unit circle and are used for estimation of the DoA. The steps involved in estimation of DoA using RM are: (i) estimate the correlation matrix R_a, (ii) carry out the SVD, i.e. $\left[\text{UDU}^T\right] = \text{svd}(R_a)$, (iii) partition U to obtain U_e corresponding to $m - l$ smallest singular values, (iv) compute C_r, (v) find the zeros of the polynomial, (vi) from $m - 1$ roots (that are within the unit circle) choose the l roots closest to the unit circle $\left(z_q,\ z = 1, 2, ..., l\right)$, and (vii) finally obtain the DoA as

$$\theta_q = \sin^{-1}\left(\frac{2z_q}{d\pi}\right)\frac{180}{\pi} \tag{3.120}$$

3.6.1.2 Target Tracking Algorithms

Three target tracking schemes are evaluated [24]: (i) a second-order digital filter is used to remove the noise, the measurements being in the polar coordinate, and LS estimation method is used to compute the target positions in Cartesian coordinates; (ii) KF is used to estimate the states of the target in the polar frame, where the ASA

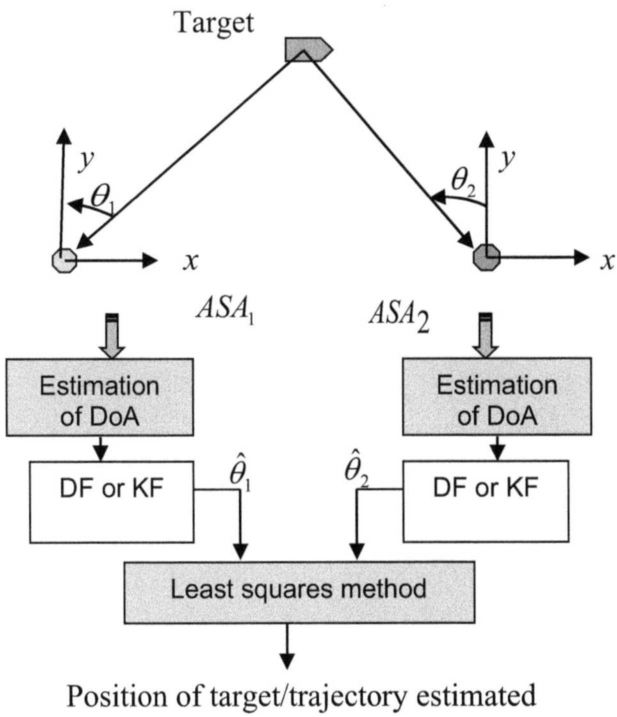

FIGURE 3.31 Target tracker schemes: 1 and 2 {Scheme 1: DF-LS method; Scheme 2: KF-LS method}.

provides measurements in the same plane, and the LS method is used to compute the position of the target; and (iii) EKF is used for estimation of the target states in the Cartesian coordinates, the measurements being in polar coordinates. These schemes are shown in Figures 3.31 and 3.32 [24].

3.6.1.2.1 *Digital Filter*

The ASA provides noisy measurements in polar coordinates. A second-order LP digital Butterworth filter with normalized cutoff frequency of 0.5 is used to remove the noise. The transfer function $H(z)$ of the filter is

$$H(z) = \frac{b_0 + b_1 z^{-1} + b_2 z^{-2}}{1 + a_1 z^{-1} + a_2 z^{-2}} \tag{3.121}$$

Here, n is the order of the filter, b_i are numerator coefficients, and a_i are denominator coefficients.

3.6.1.2.2 *Triangulation*

For Scheme 1, the measurements are in polar coordinates, and to obtain the target position in Cartesian coordinates, the LS estimation is used [22]. If the true target

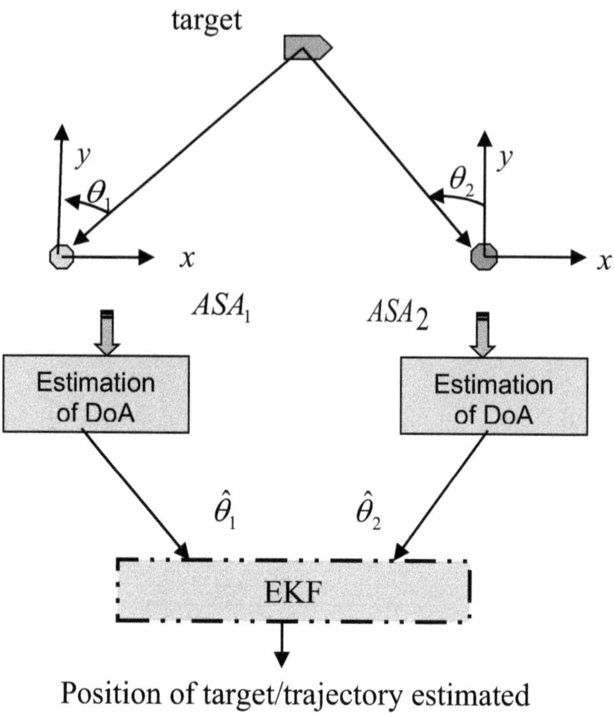

FIGURE 3.32 Target tracker – Scheme 3.

positions in Cartesian coordinates are (x_{tr}, y_{tr}), the locations of two sensors S_{r1} and S_{r2} are (S_{r1x}, S_{r1y}) and (S_{r2x}, S_{r2y}), and the target is at θ_1 and θ_2 w.r.t. S_{r1} and S_{r2}, respectively; then, we have

$$\tan(\theta_1) = \frac{y_{tr} - S_{r1y}}{x_{tr} - S_{r1x}} \text{ and } \tan(\theta_2) = \frac{y_{tr} - S_{r2y}}{x_{tr} - S_{r2x}} \tag{3.122}$$

After rearrangement, we obtain

$$\begin{bmatrix} S_{r1y} - S_{r1x}\tan(\theta_1) \\ S_{r2y} - S_{r2x}\tan(\theta_2) \end{bmatrix} = \begin{bmatrix} -\tan(\theta_1) & 1 \\ -\tan(\theta_2) & 1 \end{bmatrix} \begin{bmatrix} x_{tr} \\ y_{tr} \end{bmatrix} \tag{3.123}$$

which is represented as

$$f = Bg(11.81) \tag{3.124}$$

Here, B and f are known and g is unknown, and then the target position g is computed from $g = B^{-1}f$.

3.6.1.2.3 Results and Discussion

For performance evaluation of the DoA algorithm, the parameters given in Table 3.9, and 50 Monte Carlo simulations are used for obtaining the results. Figure 3.33(a) is obtained by varying the variance level from 0 to 0.1 and keeping the other parameters

TABLE 3.9
Parameters for Evaluation of DoA Estimation

Parameter	**Value**
No. of sensors (m)	5
No. of snapshots (N)	50
Noise variance (σ^2)	0.00001
No. of targets (l)	1
Theta (θ)	30°
Spacing between the sensors (d)	0.5λ

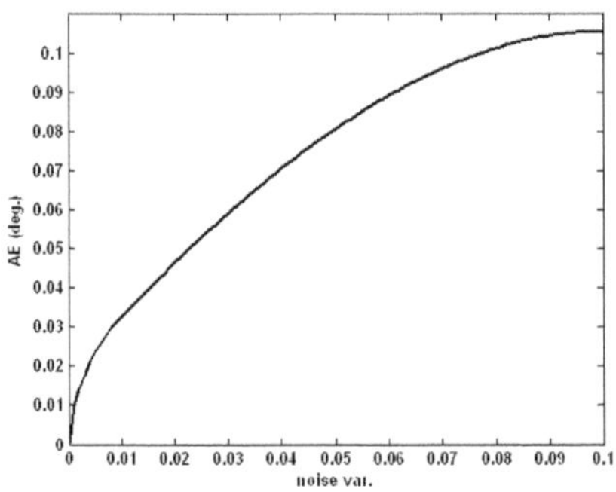

(a) Effect of noise variance on AE in DoA estimation

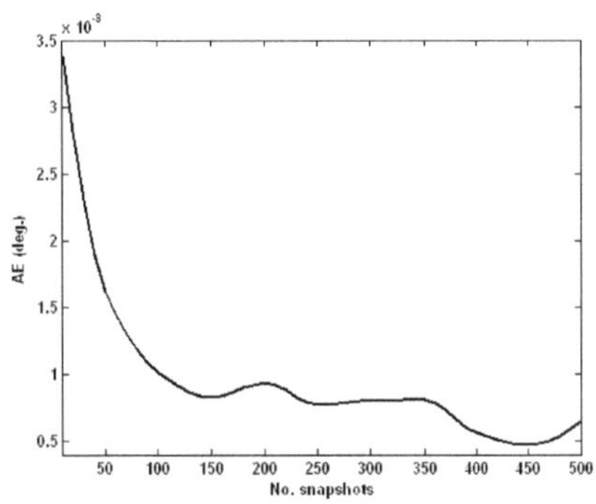

(b) Effect of snapshots on AE in DoA estimation

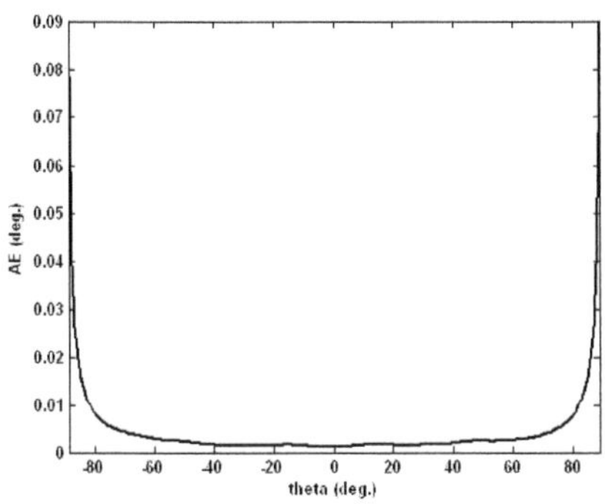

(c) Effect of theta on DoA estimation

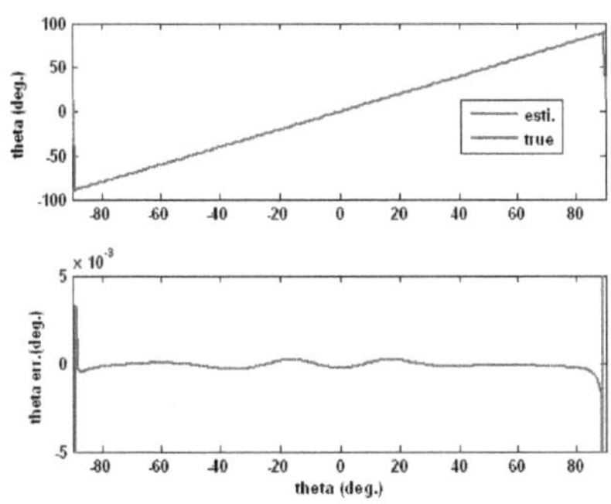

(d) Effect of theta on DoA estimation - error

FIGURE 3.33 Effect of various parameters on DoA estimation.

constant as shown in Table 3.9. It is observed that the absolute error (AE) in DoA increases with an increase in the noise level. Keeping the other parameters constant, the effect of snapshots is shown in Figure 3.33(b), and the AE is decreasing with an increase in the number of snapshots. Figure 3.33c and d shows the effect of direction of incoming signal to the ASA on DoA estimation. The error is very less except at 90°. We see from Figure 3.34 that the AE is decreasing with an increase in the number of sensors; the greater number of signals provides the average effect that would reduce the effect of noise.

3.6.1.2.4 Target Tracking

The 2DOF kinematic model in Cartesian coordinates x and y is given as

$$F = \begin{bmatrix} \Phi & 0 \\ 0 & \Phi \end{bmatrix} \text{and } G = \begin{bmatrix} \varsigma & 0 \\ 0 & \varsigma \end{bmatrix} \tag{3.125}$$

Here, $\Phi = \begin{bmatrix} 1 & T \\ 0 & 1 \end{bmatrix}$ and $\varsigma = \begin{bmatrix} T^2/2 \\ T \end{bmatrix}$

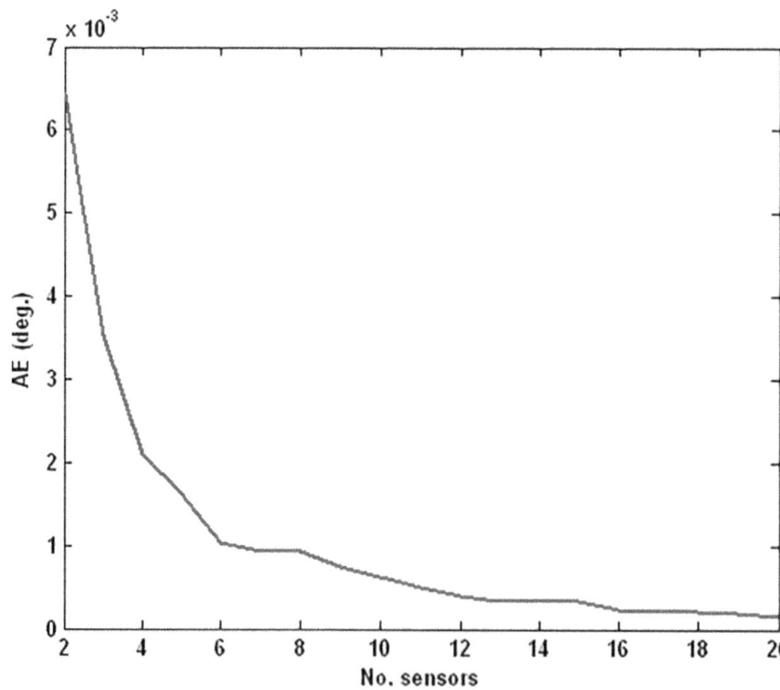

FIGURE 3.34 Effect of the number of sensors on DoA estimation.

The simulation utilizes the following parameters: (i) $T=0.1$ s, (ii) process noise variance is 0.001, (iii) simulation is run for 100 s, (iv) initial values are $\begin{bmatrix} x & \dot{x} & y & \dot{y} \end{bmatrix} = \begin{bmatrix} 0 & 1 & 10 & 0 \end{bmatrix}$, and (v) the sensor locations are $S_{1r} = \begin{bmatrix} S_{r1x} & S_{r1y} \end{bmatrix} = \begin{bmatrix} 0 & 0 \end{bmatrix}$ and $S_{r2} = \begin{bmatrix} S_{r2x} & S_{r2y} \end{bmatrix} = \begin{bmatrix} 50 & 0 \end{bmatrix}$. The target trajectory is shown in Figure 3.35a. The trajectory of the target in Cartesian coordinates is converted into polar coordinates using eqn. (3.122), and ASA output is generated for each sample. The ASA contains five sensors. It produces 100 snapshots with noise variance of 0.00001. The RM algorithm is used to estimate DoA. Measurement noise with variance of 0.001 is added to the estimated DoAs. The simulated DoAs data are shown in Figure 3.35b, and the enlarged view of a certain small portion from Figure 3.35b is shown in Figure 3.35c. The initial states for both linear and EKF filters are computed with first two measurements. Identity matrices (of the same order of the state vector) are multiplied with 0.01 and are taken as initial

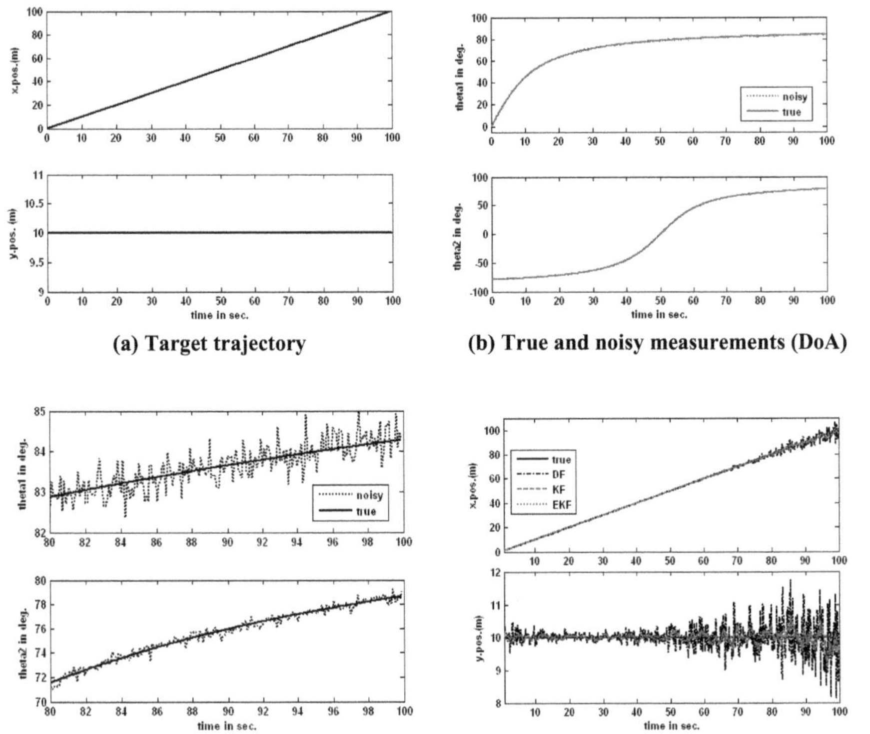

(a) Target trajectory **(b) True and noisy measurements (DoA)**

(c) Enlarged view of a portion from Figure (b) **(d) True and estimated x- and y-positions**

FIGURE 3.35 Target trajectory in Cartesian coordinates and some results.

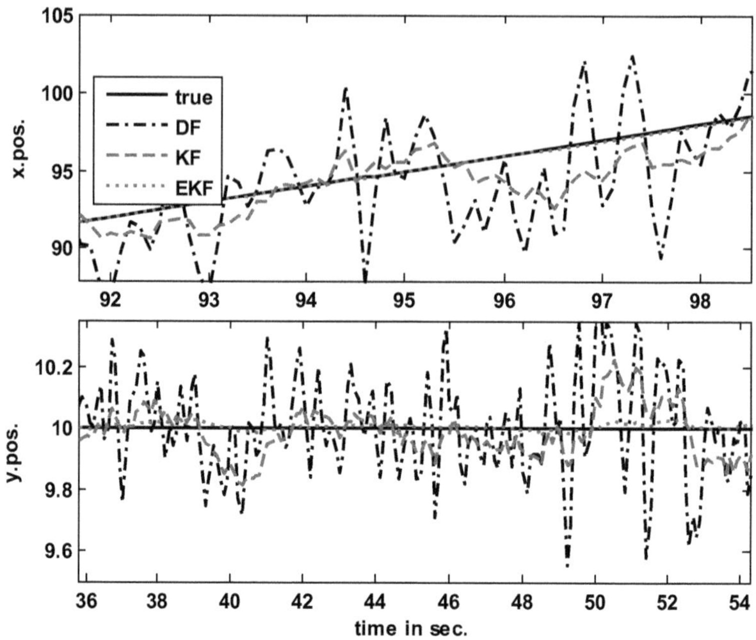

FIGURE 3.36 Enlarged view of a portion from Figure 3.35d.

state covariance matrices for both the filters. True and estimated positions are shown in Figure 3.35d, and an enlarged view of a portion from Figure 3.35d is shown in Figure 3.36. The performance metrics are given in Table 3.10. AE in positions are shown in Figure 3.37a, and an enlarged view of a certain portion from Figure 3.37a is shown in Figure 3.37b. Figure 3.37c shows the RSSPE, and the enlarged view of a certain portion is shown in Figure 3.37d. It is observed that the errors are growing when the target moves away from the sensors. From Table 3.10, it is observed that the execution time for the first scheme is the least among the three schemes; however, the performance is not that good.

3.6.1.2.5 Joint Acoustic-Image Target Tracking

The target localization and tracking (TLT) based on acoustic/audio and visual/video measurements is an important topic in various applications: (i) analysis of dynamic scenes, (ii) video conferences, and (iii) analysis of traffic situations [25]. A scheme for target tracking using joint acoustic and imaging sensor data is discussed for which a decentralized KF is used. As both the sensors provide angle measurements, a triangulation method is used to get 3D position of the target. The flow diagram of the scheme is shown in Figure 3.38. At least two sensors are required to get 3D position, since these sensors provide only 2D measurements.

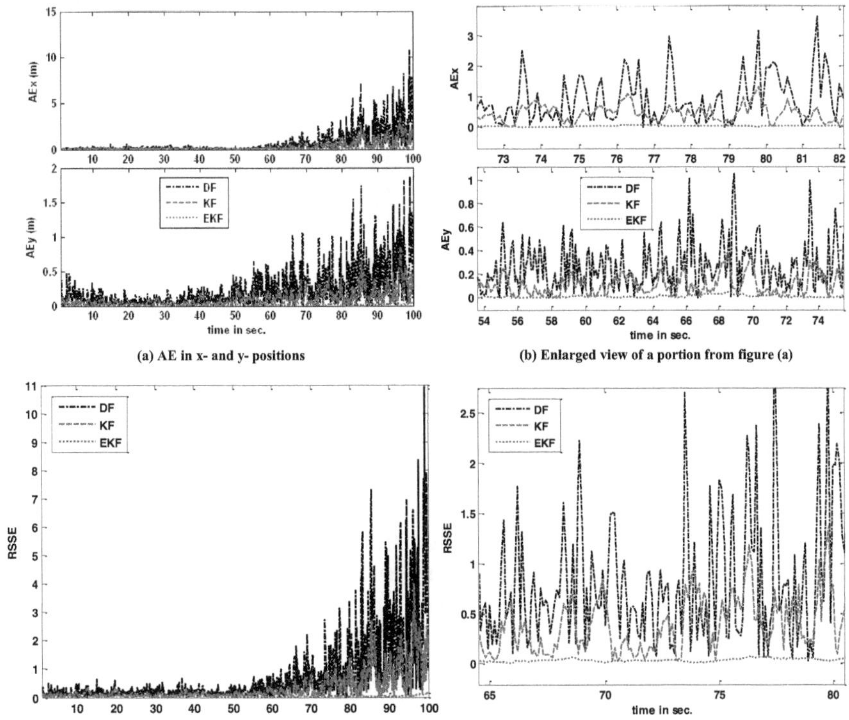

(a) AE in x- and y- positions

(b) Enlarged view of a portion from figure (a)

(c) RSS position errors

(d) Enlarged view of a portion from figure (c)

FIGURE 3.37 AE and RSSPE.

TABLE 3.10
Performance Metrics for Different Schemes

	PFEx	PFEy	MAEx	MAEy	RMSPE	Execution Time (s)
DF	2.43	5.349	0.67	0.251	1.06	0.04
KF	0.98	1.503	0.29	0.101	0.41	0.15
EKF	0.06	0.173	0.03	0.013	0.03	0.18

3.6.1.2.6 Decentralized Kalman Filter

The two local KFs (LKFs) produce track state estimates based on a respective local sensor and a fusion center; i.e., global KF (GKF) combines these local target state estimates with a global target state estimate. The target dynamics are given as

$$X(k+1) = FX(k) + Gw(k) \tag{3.126}$$

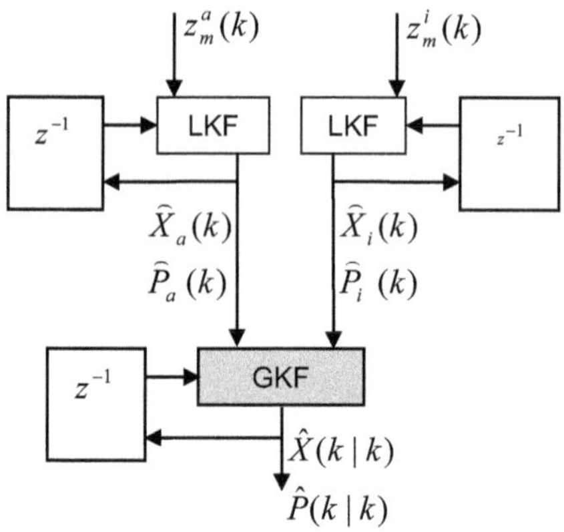

FIGURE 3.38 Decentralized KF scheme.

Here, $X(k)$ is the state vector of the target at time instant k, F is time-invariant state transition matrix, G is process noise gain matrix, and $w(k)$ is process noise sequence with covariance matrix Q. The linear measurement models are given by

$$z_m^a(k+1) = HX(k+1) + v_a(k+1) \tag{3.127}$$

$$z_m^i(k+1) = HX(k+1) + v_i(k+1) \tag{3.128}$$

Here, $z_m^a(k)$ is the measurement vector of the acoustic sensor, $z_m^i(k)$ is the measurement vector of the imaging sensor, H is an observation matrix, $v_a(k)$ is the measurement noise sequence with a covariance matrix R_a, and $v_i(k)$ is the measurement noise sequence with a covariance matrix R_i. In the first step where the LKF is working, it would take the measurements from the sensors and estimate the target states. In second step, the GKF would take the local estimates with associated covariance and fuse them to get global target state estimate.

1st step: at LKF

It is assumed that there are two sensors, viz., acoustic and imaging sensor, and these sensors provide measurements synchronously. The processing steps followed at LKF where the standard KF works are time prediction and measurement updation.

Time prediction: (a) at acoustic sensor:

$$\tilde{X}_a(k|k-1) = F\hat{X}_a(k-1|k-1) \tag{3.129}$$

$$\tilde{P}_a(k|k-1) = F\hat{P}_a(k-1|k-1)F^T + \mathrm{GQG}^T \tag{3.130}$$

(b) at imaging sensor:

$$\tilde{X}_i(k|k-1) = F\hat{X}_i(k-1|k-1) \tag{3.131}$$

$$\tilde{P}_i(k|k-1) = F\hat{P}_i(k-1|k-1)F^T + \text{GQG}^T \tag{3.132}$$

Measurement update: (a) at acoustic sensor:

$$\vartheta = z_m^a(k) - H\tilde{X}_a(k|k-1) \tag{3.133}$$

$$S = H\tilde{P}_a(k|k-1)H^T + R_a \tag{3.134}$$

$$K = \tilde{P}_a(k|k-1)H^T S^{-1} \tag{3.135}$$

$$\hat{X}_a(k|k) = \tilde{X}_a(k|k-1) + K\vartheta \tag{3.136}$$

$$\hat{P}_a(k|k) = [I - KH]\tilde{P}_a(k|k-1) \tag{3.137}$$

$$\widehat{X}_a(k) = \hat{P}_a^{-1}(k|k)\hat{X}_a(k|k) - \tilde{P}_a^{-1}(k|k-1)\tilde{X}_a(k|k-1) \tag{3.138}$$

$$\widehat{P}_a(k) = \hat{P}_a^{-1}(k|k) - \widehat{P}_a^{-1}(k|k-1) \tag{3.139}$$

(b) at imaging sensor:

$$\vartheta = z_m^i(k) - H\tilde{X}_i(k|k-1) \tag{3.140}$$

$$S = H\tilde{P}_i(k|k-1)H^T + R_i \tag{3.141}$$

$$K = \tilde{P}_i(k|k-1)H^T S^{-1} \tag{3.142}$$

$$\hat{X}_i(k|k) = \tilde{X}_i(k|k-1) + K\vartheta \tag{3.143}$$

$$\hat{P}_i(k|k) = [I - KH]\tilde{P}_i(k|k-1) \tag{3.144}$$

$$\widehat{X}_i(k) = \hat{P}_i^{-1}(k|k)\hat{X}_i(k|k) - \tilde{P}_i^{-1}(k|k-1)\tilde{X}_i(k|k-1) \tag{3.145}$$

$$\widehat{P}_i(k) = \hat{P}_i^{-1}(k|k) - \widehat{P}_i^{-1}(k|k-1) \tag{3.146}$$

The processing steps followed at fusion center where the GKFs are as follows:

Time prediction:

$$\tilde{X}(k|k-1) = F\hat{X}(k-1|k-1) \tag{3.147}$$

$$\tilde{P}(k|k-1) = F\hat{P}(k-1|k-1)F^T + \text{GQG}^T \tag{3.148}$$

Estimate correction:

$$\hat{P}^{-1}(k|k) = \tilde{P}^{-1}(k|k-1) + \widehat{P}_a(k) + \widehat{P}_i(k) \tag{3.149}$$

$$\hat{X}(k|k) = \hat{P}(k|k)\left[\tilde{P}^{-1}(k|k-1)\tilde{X}(k|k-1) + \hat{X}_a(k) + \hat{X}_i(k)\right] \tag{3.150}$$

3.6.2.2.7 3D Target Tracking

Since the ASA provides only angle measurements, triangulation is used to obtain the target's 3D position. The two sensors are placed apart: at two different locations, and each set contains acoustic and imaging sensors. The scheme is shown in Figure 3.39. The estimates from the two fusion centers are used to determine the 3D target position using triangulation. Let the target state estimate from first and second fusion centers be

$$\hat{X}_1(k|k) = \begin{bmatrix} \theta_1 & \dot{\theta}_1 & \ddot{\theta}_1 & \phi_1 & \dot{\phi}_1 & \ddot{\phi}_1 \end{bmatrix}^T \tag{3.151}$$

$$\hat{X}_2(k|k) = \begin{bmatrix} \theta_2 & \dot{\theta}_2 & \ddot{\theta}_2 & \phi_2 & \dot{\phi}_2 & \ddot{\phi}_2 \end{bmatrix}^T \tag{3.152}$$

Similarly, let the locations of two sensors sets S_{r1} and S_{r2} be $(S_{r1x}, S_{r1y}, S_{r1z})$ and $(S_{r2x}, S_{r2y}, S_{r2z})$. Let the target be at an angle of (θ_1, φ_1) and (θ_2, φ_2) with respect to S_{r1} and S_{r2}, respectively. Let the target position in Cartesian coordinates be (x_{tr}, y_{tr}, z_{tr}).

$$\tan(\theta_1) = \frac{y_{tr} - S_{r1y}}{x_{tr} - S_{r1x}} \tag{3.153}$$

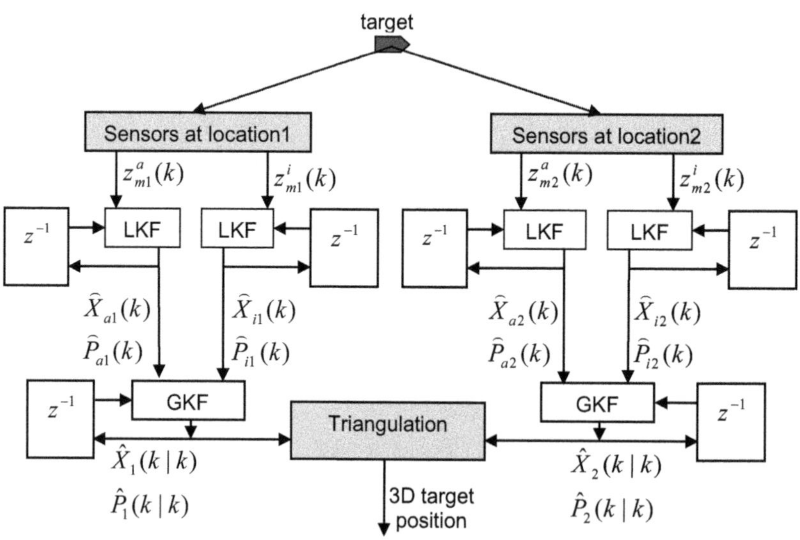

FIGURE 3.39 Decentralized KF with triangulation.

$$\tan(\theta_2) = \frac{y_{tr} - S_{r2y}}{x_{tr} - S_{r2x}} \tag{3.154}$$

$$\tan(\phi_1) = \frac{z_{tr} - S_{r1z}}{\sqrt{(x_{tr} - S_{r1x})^2 + (y_{tr} - S_{r1y})^2}} \text{ and} \tag{3.155}$$

$$\tan(\varphi_2) = \frac{z_{tr} - S_{r2z}}{\sqrt{(x_{tr} - S_{r2x})^2 + (y_{tr} - S_{r2y})^2}} \tag{3.156}$$

By arranging the above equations, we can get

$$\begin{bmatrix} S_{r1y} - S_{r1x}\tan(\theta_1) \\ S_{r2y} - S_{r2x}\tan(\theta_2) \\ S_{r1z} - S_{r1x}\tan(\varphi_1)\sqrt{1+\tan^2(\theta_1)} \end{bmatrix} = \begin{bmatrix} -\tan(\theta_1) & 1 & 0 \\ -\tan(\theta_1) & 1 & 0 \\ -\tan(\varphi_1)\sqrt{1+\tan^2(\theta_1)} & 0 & 1 \end{bmatrix} \begin{bmatrix} x_{tr} \\ y_{tr} \\ z_{tr} \end{bmatrix} \tag{3.157}$$

The above equation is represented as $f = Bg$, where B and f are known, and g is unknown. The vector g of the target position is computed as $g = B^{-1}f$.

3.6.1.3 Simulation Results

The 3DOF kinematic model with position, velocity, and acceleration components in the Cartesian coordinate x-, y-, and z-axis is given as

$$F = \text{diag}\begin{bmatrix} \Phi & \Phi & \Phi \end{bmatrix} \text{ and } G = \text{diag}\begin{bmatrix} \varsigma & \varsigma & \varsigma \end{bmatrix} \tag{3.158}$$

$$\text{where } \Phi = \begin{bmatrix} 1 & T & T^2/2 \\ 0 & 1 & T \\ 0 & 0 & 1 \end{bmatrix} \text{ and } \varsigma = \begin{bmatrix} T^3/6 \\ T^2/2 \\ T \end{bmatrix}$$

The simulation utilizes the following parameters: sampling interval is $T = 0.25$, process noise variance as $\sigma_x^2 = \sigma_y^2 = \sigma_z^2 = 10^{-6}$, duration of simulation is 125 seconds, and initial state vector is $X_i = \begin{bmatrix} 100 & 10 & 0.01 & 100 & -10 & 0.01 & 125 & -1 & 0 \end{bmatrix}^T$. The trajectory of the target in Cartesian coordinates is converted into polar coordinates using equation (3.157), and the following measurement noises are added to obtain the realistic measurements. The two sensor locations are $S_{r1} = (0,0,0)$ and $S_{r1} = (1000,-1000,0)$. Noise with variance of $(\sigma_\theta^a)^2 = (\sigma_\varphi^a)^2 = 10^{-4}$ is added to get the acoustic measurements, and noise with variance of $(\sigma_\theta^i)^2 = (\sigma_\varphi^i)^2 = 10^{-6}$ is added to obtain imaging sensor measurements. The acoustic and imaging sensor data at locations 1 and 2 are shown in Figures 3.40 and 3.41, respectively. For each sensor

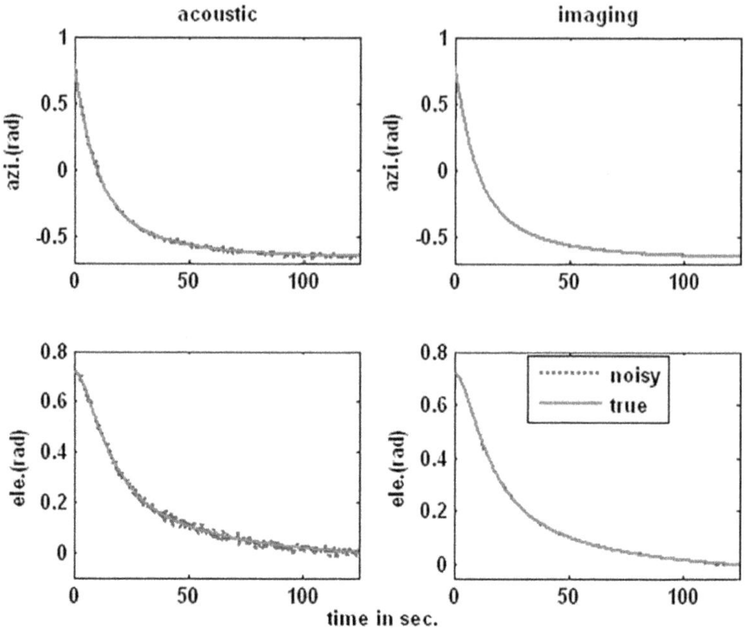

FIGURE 3.40 True/noisy measurements from acoustic and imaging sensors from sensor location 1.

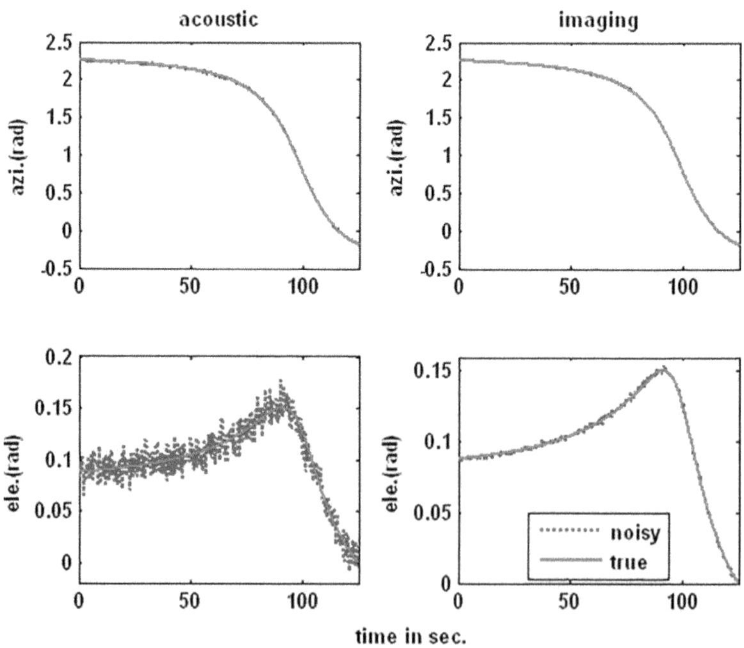

FIGURE 3.41 True/noisy measurements from acoustic and imaging sensors from sensor location 2.

location, the 3DOF kinematic model, with angular position, angler velocity, and angular acceleration components in each of the two polar coordinates azimuth and elevation, has the following transition and process noise gain matrices:

$$F = \begin{bmatrix} \Phi & 0 \\ 0 & \Phi \end{bmatrix} \text{and } G = \begin{bmatrix} \varsigma & 0 \\ 0 & \varsigma \end{bmatrix} \tag{3.159}$$

Here, $\Phi = \begin{bmatrix} 1 & T & T^2/2 \\ 0 & 1 & T \\ 0 & 0 & 1 \end{bmatrix}$ and $\varsigma = \begin{bmatrix} T^3/6 \\ T^2/2 \\ T \end{bmatrix}$

Initial true state vector is $X_t = \begin{bmatrix} \theta & \dot{\theta} & \ddot{\theta} & \phi & \dot{\phi} & \ddot{\phi} \end{bmatrix}^T$ computed using the first two true measurements. The state errors for each sensor location were found to be within the bounds. The innovation sequences are shown in Figures 3.42 and 3.43. RSS errors in angler position are shown in Figure 3.44. The PFE and AE in angler positions are given Table 3.11. The RMS error in angular position is given in Table 3.12. The true and estimated target position from triangulation is shown in Figure 3.45. Percentage fit error and mean AE in all three x-, y-, and z-positions are shown in Table 3.13. It is inferred that EKF gives better results than LKF along with LS estimation and digital filter with LS estimation at the cost of execution time.

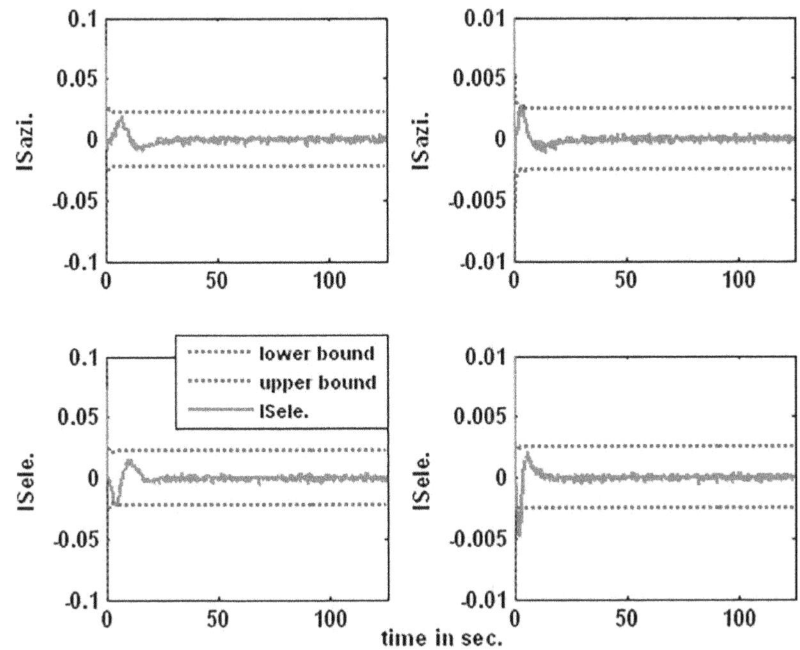

FIGURE 3.42 Innovation sequences (left: acoustic sensor; right: imaging sensor obtained from sensor location 1).

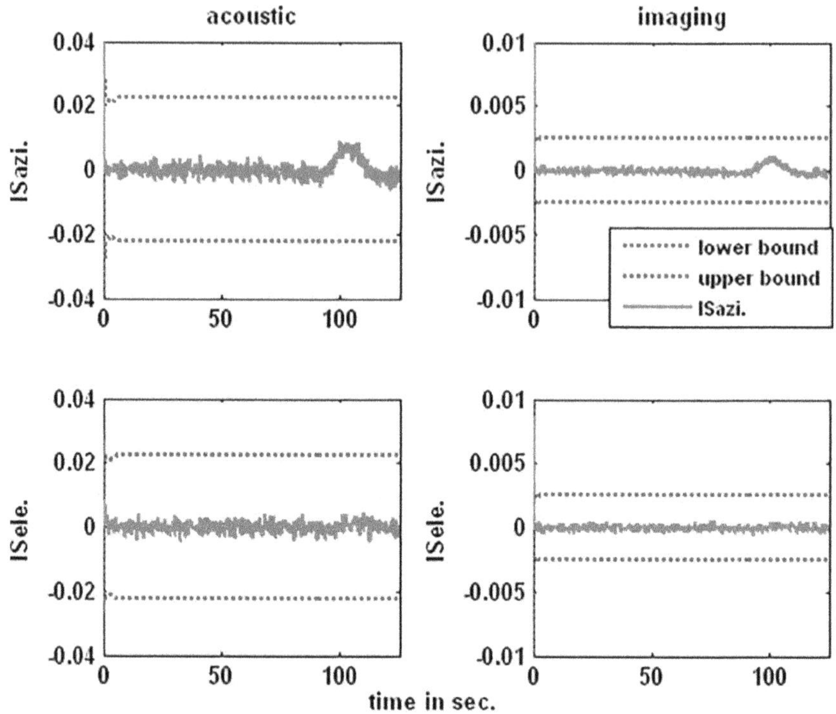

FIGURE 3.43 Innovation sequences (left: acoustic sensor; right: imaging sensor obtained from sensor location 2).

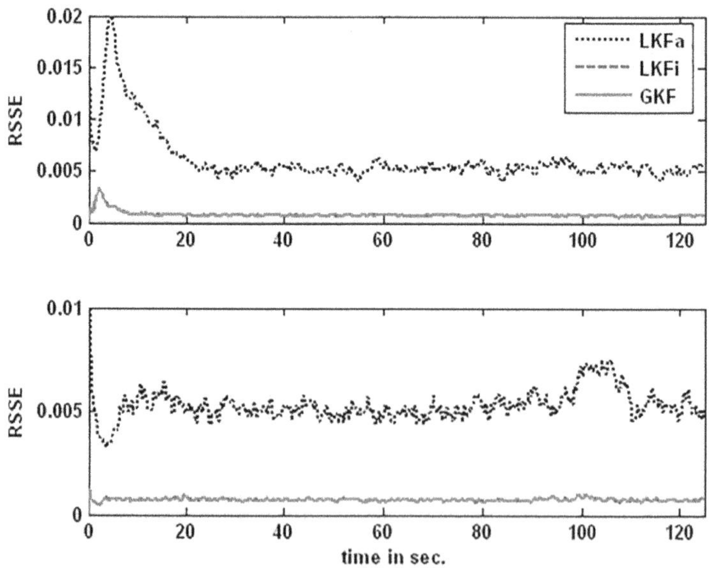

FIGURE 3.44 RSSE in angle (upper: location 1; and lower: location 2).

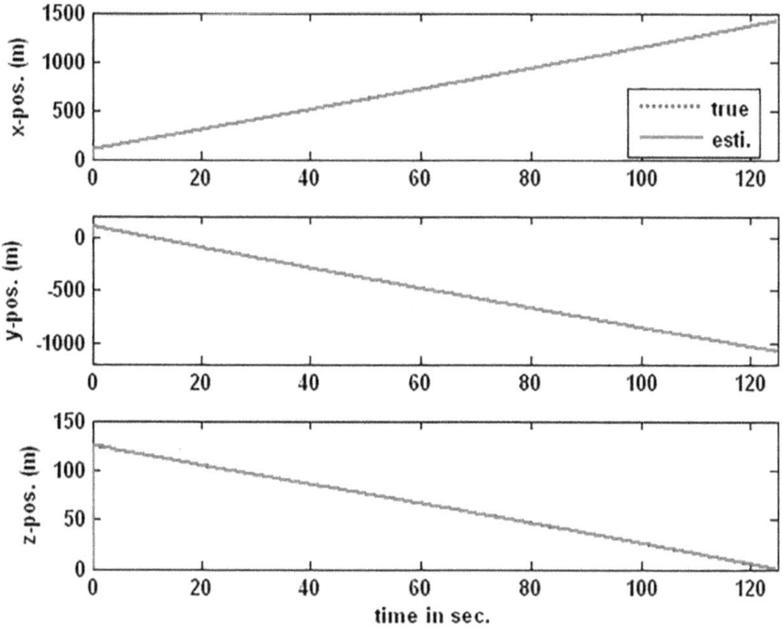

FIGURE 3.45 True and estimated target positions.

TABLE 3.11
PFE and MAE in Angular Positions

	Sensor Location 1				Sensor Location 1			
	PFEθ_1	PFEϕ_1	MAEθ_1	MAEϕ_1	PFEθ_1	PFEϕ_1	MAEθ_1	MAEϕ_1
LKFa	0.8824	2.2399	0.0038	0.0041	0.242	3.8994	0.0035	0.0033
LKFi	0.1135	0.2887	0.0005	0.0005	0.0328	0.5538	0.0005	0.0005
DKF	0.1129	0.2875	0.0005	0.0005	0.0326	0.5514	0.0005	0.0005

TABLE 3.12
RMSE in Angular Position

	Sensor Location 1	Sensor Location 2
LKFa	0.0052	0.0042
LKFi	0.0007	0.0006
GKF	0.0007	0.0006

TABLE 3.13
PFE and MAE in *x*-, *y*-, and *z*-Positions

PFE*x*	PFE*y*	PFE*z*	MAE*x*	MAE*y*	MAE*z*
0.1194	0.1231	0.9217	0.7175	0.5427	0.5082

EXERCISES

3.1 How a model transition matrix for the IMM-based filter for tracking a maneuvering target is decided?
3.2 What are the major differences between the NNKF, PDAF, and the strongest neighbor KF (SNKF) data association algorithms?
3.3 What is the major problem with NNKF?
3.4 How PDAF tries to circumvent this problem of NNKF (Exercise I.49)?
3.5 Explain the difference between decision fusion and data fusion in the context of image fusion.
3.6 In the context of, say, acoustic- and vision-based information (for speech recognition), how would you use the process of data/feature fusion and the decision fusion?
3.7 In the context of visual sensors what is the distinction between active and passive sensors?
3.8 What is an "active vision"?
3.9 What are passive imaging sensors/systems?
3.10 What are active sensing systems/techniques?
3.11 In the context of image/vision systems, give examples of some fusion configurations (basically sensor NW configurations).
3.12 Why image synchronization and registration are important for image fusion process?
3.13 Why segmentation is required in image analysis/fusion process?
3.14 What is the feature available from a thermal imager?
3.15 What type of sensory information do we get/obtain/use for pixel-level and feature-level fusion processes?
3.16 What is the fusion process for pixel- and feature-level image fusion?
3.17 What are the common noise sources that an image can get contaminated with?
3.18 What are possible methods of reducing the effect of noise processes in the acquired image?
3.19 How the image registration algorithm is used to help target tracking?
3.20 What are the fundamental requirements for the fusion results in case of image fusion?

REFERENCES

1. Hall, D. L. *Mathematical Techniques in Multi-Sensor Data Fusion.* Artech House, Norwood, MA, 1992.
2. Shalom, B. *Multi Target Multi Sensor Tracking: Principles and Techniques.* Academic Press, 1990.
3. Blackman, S. S. *Multiple-Target Tracking with Radar Applications.* Artech House, Inc, Norwood, MA, 1986.
4. Raol, J. R., Girija, G., and Singh, J. *Modelling and Parameter Estimation of Dynamic Systems.* IEE Control Engg. Series Book, Vol. 65, IEEE, London, 2004.
5. Blom, H. A. P., and Bar-Shalom, Y. The interacting multiple model algorithm for systems with Markovian switching coefficients. *IEEE Trans. Aut. Con.*, AC-33, 780–783, 1988.
6. Mazor, A. A., Bae-Shalom, Y., and Dayan, J. Interacting multiple model methods in target tracking: a survey. *IEEE Trans. Aerosp. Electron. Syst.*, 34, 103–23, 1998.
7. Naidu, V. P. S., Girija, G., and Santhakumar, N. Three Model IMM-EKF for Tracking Targets Executing Evasive Maneuvers, AIAA-66928, 45th *AIAA Conference on Aerospace Sciences*, 2007, Reno, NV.
8. Bar-Shalom, Y. and Li, X.-R. *Multitarget-Multisensor Tracking: Principles and Techniques.* YBS, Storrs, CT, 1995.
9. Simeonova, L. and Semerdjiev, T. Specific features of IMM tracking filter design. *Int. J. Inf. Secur.*, 9, 154–165, 2002.
10. Naidu, V. P. S., Girija G., and Santhakumar, N. Three Model IMM-EKF for Tracking Targets Executing Evasive Maneuvers, AIAA-66928, *45th AIAA Conference on Aerospace Sciences*, 2007, Reno, NV.
11. Bar-Shalom, Y. and Chang, K. C. Tracking a maneuvering target using input estimation versus the interacting multiple model algorithm. *IEEE Trans. Aerosp. Electron. Syst.*, 26(2), 1989.
12. Mehrotra, K. and Mahapatra, P. A jerk model for tracking highly maneuvering targets. *IEEE Trans. Aerosp. Electron. Syst.*, 33(4), 1094–1105, 1997.
13. Ananthasayanam, M. R., Sarkar, A. K., Bhattacharya, A., Tiwari, P., and Vorha, P. *Nonlinear Observer State Estimation From Seeker Measurements and Seeker-Radar Measurements Fusion*, Paper No AIAA-2005-6066-CP, AIAA Guidance Navigation, and Control Conference, and Exhibit, 15–18 August 2005, San Francisco, California, USA.
14. Vora, P. Abhijit Bhattacharyya, M., Jyothi, P. K. T., and Bhattacharjee, R. N. *RF Seeker Modeling and Seeker Filter Design, National Workshop on Tactical Missile Guidance (NWTMG), DRDL*, 2004, Hyderabad, India.
15. Kashyapl, S.K., Shanthakumar, N., Naidu, V. P. S., Girija, G., and Raol, J. R. State *Estimation for Pursuer Guidance Using Interacting Multiple Model Based Augmented Extended Kalman Filter.* International Radar Symposium, India, (IRSI), Bangalore, 2007.
16. Kashyap, S. K., Naidu, V. P. S., Singh, J., Girija, G., and Raol, J. R. Tracking of multiple targets using interactive multiple model and data association filter. *JOAST*, 58(1), 65–74, 2006
17. Romine, J. B., and Kamen, E. W. Modeling and fusion of radar and imaging sensor data for target tracking. *Opt. Eng.*, 35(3), 659–672, 1996.
18. Naidu, V. P. S., Girija, G., and Raol, J. R. Target tracking and fusion using imaging sensor and ground based radar data, *AIAA Guidance, Navigation and Control Conference and Exhibit, Paper No. AIAA-2005-5842*, 15–18 August 2005, San Francisco, CA.
19. Dogancay, K., and Hashemi-Sakhtsari, A. Target tracking by time difference of arrival using recursive smoothing. *Signal Process.*, 85 (40), 667–679, 2005.

20. Stoica, P., and Nehorai, A. MUSIC, Maximum likelihood, and Cramer-Rao bound. *IEEE Trans. Signal Process.*, 37(5), 720–741, 1989.
21. Johnson, D. H., and Dudgeon, D. E. *Array Signal Processing: Concepts and Techniques.* Prentice Hall Signal Processing Series, Upper Saddle River, New Jersey, USA, 1993.
22. Wijk, O., and Christensen, H. I. Triangulation-based fusion of sonar data with application in robot pose tracking. *IEEE Trans. Robot.*, 16(6), 740–752, 2000.
23. Dufour, F., and Mariton, M. Tracking a 3D maneuvering target with passive sensors. *IEEE Trans. Aerosp. Electron. Syst.*, 27(4), 725–738, 1991.
24. Naidu, V. P. S., and Raol, J. R. Target tracking with multi acoustic array sensors data. *Def. Sci. J.*, 57(3), 289–303, 2007.
25. Strobel, N., Spors, S., and Rabenstein, R. Joint audio-video object localization and tracking. *IEEE Signal Process. Mag.*, 18(1), 22–31, 2001.

4 Mathematical Modelling for Two-Vehicle Coordination

Sudesh K. Kashyap and Jitendra R. Raol

4.1 INTRODUCTION

Unmanned aerial vehicles (UAVs) are often deployed for: (i) search and rescue, (ii) mapping, and (iii) real-time monitoring. Here, a problem of optimally coordinating the motions of two UAVs for the purpose of tracking a moving target on the ground is considered. Each UAV flies at a fixed altitude and is equipped with a GPS, an INS, and a gimbaled video camera (GVC). To track objects on the ground, video processing algorithms (VPAs) are used to determine the centroid pixel coordinates of targets that are moving in the image frame [1]. With these pixel coordinates, the camera parameters, and the terrain data, the 3D location of the target in inertial coordinates is estimated, and the error covariance is computed; this is the process of geolocation for video cameras. A good practice is to have the UAVs orbit the target at a nominal standoff distance and maintain an angular separation of 90°; this minimizes the instantaneous geolocation error covariance as the error ellipses are orthogonal. In the optimization-based approach, there is no restriction on the motion of the UAVs, other than kinematics. Here, the optimal (with respect to the fused geolocation error covariance) joint control policy for a long horizon using dynamic programming is determined. The optimal coordination has the UAVs taking turns passing over the target; these results may guide the design of more practical heuristic controllers that are near optimal for tracking.

The problem of formation control and coordination is very important for multi-robot tasks [2–4]. The main reason is that many cooperative tasks like surveillance, transportation, and search and rescue missions can be performed very efficiently with more than one robot. In such cases, a desired formation and coordinated control are required to be maintained by the participant robots [5]; this is illustrated with an example implemented in MATLAB.

4.2 PROBLEM FORMULATION FOR UAV COORDINATION

A group of two UAVs tasked with estimating the state of a moving target vehicle is considered. The UAVs fly at a fixed forward speed with a constant altitude. The ground-target vehicle has a forward speed that can be significantly slower than that

DOI: 10.1201/9781003429869-4

of the UAVs. The main aim is to optimize the coordination of the UAVs with respect to the joint estimation error covariance.

4.2.1 Math Model of the Measurements

Each UAV has a video sensor that makes image plane measurements of the target, which are related to the topographical coordinates. Video tracking uses main coordinate frames: (i) the topographic coordinate frame (T frame), which is the primary coordinate frame, and its *X*, *Y*, and *Z* axes are along the East, North, and upward directions, respectively; and (ii) the sensor coordinate frame (S frame), which is the secondary frame and has its origin at the optical center of the camera; the *Z* axis is along the optical axis and is pointed in the general downward direction toward the Earth.

To perform the coordinate transformation from the T frame to the S frame, we use the orthogonal attitude matrix $T_T^S(\theta)$, which is a nonlinear function of the Euler angles $\left(\theta \in \mathbb{R}^3\right)$. We note that the Euler angles are those of the camera sensor and not the aircraft and that its inverse is denoted by $T_S^T(\theta)$. Moreover, we assume that image tracking software is available to control the camera's gimbal platform and keep the target in the camera's field of view.

Let the true sensor and ground-target positions measured from the T frame origin be given as $s := \begin{bmatrix} s_x, & s_y, & s_z \end{bmatrix}'$ and $o := \begin{bmatrix} o_x, & o_y, & o_z \end{bmatrix}'$, respectively.

The relationship between the object and the image point is given as

$$r^S := o^S - s^S = T_T^S(\theta)(o - s) \tag{4.1}$$

and we let $r := r^S$ and $u^S := r^S / r$. Here, r is the range from the S frame origin to the target position, and u^S is the unit vector along the camera's line-of-sight (LOS).

The relationship between the sensor and target positions can be estimated as

$$\hat{o} = \hat{s} + \hat{r} T_S^T\left(\hat{\theta}\right)\hat{u}^S \tag{4.2}$$

The quantity $\hat{r}$ is computed using the Flat-Earth approximation. To determine $\hat{u}^S$, we need a model that relates the video sensor measurements to the world; the perspective transformation accounting for the nonlinearities in this mapping is used. Let $p := \begin{bmatrix} p_u & p_v \end{bmatrix}'$ denote the noisy pixel location corresponding to the 3D object point 'o'. The noisy video measurement is modeled as

$$p = h(o, \eta) + n \tag{4.3}$$

Here, h is the nonlinear measurement function, η is the video camera parameters' vector (including the extrinsic and intrinsic camera parameters), and n is a Gaussian measurement noise $n \sim N(0, R)$. $\hat{u}^S$ can be determined from the observed pixel coordinate and the camera parameters.

4.2.2 Covariance of Geolocation Error

The uncorrelated sources of error arise from the estimates for sensor position ($\hat{s}$), sensor attitude angles ($\hat{\theta}$), and terrain height ($\hat{h}_0$); the error models are given as

$$\hat{s} = s + \tilde{s}, \qquad \tilde{s} \sim N(0, R_{\tilde{s}}) \tag{4.4}$$

$$\hat{\theta} = \theta + \tilde{\theta}, \qquad \tilde{\theta} \sim N(0, R_{\tilde{\theta}}) \tag{4.5}$$

$$\hat{h} = h_0 + \tilde{h}_0, \qquad \tilde{h}_0 \sim N(0, \sigma_{\text{ter}}^2) \tag{4.6}$$

The corresponding geolocation error is computed as

$$\tilde{o} := \hat{o} - o = \tilde{s} + \hat{u}\tilde{r} + \hat{r}\tilde{u} \tag{4.7}$$

Here, $\hat{u} = T_S^T\left(\hat{\theta}\right)\hat{u}^S$. Linearizing the attitude matrix $T_S^T(\theta)$ about the operating point, one obtains a relation for $\tilde{u}$ as

$$\tilde{u} = T_S^T(\theta)\tilde{u}^S + C\left(u^S, \theta\right)\tilde{\theta} + C\left(\tilde{u}^S, \theta\right)\tilde{\theta} \tag{4.8}$$

where

$$C\left(u^S, \theta\right) = \frac{\partial\left(T_S^T(\theta)\hat{u}^S\right)}{\partial\theta} \tag{4.9}$$

is the linearization of the attitude matrix. Ignoring the higher order term $C\left(\tilde{u}^S, \theta\right)\tilde{\theta}$ and the effects of the pixel location error (manifested in $\tilde{u}^S$), we get the approximation

$$\tilde{u} \approx C\left(u^S, \theta\right)\tilde{\theta} \tag{4.10}$$

Next, starting from the definition for the estimated object position (2) and using the Flat-Earth approximation, we get '$\hat{r}$':

$$\hat{r} = \frac{\hat{h}_0 - \hat{s}_z}{\hat{u}_z} \tag{4.11}$$

By substituting $h_0 + \tilde{h}_0$ for $\hat{h}_0$, $s_z + \tilde{s}_z$ for $\hat{s}_z$, and $u_z + \tilde{u}_z$ for $\hat{u}_z$, we obtain

$$\tilde{r} = \frac{\tilde{h}_0 - \tilde{s}_z}{u_z + \tilde{u}_z} - \left(\frac{h_0 - s_z}{u_z}\right)\left(\frac{\tilde{u}_z}{u_z + \tilde{u}_z}\right) \tag{4.12}$$

Considering $\tilde{r}$ as a function of $\zeta = \begin{bmatrix} \tilde{h}_0 & \tilde{s}_z & \tilde{u}_z \end{bmatrix}'$ and linearizing about $\zeta = 0$, we obtain

$$\tilde{r} = \frac{\tilde{h}_0 - \tilde{s}_z}{u_z} - \frac{h_0 - s_z}{u_z^2}\tilde{u}_z \tag{4.13}$$

By substituting eqns. (4.10) and (4.13) into eqn. (4.7), we obtain

$$\tilde{o} = B\tilde{s} + \frac{u}{u_z}\tilde{h}_0 + rBC\left(u^S, \theta\right)\tilde{\theta} \tag{4.14}$$

Here, B is a matrix defined by

$$B = I_{3\times3} - u\begin{bmatrix} 0 & 0 & 1/u_z \end{bmatrix} \tag{4.15}$$

By computing the covariance of eqn. (4.14), we obtain the geolocation error covariance,

$$P_{\tilde{o}} = BR_{\tilde{s}}B' + \frac{1}{u_z^2}uu'\sigma_{\text{ter}}^2 + r^2BC\left(u^S, \theta\right)R_{\tilde{\theta}}C'\left(u^S, \theta\right)B' \tag{4.16}$$

Here, $R_{\tilde{s}}$, σ_{ter}^2, and $R_{\tilde{\theta}}$ are the covariances for the sensor position, the terrain height, and the Euler angles, respectively.

Uncertainty in the sensor attitude angles is the main source of error. By considering $R_{\tilde{s}} = 0_{3\times3}$ and $\sigma_{\text{ter}}^2 = 0$, the geolocation error covariance is simplified to

$$P_{\tilde{o}} = r^2BC\left(u^S, \theta\right)R_{\tilde{\theta}}C'\left(u^S, \theta\right)B' \tag{4.17}$$

Since tracking will be done in the ground plane, only the 2 x 2 upper-left submatrix of $P_{\tilde{o}}(k)$ is relevant,

$$P_i = \begin{bmatrix} P_{\tilde{o}:1,1} & P_{\tilde{o}:1,2} \\ P_{\tilde{o}:2,1} & P_{\tilde{o}:2,2} \end{bmatrix} \tag{4.18}$$

As the independent measurements of the target are collected by the two UAVs, the fused geolocation error covariance can be computed as

$$P = \left(P_1^{-1} + P_2^{-1}\right)^{-1} \tag{4.19}$$

4.2.3 UAV Modelling

The Dubins vehicle is a planar vehicle that moves forward at a fixed velocity and has a bounded turning radius. Generally, it is used to provide a simple model for UAVs, flying at a fixed altitude.

We assume that all UAVs are flying at a constant speed of v_a and have a bounded turning rate, $w_{\max} > 0$. For an individual agent i, let $(x_{i,}\ y_{i,}\ h)$ denote its absolute position in the T frame, where h is fixed, and let $\dot{\psi}_i$ denote its heading. Then, the kinematics of the UAV are described by

$$\dot{x}_i(t) = v_a \cos(\psi_i(t))$$

$$\dot{y}_i(t) = v_a \sin(\psi_i(t))$$

$$\dot{\psi}_i = u_i(t),\ |u_i| \leq w_{\max} \quad (4.20)$$

We assume that the ith UAV and its camera sensor have the same position, $s_i = \begin{bmatrix} x_i & y_i & h \end{bmatrix}'$. Generally, for the path planning problem, the solutions are obtained in continuous time. However, in coordinated target tracking, discrete time dynamics are appropriate because of the following:

1. Communication constraints, sensor sampling rates, and on-board computing power limit the frequency of target geolocation updates.
2. Discrete motion primitives are an efficient solution to computing optimal trajectories for Dubins Vehicles.

A zero-order hold (ZOH) is applied on the control input, with a sampling time of T_s to discretize the Dubins vehicle dynamics. The equations of motion for the nonzero input become

$$x_i^+ = \frac{v_a}{u_i}\left[\sin(u_i T_s + \psi_i) - \sin\psi_i\right] + x_i$$

$$y_i^+ = \frac{v_a}{u_i}\left[\cos(\psi_i) - \cos(u_i T_s + \psi_i)\right] + y_i$$

$$\psi_i^+ = u_i T_s + \psi_i \quad (4.21)$$

Whereas the equations of motion for zero input is

$$x_i^+ = v_a T_s \cos(\psi_i) + x_i$$

$$y_i^+ = v_a T_s \sin(\psi_i) + y_i$$

$$\psi_i^+ = \psi_i \quad (4.22)$$

4.3 DYNAMIC PROGRAMMING

Dynamic programming can address a wide class of applications comprised of many different physical systems described by dynamic equations of motion that require optimized trajectories. Here, for a pair of UAVs tracking a target, the technique of dynamic programming is applied to produce an optimal joint control policy [1]. We first present the state space and quantization required for tractability and then describe the cost function.

4.3.1 STATE SPACE AND QUANTIZATION

Each UAV requires three states, and the target requires additional states. While solving a dynamic programming problem, the computational cost involved is highly sensitive to the dimension of the state space; thus, the dimension of the state space is reduced by considering the UAV position relative to the target position,

$$z_{i(k)} = \begin{bmatrix} x_r(k) - x_i(k) \\ y_r(k) - y_i(k) \\ \dot{\psi}_i \end{bmatrix}, \; i = 1,2 \tag{4.23}$$

where $(x_r, y_r, 0)$ is the absolute position of the target (or reference) in the T frame, and k is the discrete time index. The individual relative coordinates can be arranged to form the state

$$z(k) = \begin{bmatrix} z_1'(k), & z_2'(k) \end{bmatrix}'.$$

Correspondingly, we get the planar distance of the ith agent to the target as

$$d_i(k) := \sqrt{z_{i,1}^2(k) + z_{i,2}^2(k)} \tag{4.24}$$

which plays a major role in the UAV control policy. Here, $z_{i,1}(k)$ and $z_{i,2}(k)$ are the first two entries of $z_{i(k)}$. Moreover, $d_i(k) = 0$ indicates that agent i is directly above the target at an altitude of h.

The UAV platform under consideration travels at a fixed speed of $v_a = 15$ m/s, and the UAVs communicate to a fusion center at a frequency of 1 Hz. Accordingly, the sampling time (T_s) is 1 second. We quantize the control and state spaces to make the optimization feasible. Optimal Dubins paths use only three inputs, accordingly we consider

$$\mathcal{U} := \{-w_{\max}, 0, w_{\max}\}$$

$$u_i(k) \in \mathcal{U}, \; i = 1,2$$

with $w_{max} = 0.5$. In quantizing the state space, note that with $T_s = 1$ and $w_{max} = 0.5$, the UAV headings will be well approximated by

$$\Psi := \{0.5n\}_{n=0}^{12}$$

$$\psi_i(k) \in \Psi,\ i = 1,2$$

To quantize the relative positions, let $z_{max} > 0$ denote the boundary of the relative spatial coordinates and $q > 0$ denote the quantization step between each of the coordinates,

$$Z := \{-z_{max} + qn\}_{n=0}^{2z_{max}/q}$$

$$z_{i,1}(k) \in Z, \forall_i = 1,2$$

$$z_{i,2}(k) \in Z, \forall_i = 1,2$$

The parameters z_{max} and q should be selected to balance the fidelity of spatial resolution with practical considerations such as time and computer memory.

4.3.2 Cost Function

The purpose of the cost function, which is used for dynamic programming, is to produce UAV trajectories that optimize the estimation performance over a long planning horizon. The estimation performance is quantified by the trace of the fused geolocation error covariance as shown in eqn. (4.19). The resulting cost function is evaluated over a planning horizon of N_s stages (time instances) according to

$$V_0(z) = \frac{1}{N_s}\sum_{k=0}^{N_s-1} L(z(k)) \tag{4.25}$$

where

$$L(z(k)) = \text{trace}\big(P(z(k))\big) \tag{4.26}$$

is the instantaneous cost in which the fused covariance is evaluated with respect to the relative coordinates.

The cost function in eqn. (4.25) is based entirely on the objective of gathering the "best" joint measurements. Since the target state estimate error covariance would yield a problem incompatible with dynamic programming, the joint measurement error covariance eqn. (4.19) is incorporated into the cost. However, optimizing measurements can be interpreted as optimizing state estimates for the case in which process noise dominates measurement noise.

Dynamic programming uses backward induction to find the optimal joint control policy, $\Pi_k(z)$, which maps each relative state z at time k to the optimal steering input for each UAV. We apply a standard technique known as value iteration in which the optimal cost to go is computed using the recursion

$$V_k(z) = \min_{u \in \mathcal{U}^2} \left(L(z(k)) + V_{k+1}(f(z,u)) \right), \ \forall z \tag{4.27}$$

from $k = N_s - 1$ to 0 with $V_{N_s}(z) = 0, \ \forall z$. At time k, the input obtaining the minimum in eqn. (4.27) for each z is stored in $\Pi_k(z)$. Here, $f(z,u)$ is the relative dynamic derived from eqn. (4.21) by assuming a particular (known) target motion.

4.4 SIMULATION RESULTS

The efficiency of dynamic programming applied to the optimal coordinated target tracking problem is demonstrated [1]. Two scenarios, namely, Scenario 1 and Scenario 2, are considered. In Scenario 1, the target has a slow speed compared to that of the UAVs, whereas in Scenario 2, the target is moving faster. In both the scenarios, the target has a constant velocity along the X-axis at zero altitude. Scenario 2 is used to compare the performance of the optimal controller to Kingston's splay state controller [1]. The simulation parameters are summarized in Table 4.1. By using the

TABLE 4.1
Simulation parameters

Parameter	Description	Value	Units
h	UAV Altitude (AGL)	100	m
v_a	UAV Speed	15	m/s
v_r	Target Speed	{5,7}	m/s
N_s	Planning Horizon	{90,60}	s
T_s	Sampling Time	1	s
$R_{\tilde{\theta}}$	Sensor Attitude Covariance	$9I_{3X3}$	deg^2
w_{max}	Max. Turning Rate	0.5	rad/sec
z_{max}	Max. UAV-Target Distance	75	m
q	Quantization Step	5	m

Source: Quintero, S. A. P., Papi, F., Klein, D. J., Chisci, L., and Hespanha, J. P. Optimal UAV coordination for target tracking using dynamic programming, *49th IEEE Conference on Decision and Control* (CDC), Atlanta, GA, USA, 2010, pp. 4541–4546, doi: 10.1109/CDC.2010.57179331. Accessed February 2024.

N_s-horizon optimal controller $\Pi_0(z)$, one can simulate continuous UAV trajectories. The input for the nearest quantization point is applied and held over a single sampling period, after which the control is recomputed. Note that with dynamic programming, the optimal control policy is independent from the initial conditions.

4.4.1 Scenario 1

In Scenario 1, the target travels at a speed of 5 m/s, over an optimized horizon of $N_s = 90$ seconds. Figure 4.1 shows the UAV trajectories, which are obtained as a result, from an arbitrarily chosen initial state. Figure 4.2 shows the corresponding planar distances to the target, d_1 and d_2, the relative angle γ, and the instantaneous cost, $L(z)$. The UAVs making alternating passes over the target is shown in Figure 4.1. It is apparent that the two UAVs do not maintain a 90° relative angle to the target from Figure 4.2. Although the UAVs do occasionally pass through a 90° relative angle, the trace of the fused error covariance has its minimum values when at least one UAV is on top of the target, i.e., $d_1(t) = 0$ or $d_2(t) = 0$. For example, at the times of 20 and 40 seconds, the values of the trace are comparatively lower than at other time instants. Therefore, one can conclude that independent of the relative viewing angles, the optimal controller maneuvers the UAVs so that one is always close to the target. From any initial condition, similar behavior is observed.

4.4.2 Scenario 2

In Scenario 2, the target travels at a speed of 7 m/s, and over a horizon of $Ns = 60$ seconds. The purpose of Scenario 2 is twofold. First, we want to understand if the UAV behavior is dominated by angle or distance coordination for a target moving faster

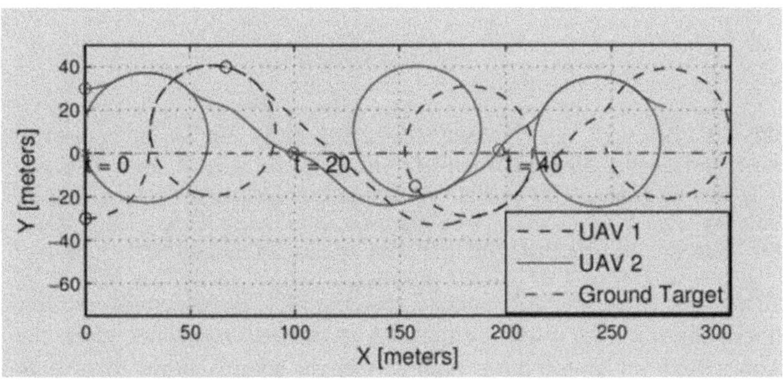

FIGURE 4.1 Trajectories for two unmanned aerial vehicles (UAVs) tracking a single target with a speed of 5 m/s. Adopted/Modified from: Quintero, S. A. P., Papi, F., Klein, D. J., Chisci, L., and Hespanha, J. P. Optimal UAV coordination for target tracking using dynamic programming, *49th IEEE Conference on Decision and Control (CDC)*, 2010, Atlanta, GA, pp. 4541–4546, doi: 10.1109/CDC.2010.57179331. Accessed February 2024.

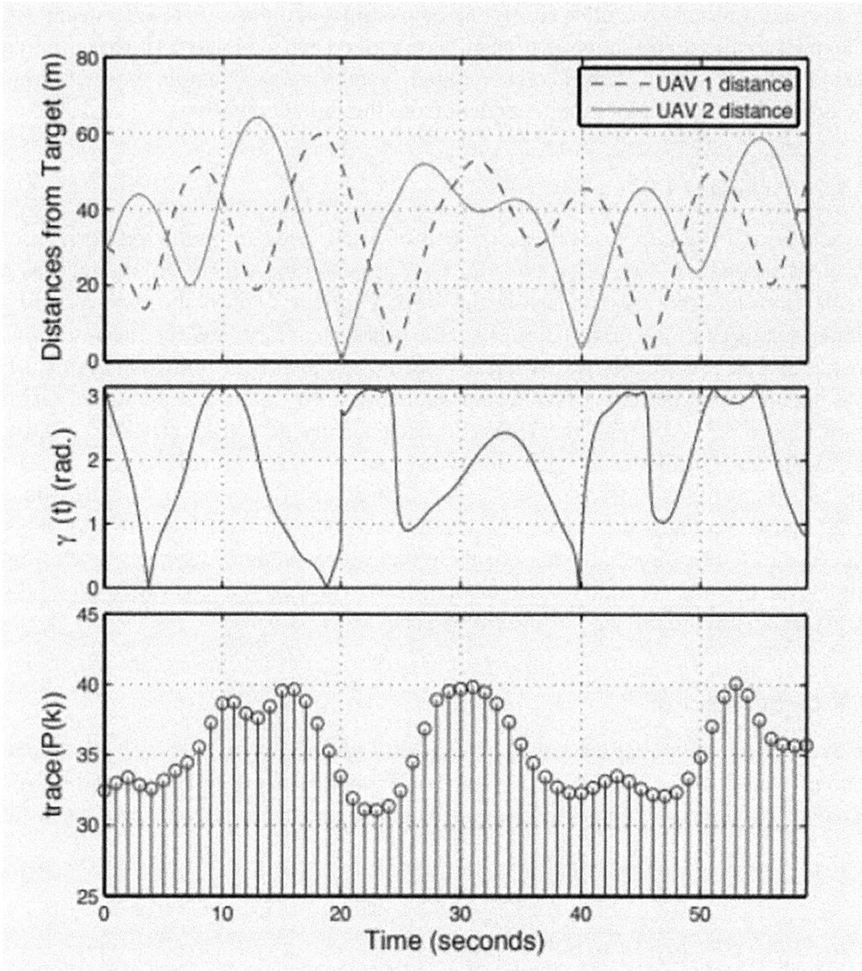

FIGURE 4.2 Distances, relative angle, and instantaneous cost for the optimal trajectories shown in Figure 1. Adopted/Modified from: Quintero, S. A. P., Papi, F., Klein, D. J., Chisci, L., and Hespanha, J. P. Optimal UAV coordination for target tracking using dynamic programming, *49th IEEE Conference on Decision and Control (CDC)*, 2010, Atlanta, GA, pp. 4541–4546, doi: 10.1109/CDC.2010.57179331. Accessed February 2024.

than in the slow target scenario. Second, the optimal UAV trajectories are compared with those coming from a well-established coordinated controller. Here, the splay state controller from Kingston is compared with the nominal orbit radius set to 65 m. This orbit radius is the distance the UAVs should maintain from the target. Figure 4.3 shows the typical behavior of the UAV trajectories, resulting from the application of these two controllers. The corresponding data are shown for the optimal controller in Figure 4.4 and for the splay state controller in Figure 4.5. For Kingston's controller, the distance to the target is not shown in Figure 4.5 as it stays within 3 m of the set point. The first observation regarding the optimal UAV trajectories is that they now

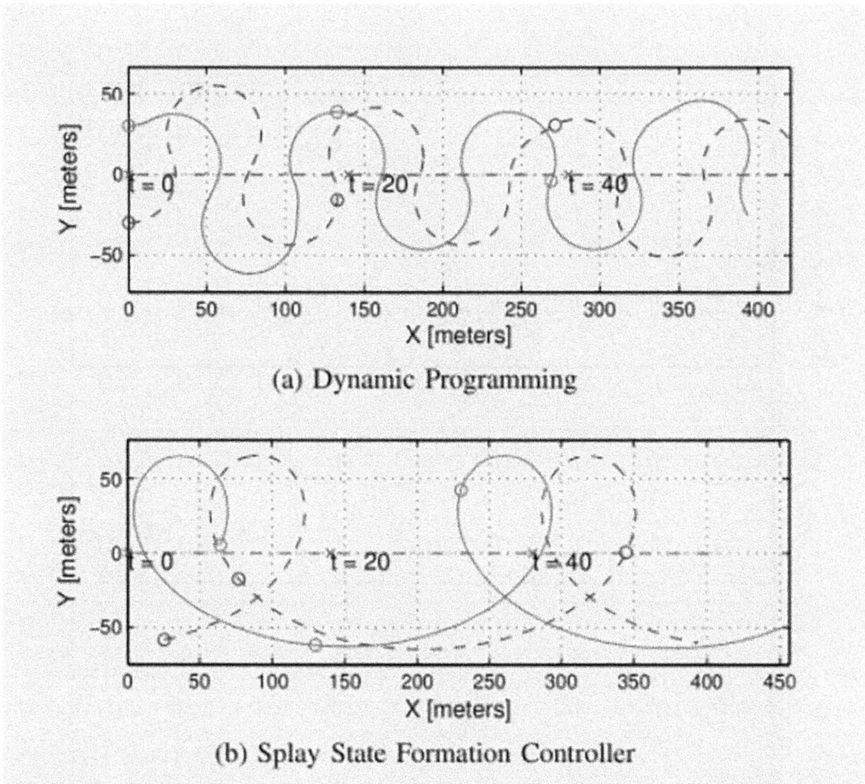

FIGURE 4.3 Comparison of trajectories generated by (a) dynamic programming and (b) the splay state controller. In this scenario, the target travels at 7 m/s. Adopted/Modified from: Quintero, S. A. P., Papi, F., Klein, D. J., Chisci, L., and Hespanha, J. P. Optimal UAV coordination for target tracking using dynamic programming, *49th IEEE Conference on Decision and Control (CDC)*, 2010, Atlanta, GA, pp. 4541–4546, doi: 10.1109/CDC.2010.57179331. Accessed February 2024.

have a different shape. Instead of the combination of circling and straight motions observed in Figure 4.1, the UAVs now make back-and-forth motions over the target. In analyzing Figure 4.4, it is again apparent that the distance is the dominating factor in the optimal coordination. This point is further apparent in comparing the instantaneous cost between the optimal controller and the splay state controller. The splay state controller maintains an angle separation near 90° throughout the simulation. However, the trace of the geolocation error covariance is significantly higher than that of the optimal controller because the UAVs stay far from the target. Moreover, even though a 90° separation angle is advantageous in reducing the trace of the error covariance, its benefit is overshadowed by that of having at least one UAV as close to the target as possible. Thus, we again conclude that distance coordination dominates angle coordination.

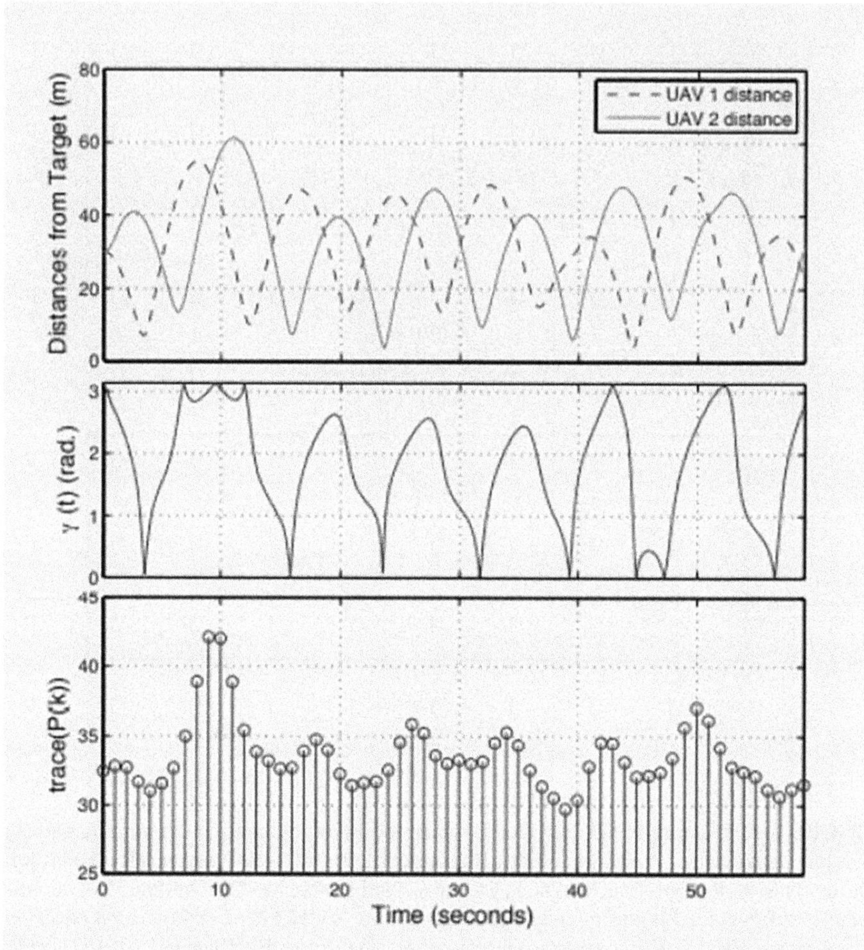

FIGURE 4.4 Distance, relative angle, and instantaneous cost for the optimal trajectories shown in Figure 4.3a. Adopted/Modified from: Quintero, S. A. P., Papi, F., Klein, D. J., Chisci, L., and Hespanha, J. P. Optimal UAV coordination for target tracking using dynamic programming, *49th IEEE Conference on Decision and Control (CDC)*, 2010, Atlanta, GA, pp. 4541–4546, doi: 10.1109/CDC.2010.57179331. Accessed February 2024.

4.4.3 Static Interpretation

To understand the performance of observed distance-dominated coordination, there are two stationary scenarios setup, in which the first UAV is fixed at a distance of $d_1 = 50$ m from the target. Second, for two different UAV altitudes, the trace of the fused geolocation error covariance is plotted as a function of d_2 and relative angle, γ, as shown in Figure 4.6. The error covariance ellipses are significantly elongated when the UAVs are at low altitude, relative to the distance to the target. In such case, orthogonality between viewing angles is advantageous but primarily when d_2 is

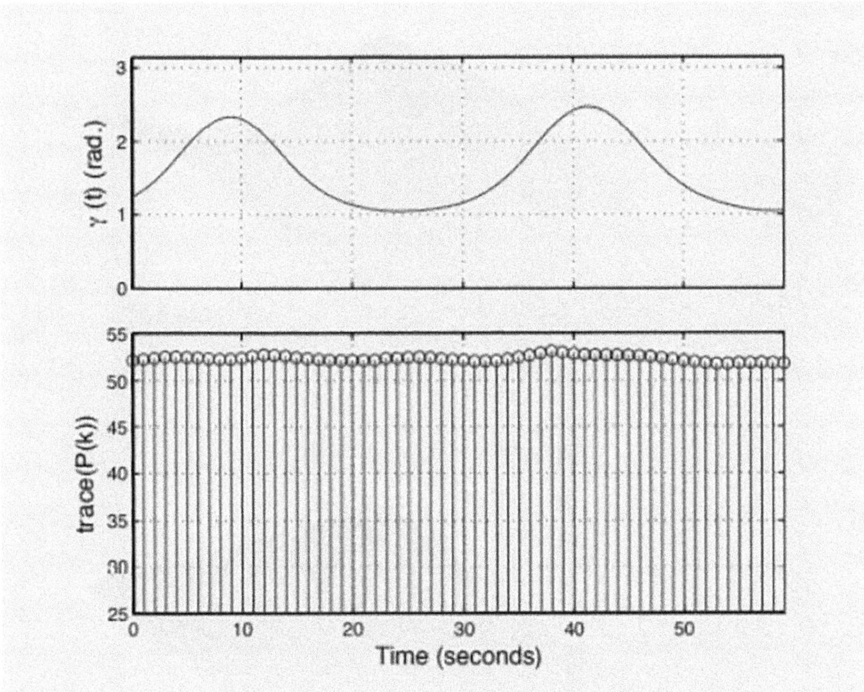

FIGURE 4.5 Relative angle and instantaneous cost for the splay state trajectories shown in Figure 3(b) The UAVs maintained a distance of 65±3 m from the target. Adopted/Modified from: Quintero, S. A. P., Papi, F., Klein, D. J., Chisci, L., and Hespanha, J. P. Optimal UAV coordination for target tracking using dynamic programming, *49th IEEE Conference on Decision and Control (CDC)*, 2010, Atlanta, GA, pp. 4541–4546, doi: 10.1109/CDC.2010.57179331. Accessed February 2024.

large, whereas the effect of orthogonal angles is unnoticeable when the UAVs fly at a more reasonable altitude. Though, these are static scenarios, the dynamic programming effectively brings the motion constraints into the UAV routing optimization.

4.5 DYNAMIC EQUATIONS OF WHEELED MOBILE ROBOT

There are various approaches: behavior-based, planned architecture, reactive-based, and the hybrid types of formation control [5]. The simple one is that of the leader-follower (L-F) mode in which a follower robot (RF) follows a leader robot (RL), by obtaining one or more inputs (data/signals) from the RL and then maintains the desired/specified distance and the angle with respect to the leader. To maintain these desired values, the RF would have a controller algorithm working in its on-board computer. This also requires the transmission of the data from the RL sensor/s, and these inputs are received by the RF. Often such data would be noisy, and if used as received, the follower controller would perform badly. Hence, it would be necessary

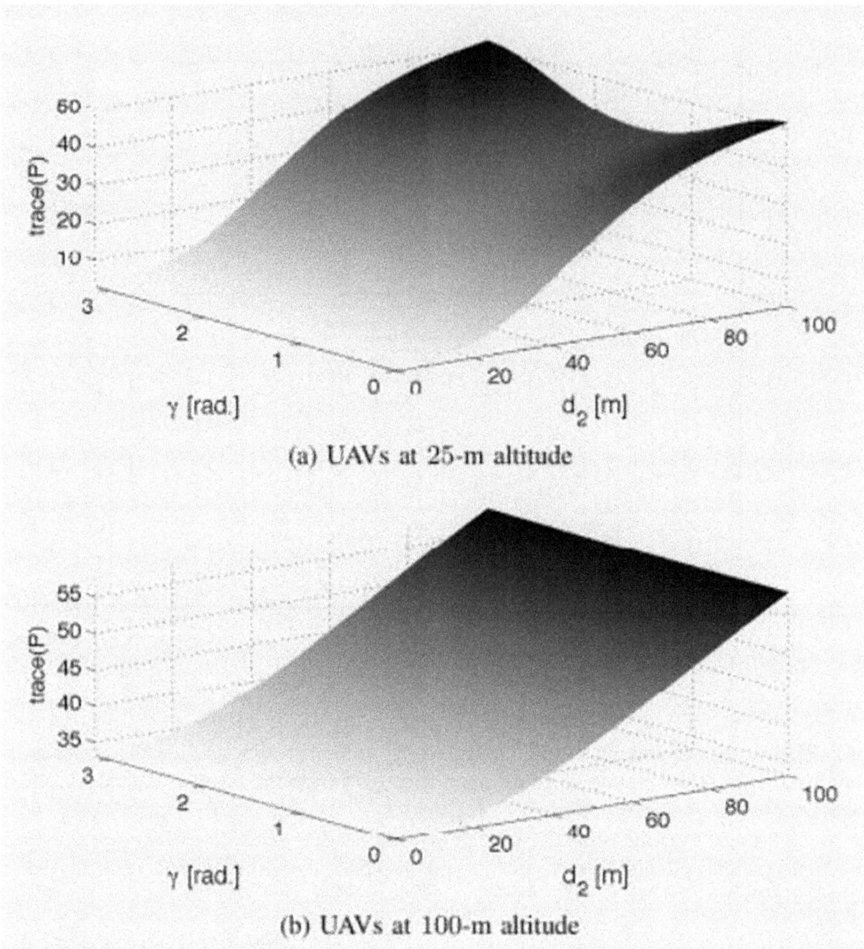

FIGURE 4.6 Effects of distance from target versus angles between UAVs. Adopted/Modified from: Quintero, S. A. P., Papi, F., Klein, D. J., Chisci, L., and Hespanha, J. P. Optimal UAV coordination for target tracking using dynamic programming, *49th IEEE Conference on Decision and Control (CDC)*, 2010, Atlanta, GA, pp. 4541–4546, doi: 10.1109/CDC.2010.5717933l. Accessed February 2024.

to filter the received data/signals before these are utilized in the follower's control laws. Often Kalman filter (KF) is used, which requires a full-fledged model of the signal that is to be filtered. A simple way is to use the degenerate mode of the KF, leading to the so called recursive least squares (RLS) filter. The RLS filter equations are then very simple to implement in the follower controller. In this case, it is then required to study and evaluate the performance of such an L-F formation control-cum-coordination and controller-cum-RLS filtering task in the presence of noisy transmitted data and the RLS-filtered inputs to the follower controller.

4.5.1 Mathematical Modelling

A situation of two-robot coordination is considered, and mathematical modelling for formation control is studied. Basically, the L-F configuration is discussed. The performance is studied without and with noise in the transmitted velocity of the RL. Then, the KF in the degenerate form is used to filter the noisy transmitted velocity, and the performance of the coordination system is evaluated using the simulated data in MATLAB where the necessary concept of local error feedback control/s has been utilized to achieve a good performance. It is found that the main controller and the local error controls work very well, and the only recursive part of the KF called RLS filter also works very well and provides very good results, which are evaluated using percentage fit errors. The dynamic equations for a nonholonomic wheeled mobile robot are given by

$$\begin{bmatrix} \dot{x}(t) \\ \dot{y}(t) \\ \dot{\theta}(t) \end{bmatrix} = \begin{bmatrix} \cos\theta(t) & 0 \\ \sin\theta(t) & 0 \\ 0 & 1 \end{bmatrix} \begin{bmatrix} v(t) \\ \omega(t) \end{bmatrix} \tag{4.28}$$

In eqn. (4.28), we have $x(.)$, and $y(.)$ as the position coordinates of the robot/s, $\theta(.)$ is the angular orientation of the robot, $v(.)$ is the linear velocity, and $\omega(.)$ is the angular velocity; eqn. (4.28) is individually valid for both the robots: RL and RF. The simplified block diagram of the L-F formation-cum-coordination controller is shown in Figure 4.7. Integration of eqn. (4.28) would give the position and orientation of any of these robots in the inertial coordinates' frame of reference. An off-the-axis point '*a*' is considered that lies on the axis of the orientation of the robot, and it is at a distance *L* from the center; it is this point where a vision camera would be located. The coordinates of this point are given as [1]

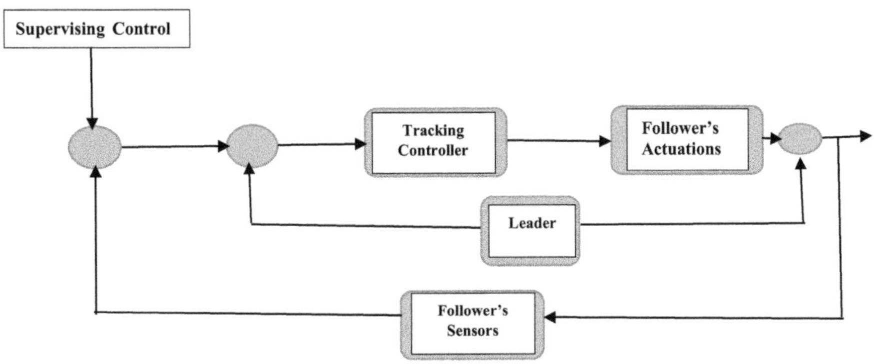

FIGURE 4.7 The leader-follower (L-F) formation coordination controller.

$$\begin{aligned} x_a &= x + L\cos(\theta) \\ y_a &= y + L\sin(\theta) \\ \theta_a &= \theta \end{aligned} \tag{4.29}$$

The time derivative of the equations in eqn. (4.29) is given by simple differentiation of eqn. (4.29), and utilizing eqn. (4.28) thereafter in the resultant equations:

$$\begin{bmatrix} \dot{x}_a \\ \dot{y}_a \\ \dot{\theta}_a \end{bmatrix} = \begin{bmatrix} \cos\theta - L\sin\theta \\ \sin\theta \quad L\cos\theta \\ 0 \quad 1 \end{bmatrix} \begin{bmatrix} v(t) \\ \omega(t) \end{bmatrix} \tag{4.30}$$

Now, using the equations and the information proivded in eqns. (4.28) and (4.29), we obtain the tracking controller equations for the L-F robot coordination for which the aim is to see that the RF, the following/RF maintains the desired distance d_d and the desired angle φ_d with respect to its RL, the leader/leading robot. For this, it is assumed that there is a virtual RL, virtual robot (RV) that is just in the expected position. Then, the main idea is to control RF such that it tracks the RV, and using the RL as the reference, we can obtain the dynamic equations of the RV as follows [5]:

$$\begin{bmatrix} \dot{x}_V \\ \dot{y}_V \\ \dot{\theta}_V \end{bmatrix} = \begin{bmatrix} \cos\theta_L & 0 \\ \sin\theta_L & 0 \\ 0 & 1 \end{bmatrix} \begin{bmatrix} v_L(t) \\ \omega_L(t) \end{bmatrix} + \begin{bmatrix} d_d \sin(\varphi_d + \theta_L)\dot{\theta}_L \\ -d_d \cos(\varphi_d + \theta_L)\dot{\theta}_L \\ 0 \end{bmatrix} \tag{4.31}$$

Then, the tracking error equations between RV and RF are obtained as follows:

$$\begin{bmatrix} \dot{x}_e \\ \dot{y}_e \\ \dot{\theta}_e \end{bmatrix} = \begin{bmatrix} \cos\theta_L & 0 \\ \sin\theta_L & 0 \\ 0 & 1 \end{bmatrix} \begin{bmatrix} v_L \\ \omega_L \end{bmatrix} - \begin{bmatrix} \cos\theta_F & 0 \\ \sin\theta_F & 0 \\ 0 & 1 \end{bmatrix} \begin{bmatrix} v_F \\ \omega_F \end{bmatrix} + \begin{bmatrix} d_d \sin(\varphi_d + \theta_L)\dot{\theta}_L \\ -d_d \cos(\varphi_d + \theta_L)\dot{\theta}_L \\ 0 \end{bmatrix} + \begin{bmatrix} L\sin(\theta_F)\dot{\theta}_F \\ -L\cos(\theta_F)\dot{\theta}_F \\ 0 \end{bmatrix} \tag{4.32}$$

In eqn. (4.32), we have, $x_e = x_V - x_F$; and so on. The error equations are then written in the new coordinates as follows:

$$\begin{aligned} \dot{z}_1 &= -v_F + v_L \cos\theta_e - z_2\omega_F + d_d\omega_L \sin(\varphi_d + \theta_e) \\ \dot{z}_2 &= -v_L \sin\theta_e + z_1\omega_F + L\dot{\theta}_F + d_d\omega_L \cos(\varphi_d + \theta_e) \\ \dot{\theta}_e &= -\omega_F + \omega_L \end{aligned} \tag{4.33}$$

In eqn. (4.33), we have used the following transformations:

$$z_1 = x_e \cos\theta_F + y_e \sin\theta_F$$
$$z_2 = x_e \sin\theta_F - y_e \cos\theta_F \tag{4.34}$$

In eqns. (4.33) and (4.34), the $z(.)$ are the position errors between RV and RF robots in the new coordinate system. In eqn. (4.33), the leader's linear and angular velocities are considered as inputs to the follower's controller. Next, we need to minimize the tracking errors between the leader and the follower, and hence, we need to determine the follower's linear and angular velocities. For this, we have the following structure of the control law:

$$v_F = v_L \cos\theta_e + K_1 z_1 + d_d \omega_L \sin(\varphi_d + \theta_e)$$
$$\omega_F = \frac{1}{L}[v_L \sin\theta_e - K_2 z_2 - d_d \omega_L \cos(\varphi_d + \theta_e)] \tag{4.35}$$

The $K(.)$ are the controller gains, and the leader's velocities are communicated/transmitted to the follower's controller as an input. Since RV is a virtual robot, $z(.)$ are not measureable, and hence, we use the following set of equations:

$$z_1 = d\cos\varphi - d_d \cos(\varphi_d + \theta_e)$$
$$z_2 = d\sin\varphi - d_d \sin(\varphi_d + \theta_e) \tag{4.36}$$

In eqn. (4.36), d and φ are the distance and angle between RF and RV and are measureable. By substituting eqn. (4.36) in eqn. (4.35), we finally obtain the following control laws, i.e., the follower's linear and angular velocities:

$$v_F = v_L \cos\theta_e + K_1(d\cos\varphi - d_d \cos(\varphi_d + \theta_e)) + d_d \omega_L \sin(\varphi_d + \theta_e)$$
$$\omega_F = \frac{1}{L}[v_L \sin\theta_e - K_2(d\sin\varphi - d_d \sin(\varphi_d + \theta_e)) - d_d \omega_L \cos(\varphi_d + \theta_e)] \tag{4.37}$$

Once the RF receives the input velocity from the RL, the control law as in eqn. (4.37) will direct the RF to follow the lead robot to maintain the desired distance d_d and the desired angle φ_d. As such, this is the part of the robot motion control and can be taken care in the on-board computer of the RF and the motor/s that drive the robot motion of RF.

4.6 KALMAN AND RLS FILTERING

Often, the transmitted signals from the RL would be contaminated by random noise, and this might be due to the transmission channel noise or the sensor noise from the RL. If such a noisy signal is used as an input to the RF control law computation, then the motion performance of the RF would be greatly affected. Hence, it would be necessary to filter the noisy signal and then use this filtered signal as an input in the RF. In general, KF can be used; however, this would require a full mathematical model of the system and might require more computation in the on-board computer of the RF. Since, we are considering to transmit only one signal, i.e., the leader velocity, we can use the degenerate form of the KF, that is often called as the RLS filter. We use the simple RLS form of the filter to remove/reduce the effect of the random noise in the transmitted velocity signal. We, for the sake of completeness, give the KF equations in the discrete form here [5,6].

4.6.1 State Propagation/Evolution

State estimate

$$\tilde{x}(k+1) = \varphi\,\hat{x}(k) \tag{4.38}$$

Covariance (a priori)

$$\tilde{P}(k+1) = \varphi\hat{P}(k)\varphi^T + GQG^T \tag{4.39}$$

4.6.2 Measurements/Data Update

Residual/innovations

$$e(k+1) = z(k+1) - H\,\tilde{x}(k+1) \tag{4.40}$$

Kalman Gain

$$K = \tilde{P}H^T(H\tilde{P}H^T + R)^{-1} \tag{4.41}$$

Filtered estimate

$$\hat{x}(k+1) = \tilde{x}(k+1) + Ke(k+1) \tag{4.42}$$

Covariance (a posteriori)

$$\hat{P} = (I - KH)\tilde{P} \tag{4.43}$$

In eqn. (4.38), $x(.)$ is state variable and would constitute, in general, the positions of the concerned robot, $P(.)$ are the respective state-error covariance matrices, and $Q(.)$, and $R(.)$ are the noise covariance matrices. Since we only consider using leader velocity as the input signal, we would need to filter if it is contaminated by random noise. We use the simplified model for the KF, and hence, we obtain the RLS filter as follows.

4.6.3 Simple RLS Filter Updates

Gain

$$K = P\,/\,(P + R) \tag{4.44}$$

Residual

$$e(k+1) = V_{Ln}(k+1) - \hat{V}_L(k) \tag{4.45}$$

Filtered velocity

$$\hat{V}_L(k+1) = \hat{V}_L(k) + Ke(k+1) \tag{4.46}$$

Covariance

$$P = (1-K)P \tag{4.47}$$

We see that the filtered-velocity updates from eqn. (4.46) can be directly used in eqn. (4.37) to compute the follower velocity.

4.7 IMPLEMENTATION AND RESULTS

The robot dynamic equations and the control laws are implemented in MATLAB for the RF. The simulated data are generated for formation/coordination control using eqns. (4.28), (4.31)–(4.33), and (4.36). The integrations of respective equations, where required, are carried out using the simple Euler approximation that would work well if the sampling time interval is reasonably and relatively small, as is the case for the present simulation. The performance of the system with and without noise is evaluated. Where necessary, the concept of local error feedback in certain variable/s (say, some angle/s and/or some positions) has been utilized, along with the main RLS filter. The desired distance and the desired orientation are specified. It is assumed that the RL transmits only velocity, vL = 100 mm/s, and that it is assumed to be received by the RF. Then, the controller is activated, and the RF follows the RL. The percentage fit errors for various cases are presented in Table 4.2; for velocity, distance, and orientation, consistently a certain number fixed initial points, due to transients, were

TABLE 4.2
PFE in Two-Robot Coordination with Noise and without RLS Filter and with the Filter

	Noise	R	Filter	X-axis	Y-axis	Velocity	Distance	Orientation
1	No	–	No	**1.159**	**0.3206**	**0.1751**	**3.95**	**10.6**
2	Yes	1	No	1.15	0.32	1.378	10.924	13.15
3	**Yes**	**1**	**Yes**	**1.159**	**0.32**	**0.168**	**3.703**	**10.2**
4	Yes	10	No	1.158	0.3204	4.335	32.307	20.91
5	Yes	10	**Yes**	**1.158**	**0.3167**	**0.1764**	**3.514**	**13.735**
6	Yes	100	*No*	*1.154*	*0.3316*	*13.59*	*101.68*	*49.51*
7	**Yes**	**100**	Yes	**1.1615**	**0.318**	**0.211**	**3.804**	**11.23**

not included in the fit error computation. It is found that when the noisy velocity is used in the follower controller, the results are quite poor; however, for the same case with the RLS filter activated along with the controller in the RF, the results certainly improve, as seen from cases 6 and 7 in Table 4.2 (nos. in bold). This ascertains that the simple RLS filter works very well. Although, the orientation error is relatively large, the purpose here has been to show the efficacy of the controller-cum-RLS filter performance in the L-F mode, and it has been found to work very satisfactorily. Various performance plots are shown in Figures 4.8–4.10, and these show that the

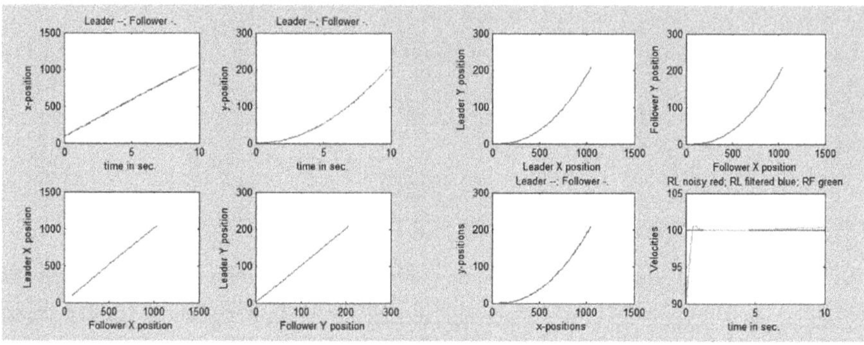

FIGURE 4.8A The L-F formation coordination: robots' positions and velocity following (no noise).

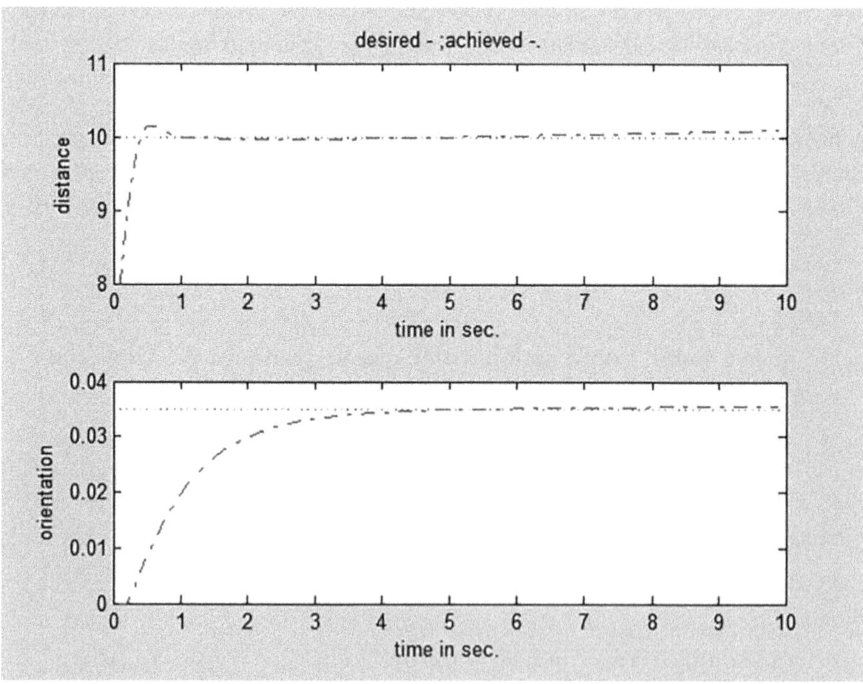

FIGURE 4.8B The L-F formation coordination: desired/achieved distances and orientations (no noise).

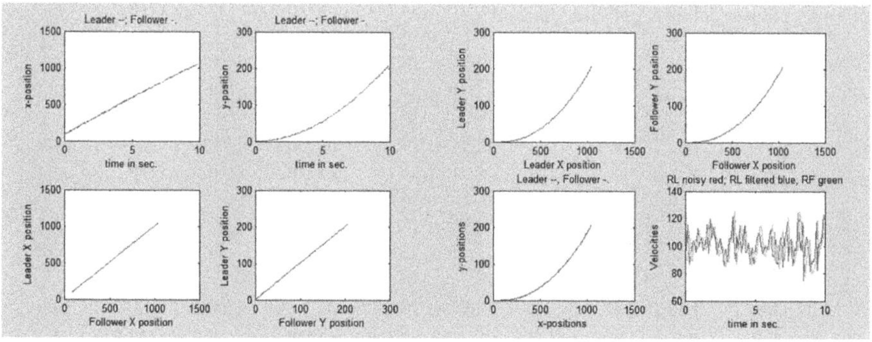

FIGURE 4.9A The L-F formation coordination: robots' positions and velocity following (with noise, $R = 100$, and no filtering).

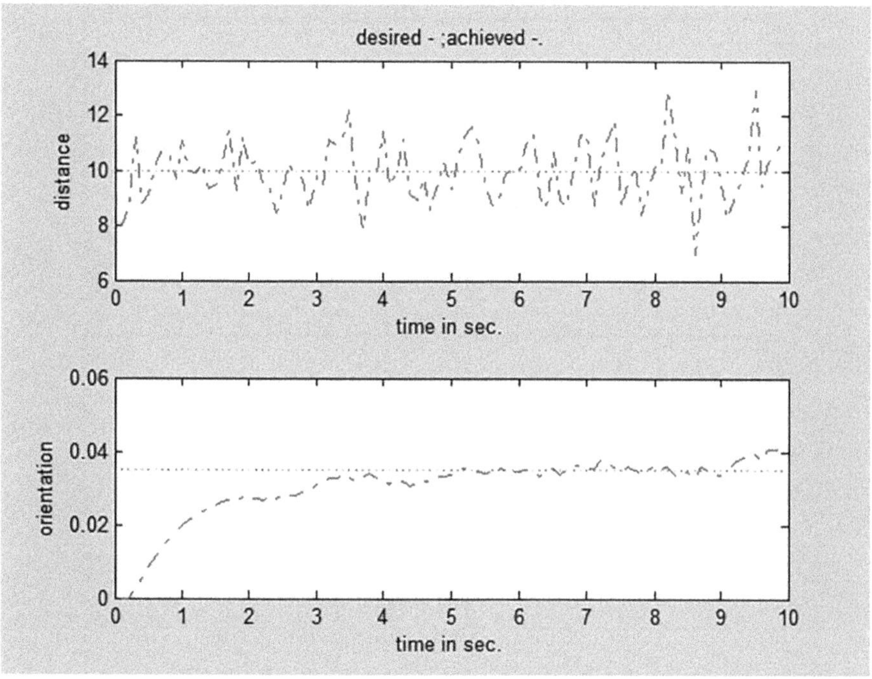

FIGURE 4.9B The L-F formation coordination: desired/achieved distances and orientations (with noise, $R = 100$, and no filtering).

follower controller-cum-RLS filter works very well when there is even noise with very high variance ($R = 100$) in the transmitted leader velocity.

The various percentage fit errors and the performance plots indicate that the scheme works very well. The scheme can also be extended to multi-robot system and for additional noisy transmitted signals to the RF; in which case, a composite RLS filter or a parallel bank of such filters could be employed; this would require some further work.

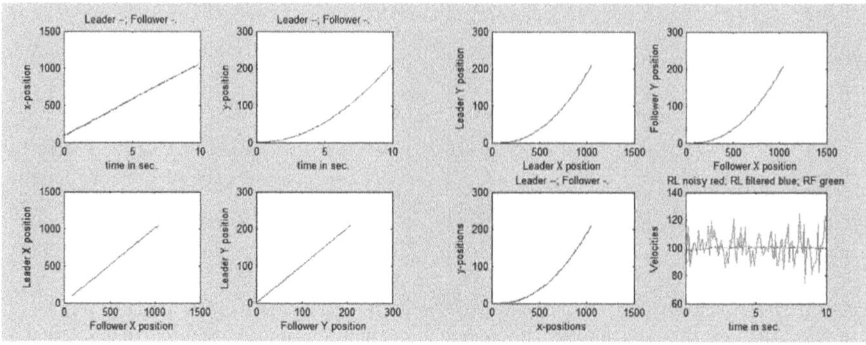

FIGURE 4.10A The L-F formation coordination: robots' positions and velocity following (with noise, $R=100$, and with RLS filter).

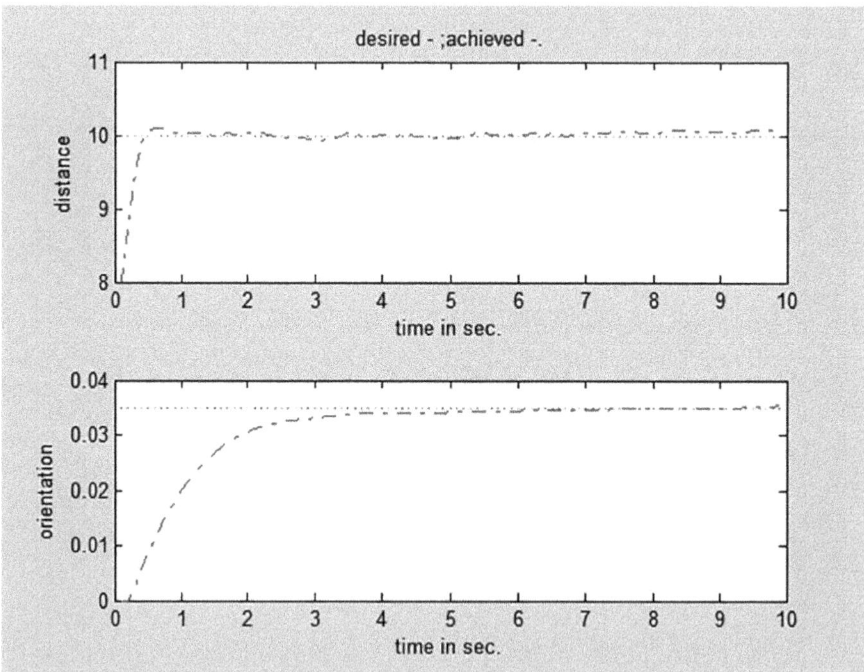

FIGURE 4.10B The L-F formation coordination: desired/achieved distances and orientations (with noise, $R=100$, and with RLS filter).

EXERCISES

1. How can the relationship between the object and the image point be defined through the attitude matrix?
2. Give the relationship between the sensor and target positions.
3. Give the measurement model for the noisy video measurements.

4. State the main uncorrelated sources of errors with their respective models.
5. State the equation of geolocation error covariance and explain how the fused geolocation error covariance for two UAVs can be computed?
6. Explain in short about the Dubins vehicle and describe the kinematic equations of the UAV.
7. How is the discretization of dynamics of Dubins vehicle achieved? Give the discretized equations of motion for nonzero input and zero input.
8. What is the parameter to which the computational cost of solving a dynamic programming is highly sensitive to?
9. Consider a UAV platform, which travels at a fixed speed of $v_a = 15\,\text{m/s}$ and the UAVs communicate to the fusion center at a frequency of 1 Hz and the sampling time is 1 second. Explain how the optimization can be made feasible by using the technique of dynamic programming.
10. Explain in brief about the cost function, which is used for dynamic programming to produce the UAV trajectories.
11. Demonstrate an example to conclude that distance coordination dominates angle coordination.
12. Describe the equations of RLS filter in the discretized form and also explain the advantages of using RLS filter over KF.
13. Explain in short about the L-F mode used to coordinate multi-robot tasks.

REFERENCES

1. Quintero, S. A. P., Papi,F., Klein, D. J., Chisci, L. and Hespanha, J. P. Optimal UAV coordination for target tracking using dynamic programming, *49th IEEE Conference on Decision and Control (CDC)*, 2010, Atlanta, GA, pp. 4541–4546, doi: 10.1109/CDC.2010.5717933l. Accessed February 2024.
2. Shao, J., Xie, G., Yu, J., and Wang, L. Leader-following formation control of multiple mobile robots. *Proceedings of the 2005 IEEE International Symposium on Intelligent Control*, 27–29 June 2005, Limassol, Cyprus, TuM01-3, pp. 809–813.
3. Kuppan Chetty, R. M., Singaperumal, M. and Nagarajan, T. Cooperative formation planning and control of multiple mobile robots. In: J. Będkowski (Ed.), *Mobile Robots - Control Architectures, Bio-Interfacing, Navigation, Multi Robot Motion Planning and Operator Training*, pp. 203–236, 2011, London, UK: InTech, ISBN: 978-953-307-842-7. Available from:https://www.intechopen.com/books/mobile-robots-control-architectures-bio-interfacing-navigation-multi-robot-motion-planning-and-operator-training/cooperative-formation-planning-and-control-of multiple-mobile-robots.
4. Dasdemir, J. and Loria, A. Robust formation-tracking control of mobile robots in a spanning-tree topology. *International Journal of Control*, 87(9), pp. 1822–1832, 2014.
5. Sreekumar, M. and Madhevan, B. Tracking algorithm using leader-follower approach for multirobots. *International Conference on Design and Manufacturing*, IconDM 2013, Procedia Engineering 64, 2013, Chennai, India. pp. 1426–1435. www.sciencedirect.com.
6. Raol, J. R., Girija, G., and Singh J. *Modelling and Parameter Estimation of Dynamic Systems*. IET/IEE Control Engineering Book Series, Vol. 65, IET, London, 2004.

5 Vision-Aided Inertial Navigation

Sudesh K. Kashyap

5.1 INTRODUCTION

Here, we present an EKF-based algorithm for real-time vision-aided inertial navigation (VAIN). A measurement model is derived to express the geometric constraints that arise when a static feature is observed from multiple camera poses. This model does not require including the 3D feature position in the state vector of the EKF and is optimal up to linearization errors. The VAIN algorithm has computational complexity, only *linear* in the number of features, and is capable of high-precision pose estimation in large-scale real-world environments. Much of the discussion in this chapter is based on refs. [1–7].

5.2 ESTIMATOR DESCRIPTION

The goal of the EKF-based estimator is to track the 3D pose of the IMU-affixed frame (I) with respect to a global frame of reference (G). The global frame is chosen as an Earth-Centered, Earth-Fixed (ECEF) frame in order to simplify the treatment of the effects of the earth's rotation on the IMU measurements [1].

5.2.1 Algorithm: Multistate Constraint Filter

Time propagation/evolution: For each IMU measurement received, propagate the filter state and covariance.

Image registration: Every time a new image is recorded: (i) augment the state and covariance matrix with a copy of the current camera pose estimate, and (ii) image processing module begins operation.

Update: When the feature measurements of a given image become available, perform an EKF measurement data update.

For propagating the EKF state and the covariance matrix, the IMU measurements are processed immediately as they become available. Conversely, each time an image is recorded, the current camera pose estimate is appended to the state vector. State augmentation is necessary for processing the feature measurements since during EKF updates, the measurements of each tracked feature are employed for imposing constraints between all camera poses from which the feature was seen. Therefore, at any time instant, the EKF state vector comprises of (i) the evolving IMU state, X_{IMU}, and (ii) a history of up to $N_{\max}$ past poses of the camera.

 DOI: 10.1201/9781003429869-5

5.2.2 State Vector Model

The evolving IMU state is described by the vector:

$$X_{\mathrm{IMU}} = \begin{bmatrix} {}^{I}_{G}\bar{q}^{T} & b_g^T & G_{V_I^T} & b_a^T & G_{P_I^T} \end{bmatrix} \quad (5.1)$$

where ${}^{I}_{G}\bar{q}^{T}$ is the unit quaternion [1, 2], describing the rotation from frame (G) to frame (I), G_{P_I} and G_{V_I} are the IMU position and velocity with respect to G, respectively, and finally, b_g and b_a are 3×1 vectors that describe the biases, affecting the gyroscope and accelerometer measurements, respectively. The IMU biases are modeled as random walk processes, driven by the white Gaussian noise vectors n_{wg} and n_{wa}, respectively. The IMU error-state is defined as follows:

$$\tilde{X}_{IMU} = \begin{bmatrix} \delta\theta_I^T & \tilde{b}_g^T & G_{\tilde{V}_I^T} & \tilde{b}_a^T & G_{\tilde{P}_I^T} \end{bmatrix}^T \quad (5.2)$$

The standard additive error definition is used for the position, velocity, and biases, (i.e., the error in the estimate $\hat{x}$ of a quantity x is defined as $\tilde{x} = x - \hat{x}$), whereas for the quaternion a different error definition is employed. In particular, if $\hat{\bar{q}}$ is the estimated value of the quaternion $\bar{q}$, then the orientation error is described by the error quaternion $\delta\bar{q}$, which is defined by the relation $\bar{q} = \delta\bar{q} \otimes \hat{\bar{q}}$; the symbol $\otimes$ denotes quaternion multiplication. The error quaternion is as follows:

$$\delta\bar{q} = \begin{bmatrix} \frac{1}{2}\delta\theta^T & 1 \end{bmatrix}^T \quad (5.3)$$

The quaternion $\delta\bar{q}$ describes the (small) rotation that causes the true and estimated attitude to coincide. Using $\delta\theta$ to describe the attitude errors, as attitude corresponds to 3 DoF, is a minimal representation. Assuming that N camera poses are included in the EKF state vector at time step k, this vector has the following form:

$$\hat{X}_k = \begin{bmatrix} \hat{X}_{IMU_k}^T & {}^{C_1}_{G}\hat{\bar{q}}^T & G_{\hat{P}_{C1}^T} & \ldots & {}^{C_N}_{G}\hat{\bar{q}}^T & G_{\hat{P}_{CN}^T} \end{bmatrix}^T \quad (5.4)$$

where ${}^{C_1}_{G}\hat{\bar{q}}$ and $G_{\hat{P}_{Ci}}$, $i = 1 \ldots$ N are the estimates of the camera attitude and position, respectively. The EKF error-state vector is defined accordingly:

$$\tilde{X}_k = \begin{bmatrix} \tilde{X}_{IMU_k}^T & \delta\theta_{C_1}^T & G_{\tilde{P}_{C1}^T} & \ldots & \delta\theta_{C_N}^T & G_{\tilde{P}_{CN}^T} \end{bmatrix}^T \quad (5.5)$$

5.2.3 State Propagation

The filter propagation equations are derived by discretization of the continuous-time IMU system model [1].

a. **Continuous-Time Modelling**

The time evolution of the IMU state is described as follows:

$$ {}^{I}_{G}\dot{\bar{q}}(t) = \frac{1}{2}\Omega(w(t))\,{}^{I}_{G}\bar{q}(t),\ \dot{b}_g(t) = n_{wg}(t)$$

$$G_{\dot{V}_I}(t) = G_a(t),\ \dot{b}_a(t) = n_{wa}(t),\ G_{\dot{P}_I}(t) = G_{V_I}(t) \tag{5.6}$$

In eqn. (5.6), G_a is the body acceleration in the global frame, $w = \begin{bmatrix} w_x\ w_y\ w_z \end{bmatrix}^T$ is the rotational velocity expressed in the IMU frame, and

$$\Omega(w) = \begin{bmatrix} -\lfloor w\times\rfloor & w \\ -w^T & 0 \end{bmatrix},\ \lfloor w\times\rfloor = \begin{bmatrix} 0 & -w_z & w_y \\ w_z & 0 & -w_x \\ -w_y & w_x & 0 \end{bmatrix} \tag{5.7}$$

The gyroscope and accelerometer measurements,sxd w_m and a_m, respectively, are given as follows:

$$w_m = w + C\left({}^{I}_{G}\bar{q}\right)w_G + b_g + n_g \tag{5.8}$$

$$a_m = C\left({}^{I}_{G}\bar{q}\right)\left(G_a - G_a + 2\lfloor w_G\times\rfloor G_{V_I} + \lfloor w_G\times\rfloor^2 G_{P_I}\right) + b_a + n_a \tag{5.9}$$

Here, C (·) denotes a rotational matrix, and n_g and n_a are zero-mean white Gaussian noise processes modelling the measurement noise. We can note that the IMU measurements incorporate the effects of the planet's rotation, w_G. Additionally, the accelerometer measurements include the gravitational acceleration, G_g, expressed in the local frame. Applying the expectation operator in the state propagation equations, we obtain the equations for propagating the estimates of the evolving IMU state as follows:

$$ {}^{I}_{G}\dot{\hat{\bar{q}}} = \frac{1}{2}\Omega(\hat{w})\,{}^{I}_{G}\hat{\bar{q}},\ \dot{\hat{b}}_g = 0_{3\times1}$$

$$G_{\dot{\hat{V}}_I} = C^T_{\hat{q}}\hat{a} - 2\lfloor w_G\times\rfloor G_{\hat{V}_I} - \lfloor w_G\times\rfloor^2 G_{\hat{P}_I} + G_g$$

$$\dot{\hat{b}}_a = 0_{3X1},\ G_{\dot{\hat{P}}_I} = G_{\hat{V}_I} \tag{5.10}$$

Here, we have denoted $C_{\hat{q}} = C\left({}^{I}_{G}\hat{\bar{q}}\right)$, $\hat{a} = a_m - \hat{b}_a$ and $\hat{w} = w_m - \hat{b}_g - C_{\hat{q}}w_G$. The linearized continuous-time model for the IMU error-state is as follows:

$$\dot{\tilde{X}}_{\text{IMU}} = F\tilde{X}_{\text{IMU}} + Gn_{\text{IMU}} \tag{5.11}$$

Here, $n_{\text{IMU}} = \left[\begin{array}{cccc} n_g^T & n_{\text{wg}}^T & n_a^T & n_{\text{wa}}^T \end{array}\right]^T$ is the system noise. The covariance matrix of n_{IMU} and Q_{IMU} depends on the IMU noise characteristics and is computed offline during sensor calibration. Finally, the matrices F and G that appear are given by:

$$F = \begin{bmatrix} -\lfloor \hat{w}\times \rfloor & -I_3 & 0_{3\times3} & 0_{3\times3} & 0_{3\times3} \\ 0_{3\times3} & 0_{3\times3} & 0_{3\times3} & 0_{3\times3} & 0_{3\times3} \\ -C_{\hat{q}}^T\left[\hat{a}\times\right] & 0_{3\times3} & -2\lfloor w_G\times \rfloor & -C_{\hat{q}}^T & -\lfloor w_G\times \rfloor \\ 0_{3\times3} & 0_{3\times3} & 0_{3\times3} & 0_{3\times3} & 0_{3\times3} \\ 0_{3\times3} & 0_{3\times3} & I_3 & 0_{3\times3} & 0_{3\times3} \end{bmatrix} \tag{5.12}$$

where I_3 is the 3×3 identity matrix, and

$$G = \begin{bmatrix} -I_3 & 0_{3\times3} & 0_{3\times3} & 0_{3\times3} \\ 0_{3\times3} & I_3 & 0_{3\times3} & 0_{3\times3} \\ 0_{3\times3} & 0_{3\times3} & -C_{\hat{q}}^T & 0_{3\times3} \\ 0_{3\times3} & 0_{3\times3} & 0_{3\times3} & I_3 \\ 0_{3\times3} & 0_{3\times3} & 0_{3\times3} & 0_{3\times3} \end{bmatrix} \tag{5.13}$$

b. **Discrete-time implementation**

The IMU samples the signals w_m and a_m with a period T, and these measurements are used for state propagation in the EKF. Every time a new IMU measurement is received, the IMU state estimate is propagated using 5th order Runge–Kutta numerical integration of eqn. (5.9). Moreover, the EKF covariance matrix has to be propagated. For this purpose, we introduce the following partitioning for the covariance:

$$P_{k|k} = \begin{bmatrix} P_{II_{k|k}} & P_{IC_{k|k}} \\ P_{IC_{k|k}} & P_{CC_{k|k}} \end{bmatrix} \tag{5.14}$$

Here, $P_{II_{k|k}}$ is the 15×15 covariance matrix of the evolving IMU state, $P_{CC_{k|k}}$ is the $6N \times 6N$ covariance matrix of the camera pose estimates, and $P_{IC_{k|k}}$ is the correlation between the errors in the IMU states and the camera pose estimates; then the covariance matrix of the propagated state is given by:

$$P_{k+1|k} = \begin{bmatrix} P_{II_{k+1|k}} & \Phi(t_k + T, t_k) P_{IC_{k|k}} \\ P_{IC_{k|k}}^T \Phi(t_k + T, t_k)^T & P_{CC_{k|k}} \end{bmatrix} \quad (5.15)$$

Here, $P_{II_{k+1|k}}$ is computed by numerical integration of the Lyapunov equation:

$$\dot{P}_{II} = FP_{II} + P_{II} F^T + GQ_{\text{IMU}} G^T \quad (5.16)$$

Numerical integration is carried out for the time interval $(t_k + T, t_k)$, with initial condition $P_{II_{k|k}}$. The state transition matrix $\Phi(t_k + T, t_k)$ is similarly computed by numerical integration of the differential equation

$$\dot{\Phi}(t_k + \tau, t_k) = F\Phi(t_k + \tau, t_k),\ \tau \in [0, T] \quad (5.17)$$

with initial condition $\tau(t_k, t_k) = I_{15}$.

5.3 STATE AUGMENTATION

Upon recording and including a new image, the camera pose estimate is computed from the IMU pose estimate as follows:

$${}_G^C\hat{\bar{q}} = {}_I^C\bar{q} \otimes {}_G^I\hat{\bar{q}} \text{ and } G_{\hat{P}_C} = G_{\hat{P}_I} + C_{\hat{q}}^T I_{P_C} \quad (5.18)$$

where ${}_G^C\bar{q}$ is the quaternion, expressing the rotation between the IMU and camera frames, and I_{P_C} is the position of the origin of the camera frame with respect to (I), both of which are known. This camera pose estimate is appended to the state vector, and the covariance matrix of the EKF is augmented accordingly:

$$P_{k|k} \leftarrow \begin{bmatrix} I_{6N+15} \\ J \end{bmatrix} P_{k|k} \begin{bmatrix} I_{6N+15} \\ J \end{bmatrix}^T \quad (5.19)$$

Here, the Jacobian J is derived as follows:

$$J = \begin{bmatrix} C\left({}_G^I\bar{q}\right) & 0_{3\times9} & 0_{3\times3} & 0_{3\times6N} \\ C_{\hat{q}}^T I_{P_C} X & 0_{3\times9} & I_3 & 0_{3\times6N} \end{bmatrix} \quad (5.20)$$

5.3.1 Measurement Model

As EKF is used for the state estimation, it is necessary to define a residual r that depends linearly on the state errors $\tilde{X}$, for constructing a measurement model according to the general form:

$$r = H\tilde{X} + \text{noise} \quad (5.21)$$

Here, H is the measurement Jacobian matrix, and the noise term must be zero-mean, white, and uncorrelated to the state error, for the EKF framework to be applied.

To derive the measurement model, one can be motivated by the fact that viewing a static feature from multiple camera poses results in constraints involving all these poses. The camera observations are grouped per tracked feature, rather than per camera pose where the measurements were recorded (the latter is the case, for example, in methods that compute pairwise constraints between). To define a constraint equation, all the measurements of the same 3D point are used, which relate all the camera poses at which the measurements occurred. One can achieve this without including the feature position in the filter state vector.

The measurement model is given by considering the case of a single feature, f_j, that has been observed from a set of M_j camera poses $({}^{C_i}_{G}\bar{q}, G_{P_{C_i}})$, $i \in S_j$. Each of the M_j observations of the feature is described by the model:

$$z_i^{(j)} = \frac{1}{C_i Z_j}\begin{bmatrix} C_i X_j \\ C_i Y_j \end{bmatrix} + n_i^{(j)},\ i \in S_j \tag{5.22}$$

Here $n_i^{(j)}$ is the 2×1 image noise vector, with covariance matrix $R_i^{(j)} = \sigma_{im}^2 I_2$. The feature position expressed in the camera frame, $C_{iP_{fj}}$, is given by:

$$C_{iP_{fj}} = \begin{bmatrix} C_i X_j \\ C_i Y_j \\ C_i Z_j \end{bmatrix} = C\left({}^{C_i}_{G}\bar{q}\right)\left(G_{P_{fj}} - G_{P_{Ci}}\right) \tag{5.23}$$

Here, $G_{P_{fi}}$ is the 3D feature position in the global frame. Since this is unknown, we have used the least-squares minimization to obtain an estimate $G_{\hat{P}_{fj}}$ of the feature position in the first step of our algorithm. This is achieved using the measurements $z_i^{(j)}$, $i \in S_j$ and the filter estimates of the camera poses at the corresponding time instants. Once the estimate of the feature position is obtained, we compute the measurement residual as follows:

$$r_i^{(j)} = z_i^{(j)} - \hat{z}_i^{(j)} \tag{5.24}$$

$$\hat{z}_i^{(j)} = \frac{1}{C_i \hat{Z}_j}\begin{bmatrix} C_i \hat{X}_j \\ C_i \hat{Y}_j \end{bmatrix}, \begin{bmatrix} C_i \hat{X}_j \\ C_i \hat{Y}_j \\ C_i \hat{Z}_j \end{bmatrix} = C\left({}^{C_i}_{G}\hat{\bar{q}}\right)\left(G_{\hat{P}_{fj}} - G_{\hat{P}_{Ci}}\right) \tag{5.25}$$

Linearizing about the estimates for the camera pose and for the feature position, the residual of eqn. (5.24) can be approximated as follows:

$$r_i^{(j)} \simeq H_{X_i}^{(j)} \tilde{X} + H_{fi}^{(j)} G_{\tilde{P}_{fj}} + n_i^{(j)} \tag{5.26}$$

Here, $H_{X_i}^{(j)}$ and $H_{fi}^{(j)}$ are the Jacobians of the measurement $z_i^{(j)}$ with respect to the state and the feature position, respectively, and $G_{\tilde{p}_{fj}}$ is the error in the position estimate of f_j. By stacking the residuals of all M_j measurements of this feature, we obtain:

$$r^{(j)} == H_X^{(j)}\tilde{X} + H_f^{(j)} G_{\tilde{p}_{fj}} + n^{(j)} \tag{5.27}$$

Here, $r^{(j)}$, $H_x^{(j)}$, $H_f^{(j)}$, and $n^{(j)}$ are block vectors or matrices with elements $r_i^{(j)}$, $H_f^{(j)}$, and $n_i^{(j)}$, respectively, for $i \in S_j$. As the feature observations in different images are independent, the covariance matrix of $n^{(j)}$ is $R^{(j)} = \sigma_{im}^2 I_{2Mj}$.

Note that since the state estimate, X, is used to compute the feature position estimate, the error $G_{\tilde{p}_{fj}}$ in eqn. (5.26) is correlated with the errors $\tilde{X}$. Therefore, the residual $r^{(j)}$is not in the form of eqn. (5.21) and cannot be directly applied for measurement updates in the EKF. Accordingly, to overcome this problem, we define a residual $r_O^{(j)}$ by projecting $r^{(j)}$ on the left nullspace of the matrix $H_f^{(j)}$. Specifically, if we let A denote the unitary matrix whose columns form the basis of the left nullspace H_f, we obtain:

$$r_o^{(j)} = A^T\left(z^{(j)} - \hat{z}^{(j)}\right) \simeq A^T H_X^{(j)}\tilde{X} + A^T n^{(j)} = H_o^{(j)}\tilde{X}^{(j)} + n_o^{(j)} \tag{5.28}$$

$$r_o^{(j)} = H_o^{(j)}\tilde{X}^{(j)} + n_o^{(j)} \tag{5.29}$$

Since the $2M_j - 3$ matrix $H_f^{(j)}$ has full column rank, its left nullspace is of dimension $2M_j - 3$. Therefore, $r_o^{(j)}$ is a $(2M_j - 3) \times 1$ vector. This residual is independent of the errors in the feature coordinates, and thus, EKF updates can be performed based on it. A linearized constraint between all the camera poses from which the feature f_j was observed is defined in eqn. (5.26). This expresses all the available information that the measurements $z_i^{(j)}$ provide for the M_j states, and thus, the resulting EKF update is (sub-)optimal, except for the inaccuracies caused by linearization.

Note that the unitary matrix A does not need to be explicitly evaluated, in order to compute the residual $r_o^{(j)}$ and the measurement matrix $H_o^{(j)}$. Instead, the projection of the vector r and the matrix $H_X^{(j)}$ on the nullspace of $H_f^{(j)}$can be computed very efficiently using Givens rotations, in $O\left(M_j^2\right)$ operations. Additionally, since the matrix A is unitary, the covariance matrix of the noise vector $n_o^{(j)}$ is given by:

$$E\left\{n_o^{(j)} n_o^{(j)T}\right\} = \sigma_{im}^2 A^T A = \sigma_{im}^2 I_{2Mj-3} \tag{5.30}$$

The residual defined in eqn. (5.28) is not the only possible expression of the geometric constraints that are induced by observing a static feature in Mj images. An alternative approach would be, for example, to employ the epipolar constraints that are defined for each of the $M_j\left(M_j - 1\right)/2$ pairs of images. However, the resulting $M_j\left(M_j - 1\right)/2$ equations would still correspond to only $2Mj - 3$ independent constraints since each measurement is used multiple times, rendering the equations statistically correlated. Our experiments have shown that employing linearization of the epipolar constraints results in a significantly more complex implementation and yields inferior results compared to the approach described above.

5.3.2 EKF Updates

In this section, we present the update phase of the EKF, in which the constraints from observing multiple features are used. EKF updates are triggered by one of the following two events: (i) when a feature that has been tracked in a number of images is no longer detected, and then all the measurements of this feature are processed using the method presented in the previous section. This case occurs most often as features move outside the camera's field of view; and (ii) every time a new image is recorded, a copy of the current camera pose estimate is included in the state vector. If the maximum allowable number of camera poses, N_{max}, has been reached, at least one of the old ones must be removed. Prior to discarding states, all the feature observations that occurred at the corresponding time instants are used, in order to utilize their localization information. In our algorithm, we choose $N_{max}/3$ poses that are evenly spaced in time, starting from the second-oldest pose. These are discarded after carrying out an EKF update using the constraints of features that are common to these poses. We have opted to always keep the oldest pose in the state vector because the geometric constraints that involve poses further back in time typically correspond to larger baseline and hence carry more valuable positioning information. This approach was shown to perform very well in practice.

Consider that at a given time step the constraints of L features, selected by the above two criteria, must be processed. Following the procedure described in the previous section, we compute a residual vector $r_o^{(j)}, j = 1... L$, as well as a corresponding measurement matrix $H_o^{(j)}, j = 1... L$ for each of these features. By stacking all residuals in a single vector, we obtain:

$$r_o = H_X \tilde{X} + n_o \tag{5.31}$$

where r_o and n_o are vectors with block elements $r_o^{(j)}$ and $n_o^{(j)}, j = 1...L$, respectively, and H_X is a matrix with block rows $H_X^{(j)}, j = 1... L$. The noise vectors $n_o^{(j)}$ are uncorrelated as the feature measurements are statistically independent. Consequently, the covariance matrix of the noise vector n_o is equal to $R_o = \sigma_{im}^2 I_d$, where $d = \sum_{j=1}^{L}(2M_j - 3)$ is the dimension of the residual r_o. One issue that arises in practice is that d can be a quite large number. For example, if 10 features are seen in 10 camera poses each, the dimension of the residual is 170. The QR decomposition of the matrix H_X is used in order to reduce the computational complexity of the EKF update. Specifically, we denote this decomposition as follows:

$$H_X = \begin{bmatrix} Q_1 & Q_2 \end{bmatrix} \begin{bmatrix} T_H \\ 0 \end{bmatrix} \tag{5.31}$$

where Q_1 and Q_2 are unitary matrices whose columns form bases for the range and nullspace of H_X, respectively, and T_H is an upper triangular matrix. With this definition we have:

$$r_o = \begin{bmatrix} Q_1 & Q_2 \end{bmatrix} \begin{bmatrix} T_H \\ 0 \end{bmatrix} \tilde{X} + n_o \tag{5.32}$$

$$\begin{bmatrix} Q_1^T r_o \\ Q_2^T r_o \end{bmatrix} = \begin{bmatrix} T_H \\ 0 \end{bmatrix} \tilde{X} + \begin{bmatrix} Q_1^T n_o \\ Q_2^T n_o \end{bmatrix} \tag{5.33}$$

From the last equation, it becomes clear that all the useful information in the measurements can be retained by projecting the residual r_o on the basis vectors of the range of H_X. The residual $Q_2^T r_o$ is only noise and can be completely discarded. For this reason, instead of the residual shown in eqn. (5.26), we employ the following residual for the EKF update:

$$r_n = Q_1^T r_o = T_H \tilde{X} + n_n \tag{5.34}$$

In this expression, $n_n = Q_1^T n_o$ is a noise vector whose covariance matrix is equal to $R_n = Q_1^T R_o Q_1 = \sigma_{im}^2 I_r$, with r being the number of columns in Q_1. The EKF update continues by computing the Kalman gain:

$$K = PT_H^T \left(T_H P T_H^T + R_n \right)^{-1} \tag{5.35}$$

while the correction to the state is given by the vector

$$\Delta X = K r_n \tag{5.36}$$

Finally, the state covariance matrix is updated according to:

$$P_{k+1|k+1} = \left(I_\xi - KT_H \right) P_{k+1|k} \left(I_\xi - KT_H \right)^T + KR_n K^T \tag{5.37}$$

where $\xi = 6N + 15$ is the dimension of the covariance matrix. Lastly, we discuss about the computational complexity of the operations needed during the EKF update. The residual r_n, as well as the matrix T_H, can be computed using Givens rotations in $O\left(r^2 d\right)$ operations, without the need to explicitly form Q_1. Conversely, eqn. (5.35) involves multiplication of square matrices of dimension ξ, an $O\left(\xi^3\right)$ operation. Hence, the cost of the EKF update is max($O\left(r^2 d\right), O\left(\xi^3\right)$. If, on the contrary, the residual vector r_o was employed, without projecting it on the range of H_X, the computational cost of computing the Kalman gain would have been $O\left(d^3\right)$. Since typically $d \gg \xi$, r, we can see that the substantial savings in computation can be achieved with the use of the residual r_n.

5.4 DISCUSSION OF EXPERIMENTAL RESULTS

This section presents the study of some of the properties of the described algorithm [1]. As shown in the previous section, the filter's computational complexity is linear in the number of observed features and at most cubic in the number of states that are included in the state vector. Thus, in determining the computational cost of the algorithm, the number of poses that are included in the state is the most significant factor. Since this number is a selectable parameter, it can be tuned according to the available computing resources and the accuracy requirements of a given application. If it is required to adjust to the varying availability of resources, the length of the filter state can also be adaptively controlled during filter operation. The major source of limitation in recursive state estimation with camera observations is the nonlinear nature of the measurement model. Vision-based motion estimation is very sensitive to noise, and, especially when the observed features are at large distances, false local minima can cause convergence to inconsistent solutions. In literature, the techniques such as Sigma-point Kalman filtering, particle filtering, and the inverse depth representation for features are used to address the problem introduced by nonlinearity. Two characteristics of the described algorithm increase its robustness to linearization inaccuracies: (i) the inverse feature depth parametrization used in the measurement model, and (ii) the delayed linearization of measurements. By the algorithm's construction, multiple observations of each feature are collected, prior to using them for EKF updates, resulting in more accurate evaluation of the measurement Jacobians. Due to the limited field of view of cameras, as well as occlusions, image noise, and viewpoint changes, it results in failures of the feature tracking algorithms, i.e., observation is that in typical image sequences, most features can only be reliably tracked over a small number of frames ("opportunistic" features), and only few can be tracked for long periods of time, or when revisiting places (persistent features). As previously discussed, if all the poses in which a feature has been seen are included in the state vector, then the proposed measurement model is optimal, except for linearization inaccuracies. Therefore, for realistic image sequences, the proposed algorithm is able to optimally use the localization information of the opportunistic features. Additionally, to improve the attainable localization accuracy, within areas with lengthy loops, we note that the state vector X_k is not required to contain only the IMU and camera poses, if desired, the persistent features can be also be included in the filter state, and used for SLAM.

The algorithm described in the preceding sections has been tested extensively both in simulation and with real data [1]. The simulation experiments have verified that the algorithm produces pose and velocity estimates that are consistent and can operate reliably over long trajectories, with varying motion profiles and density of visual features. Simulation data are not included; instead, we here present the results of the algorithm in an outdoor experiment, which demonstrates that the method is capable of long-term operation in a real-world setting. Additional datasets of real-world experiments, as well as simulation results of the use of our algorithm, can be found in ref. [5]. The experimental setup consisted of a camera/IMU system, placed on a car that was moving on the streets of a typical residential area in Minneapolis, MN. The system comprised a Point Grey FireFly camera, registering images of resolution 640 × 480 pixels at 3Hz, and an Inertial Science ISIS IMU, providing inertial measurements at a rate of 100Hz. During the experiment, all data were stored on a computer, and processing was done offline. Figure 5.1 shows some example images from the recorded sequence; whereas, a video of all 1598 images, which were recorded in about 9 minutes of driving, can be found online [5]. For the results shown here, feature extraction and matching were performed using the SIFT algorithm [6]. During this run, a maximum of 30 camera poses was maintained in the filter state vector. Since features were rarely tracked for more than 30 images, this number was sufficient for utilizing most of the available constraints between states, while attaining real-time performance. Even though images were only recorded at 3Hz due to limited hard disk space on the test system, the estimation algorithm is able to process the dataset at 14Hz, on a single core of an Intel T7200 processor (2GHz clock rate). During the experiment, a total of 142,903 features were successfully tracked and used for EKF updates, along a 3.2km-long trajectory. A GPS sensor was not available during the experiment, and therefore, no ground-truth trajectory data exists. However, the quality of the position estimates can be evaluated using a map of the area. In Figure 5.2, the estimated trajectory is plotted on a map of the neighborhood where the experiment took place. One can observe that this trajectory follows the street layout quite accurately, and additionally, the

FIGURE 5.1 Experimental dataset images. (Adopted/Modified from: A.I. Mourikis, and S. I. Roumeliotis. A multistate constraint Kalman filter for vision-aided inertial navigation. *2007 IEEE International Conference on Robotics and Automation Roma*, Italy, 10–14 April 2007.)

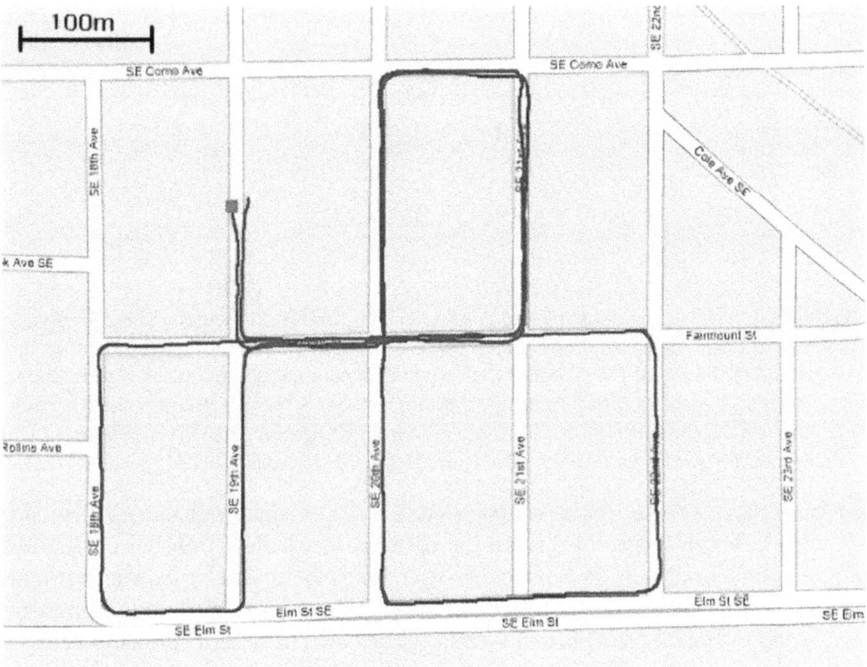

FIGURE 5.2 The estimated trajectory superimposed on a map of the region (at the experiment's site). The initial position of the car is denoted by a red square, and the scale of the map is shown on the top left corner. (Adopted/Modified from: A. I. Mourikis, and S. I. Roumeliotis. A multistate constraint Kalman filter for vision-aided inertial navigation. *2007 IEEE International Conference on Robotics and Automation Roma*, Italy, 10–14 April 2007.)

position errors that can be inferred from this plot agree with the 3σ bounds shown in Figure 5.3(a). The final position estimate, expressed with respect to the starting pose, is $\hat{X}_{\text{final}} = \begin{bmatrix} -7.92 & 13.14 & -0.78 \end{bmatrix}^T$ m. From the initial and final parking spots of the vehicle, it is known that the true final position expressed with respect to the initial pose is approximately $X_{\text{final}} = \begin{bmatrix} 0 & 7 & 0 \end{bmatrix}^T$. Thus, the final position error is approximately 10 m in a trajectory of 3.2 km, i.e., an error of 0.31% of the traveled distance. This is remarkable, given that the algorithm does not utilize loop closing, and uses no prior information (for example, non-holonomic constraints or a street map) about the car motion. Moreover, it is worth pointing out that the camera motion is almost parallel to the optical axis, a condition which is particularly adverse for image-based motion estimation algorithms [7]. Figures 5.3(b) and (c) show the 3σ bounds for the errors in the IMU attitude and velocity along the three axes. From these, it can be observed that in this particular experiment, the algorithm obtains accuracy (3σ) better than 1° for attitude and better than 0.35 m/s for velocity. The results shown here demonstrate that the proposed algorithm is capable of operating in a real-world environment and producing very accurate pose estimates in real-time. From the dataset presented here, one can observe several moving objects, such as cars, pedestrians,

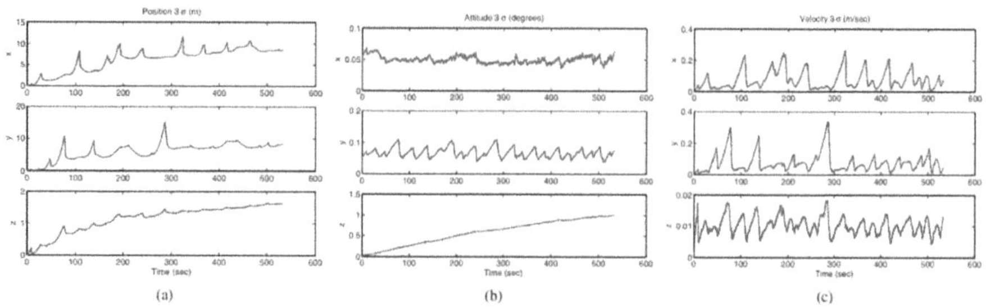

FIGURE 5.3 The errors in the position, attitude, and velocity: 3σ bounds. The EKF state is expressed in Earth-Centered, Earth-Fixed (ECEF) frame, but for plotting, all quantities are transformed in the initial IMU frame, whose x-axis is pointing approximately south, and its y-axis east. (Adopted/Modified from: A. I. Mourikis, and Stergios I. Roumeliotis. A multi-state constraint Kalman filter for vision-aided inertial navigation. *2007 IEEE International Conference on Robotics and Automation* Roma, Italy, 10–14 April 2007.)

and trees, whose leaves move in the wind. Using a simple Mahalanobis distance test, the algorithm is able to discard the outliers, which arise from visual features detected on these objects. Robust outlier rejection is facilitated by the fact that multiple observations of each feature are available, and thus, visual features that do not correspond to static objects become easier to detect. The described method can be used either as a stand-alone pose estimation algorithm or combined with additional sensing modalities to provide increased accuracy. For example, if a GPS sensor was available during this experiment, its measurements could be used to compensate for position drift.

EXERCISES

1. Explain in brief the overview of algorithm as discussed in this chapter.
2. Explain during which events the EKF update is triggered?
3. Explain the computational complexity of the operation involved during the EKF update.
4. Give the equation of linearized continuous-time model for the IMU error-state.
5. Give the covariance matrix of the propagated state and explain how it is achieved in discrete-time implementation.
6. Explain the state augmentation as discussed in this chapter.
7. Derive the measurement model that is able to express the geometric constraints that arise when a static feature is observed from multiple camera poses as discussed in this chapter.

REFERENCES

1. Mourikis, A. I., and Roumeliotis, S. I. A multistate constraint Kalman filter for vision-aided inertial navigation. *2007 IEEE International Conference on Robotics and Automation Roma*, Italy, 10–14 April 2007.
2. Breckenridge, W. G. Quaternions proposed standard conventions, JPL, Tech. Rep. Inter office memorandum, IOM 343–79–1199, Pasadena, USA, 1999.

3. Roumeliotis, S., Johnson, A., and Montgomery, J. Augmenting inertial navigation with image-based motion estimation. *IEEE International Conference on Robotics and Automation*, Washington D.C., USA, 2002, pp. 4326–4333.
4. Montiel, A. D. J. and Civera, J. Unified inverse depth parametrization for monocular slam, *Proceedings of Robotics: Science and Systems*, USA, 2006.
5. https://www.cs.umn.edu/~mourikis/icra07video.htm.
6. Lowe, D. G. Distinctive image features from scale-invariant key points. *Int. J. Comput. Vis.*, 60(2), 91–100, 2004.
7. Oliensis, J. A new structure-from-motion ambiguity. *IEEE Trans Pattern Anal Mach Intell*, 22(7), 685–700, 2000.

6 Terrain-Assisted Underwater Vehicle Navigation

6.1 TERRAIN NAVIGATION MODELS

In autonomous underwater vehicles (AUVs), terrain navigation models use specific types of methods for self-positioning and guidance. As we know, GPS is not available underwater; these models will help to analyse the surrounding underwater terrain to navigate effectively and to understand their location. This type of navigation uses pre-existing high-resolution bathymetric maps to get accurate positioning, so the accuracy depends on the quality of the reference map and the sensor data. In this section, two models are described in detail.

6.1.1 State Space Model

A state space model is a mathematical framework that represents the dynamic behaviour of the AUV and its environment, specifically focusing on the relationship between the AUV's state, i.e. position, velocity, attitude, etc., and the sensor measurements like bathymetric data from sonar, which plays a crucial role in estimating the AUV's position by providing a statistical framework for data analysis and filtering.

The position of the vehicle is usually estimated in a global coordinate system. Here, map frame {m} is considered to be in global coordinate system. The North-East-Down (NED) frame is used as a navigation coordinate system {n}, where the origin is selected as the centre of the vehicle, and the vehicle body coordinate system {b} is in Cartesian frame [1].

$$X_t = \begin{bmatrix} y_t \\ z_t \\ u_t \\ v_t \\ w_t \end{bmatrix} \tag{6.1}$$

According to the vehicle motion, the state vector is X_t at time step t where 'x' represents the east position, 'y' represents the north position, 'z' represents the depth position, 'u' represents the east velocity, 'v' represents the north velocity, and 'w'

DOI: 10.1201/9781003429869-6

represents the depth velocity. The equation below can describe the state space model of terrain-aided navigation (TAN).

$$X_t = X_{t-1} + S_{t,t-1} + n_t, \tag{6.2}$$

where X_{t-1} indicates the position coordinates of inertial navigation system (INS) at t time, $S_{t,t-1}$ represents the control input, and n_t process noise.

As we know, the tidal depth bias exists for the AUVs; and this arises due to the periodic rise and fall of water levels caused by tides. It affects the measured depth from sonar and is unpredictable or mispredicted in some natural scenes. This needs to be added to the state vector.

The three-dimensional state, control input, and process noise are given by

$$X_t = \begin{bmatrix} X_{t,n} & X_{t,e} & \delta z_t \end{bmatrix}^T, \tag{6.3}$$

$$S_{t,t-1} = \begin{bmatrix} \Delta X_{t,n} & \Delta x_{t,e} & 0 \end{bmatrix}^T, \tag{6.4}$$

$$n_t = \begin{bmatrix} n_{t,n} & n_{t,e} & n_{t,d} \end{bmatrix}^T, \tag{6.5}$$

where $X_{t,n}$ and $X_{t,e}$ represent the vehicle north and east positions at time t with a δz_t as the tidal depth bias, respectively. $S_{t,t-1}$ represents the position changes in INS measurement, and n_t represents the INS position drift error along with a random tidal depth bias error.

6.1.2 Measurement Model

The derivation of the measurement from the state is represented by the measurement model, also known as the observation model. It is important to note that the model illustrates how the measurements are obtained from the state and the way the sensor is actually operating. The procedure that transforms a measurement into a state estimate is frequently taken into consideration in various applications. But, this concept is more of a straightforward and inverted process of self-sensing. It is crucial to remember this, since it allows the filter to employ underdetermined state measurements.

The measurement model has advantages and limitations, and the choice depends on factors such as the complexity of the environments, the availability of sensor data, computational resources, and the desired level of accuracy and robustness in navigation for the AUV. Several types of measurement models can be used, depending on the specific sensors available like sonars, DVL, etc., and also based on the requirements of navigation tasks. Rangeand projection-based are such two types of measurement models.

The range-based measurement model estimates the AUV position by measuring the distance (ranges) to the surrounding objects or features in the environment; for example, sonar sensors are commonly used in AUV navigation to measure the distance of objects or the seafloor. These measurements are then used to estimate the AUV position relative to these objects or the seafloor. The projection-based measurement models estimate the AUV position by projecting sensor measurements onto a known or estimated map of the environment; for example, in terrain navigation using optical cameras or LiDAR sensors, the sensor measurements (e.g. visual features or point clouds) are projected onto a 3D environment map. These projections are then used to estimate the AUV position relative to the map. A projection-based measurement approach provides fast access to a digital map for matching computations, unlike a range-based measurement model that necessitates computationally demanding ray tracing. Thus, a projection-based measuring paradigm is used in this chapter.

For TAN using a multi-beam sonar, a projection-based measurement model relates the range measurements from each beam to the vehicle's position and orientation. This model avoids computationally expensive ray tracing by utilizing a pre-existing Digital Elevation Model (DEM) of the terrain. Here is the equation:

$$y_t = z_t \bullet I + d, \tag{6.6}$$

where,

y_t is the entire depth from the seafloor to the sea level represented as a vector, z_t is a combination of the depth measurement from a depth pressure sensor and the altitude data from the MBS.

$$y_t = h(x_t) + H \bullet c_b^n \bullet q_t + \delta z_t \bullet I \bullet v_t, \tag{6.7}$$

where $h(x_t)$: Represents the vehicle's position and orientation onto the DEM, providing an expected range for each beam in the absence of terrain irregularities or errors.

H: Beam geometry matrix (constant matrix) that translates the projected vehicle position range measurements for each beam.

c_b^n: Calibration matrix for beam n that accounts for any misalignment or scaling errors specific to each beam within the sonar and ensuring the accurate mapping of beam directions; for more details, refer to [1].

q_t: The terrain height vector represents the height of the terrain directly beneath each beam and is extracted from DEM at the projected points.

δz_t: This is the tidal depth bias that gives the scalar value correction for the difference in reference depth between the sonar and the DEM.

I: The identity matrix ensures the bias is applied equally to all beam measurements.

v_t: The inherent noise associated with each range measurement due to various factors like sensor limitations, environmental conditions, and signal processing algorithms.

Finally, the dot product of ($H \bullet c_b^n \bullet q_t$) essentially combines the geometry, calibration, and terrain information to calculate the expected range measurements for each beam.

$$x_t = \begin{bmatrix} x_{t,N} & x_{t,E} \end{bmatrix}^T \tag{6.8}$$

$$h(x_t) = h\left(\begin{bmatrix} x_{t,N} & x_{t,E} \end{bmatrix}^T\right) \tag{6.9}$$

$h(\cdot)$ denotes the terrain interpolation function where $h(x_t)$ denotes the depth (estimated) at the (x_N, x_E) interpolation point.

6.2 RECURSIVE BAYESIAN ESTIMATION METHOD

The Bayesian estimation offers a powerful framework for TAN by incorporating prior information and updating estimates with new sensor data in real-time. In the context of AUV navigation, the state refers to the vehicle's position, velocity, orientation, and possibly other relevant parameters. This approach means that the Bayes formula incorporates how the system's state evolves over time, given the sensor measurements and prior knowledge about the system. In most cases, this also means that a numerical method has to be used to establish a position. This can be computationally demanding. With the low order of the state vector, the method is easily illustrated, which can give extra insight into the estimation problem. The idea behind this concept of a likelihood function for the random variable Y to have the value y when random variable X has the value x is in the discrete case, and the most common notation is L $(x; y)$.

6.2.1 Bayesian Update Equations

Bayesian estimation is a statistical method for estimating unknown parameters based on probability distributions. Navigation involves updating the probability distribution of the vehicle's state based on new sensors, it's measurements, and prior information. Where the recursive Bayesian estimation extends Bayesian estimation by iteratively updating the probability distribution as new sensor measurements become available. This allows for real-time estimation of the vehicle state.

The Bayesian update equations are fundamental to the process of recursively refining the estimate of the system state as new measurements become available. These equations are derived from Bayes' theorem and are used in the update step of recursive Bayesian estimation. At its core, Bayesian estimation is a method for updating beliefs about the state of a system in light of new evidence. It utilizes Bayes' theorem to compute the posterior probability distribution of the state, given the observed measurements. The theorem states:

$$P(x / z) = \frac{P(z / x) \bullet P(x)}{P(z)}, \tag{6.10}$$

where the $P(x / z)$ is the posterior probability distribution of the state given the measurements. $P(z / x)$ is the likelihood function, representing the probability of observing the measurements z given the state x; depending on the sensor characteristics and measurement noise, different likelihood models can be employed. $P(x)$ is the prior probability distribution of the state, representing our belief about the state before observing the measurements. $P(z)$ is the normalization constant, ensuring that the posterior distribution integrates to 1. It acts as a scaling factor and is computed by marginalizing all possible states:

$$P(z) = \int P(z / x) \bullet P(x)\, dx \tag{6.11}$$

This integral can be difficult to compute directly, especially in high-dimensional state spaces. Techniques such as Monte Carlo methods or particle filters are often employed to approximate the normalization constant.

The non-linearity of adopting a comprehensive Bayesian estimate method is motivated by the provided measurement model. Assume that the posterior probability density of the estimated state, $p(x_t|y_{1:t})$, is conditioned on a measurement history $y_{1:t}$ of state x_t. It's the updated belief of the state after incorporating all available measurements up to time t. According to eqn. (6.10), the Bayesian formula states that it can be computed by

$$p(x_t|y_{1:t}) = \frac{p(y_t|x_t)\; p(x_t|y_{1:t-1})}{p(y_t|y_{1:t-1})}, \tag{6.12}$$

where $p(y_t|x_t)$ is the likelihood function, representing the probability of observing measurements y_t at a given state x_t. $p(x_t|y_{1:t-1})$ is the prior probability density of the state x_t given the measurements up to time t-1; this is the posterior from the previous time step.

$$p(y_t|y_{1:t-1}) = \int p(y_t|x_t)\; p(x_t|y_{1:t-1})\; dx_t \tag{6.13}$$

From eqns (6.12) and (6.13):

$$p(x_t|y_{1:t}) = \frac{p(y_t|x_t)\; p(x_t|y_{1:t-1})}{\int p(y_t|x_t)\; p(x_t|y_{1:t-1})\; dx_t} \tag{6.14}$$

According to the Bayes theorem, from eqn (6.14), the $p(y_t|y_{1:t-1})$ is the normalization constant ensuring that the posterior distribution integrates to 1. It represents the probability of observing measurements y_t, given the measurements up to time t–1, averaged over all possible states.

$$p(x_t|y_{1:t-1}) = \int p(x_t|x_{t-1})\, p(x_t|y_{1:t-1})\; dx_{t-1}, \tag{6.15}$$

where the probability density $p(x_t|y_{1:t-1})$, representing the prior belief about the state x_t given the measurements up to time $t-1$, can be calculated recursively using the Bayesian updated equation. The integral sums over all possible values of x_{t-1}, integrating the transition model $p(x_t|x_{t-1})$ weighted by the posterior probability density $p(x_t|y_{1:t-1})$ over the entire state space.

In the minimum mean square error (MMSE) criterion, the estimated state $\hat{x}_t$ and the covariance matrix P_t are calculated by computing the mean and second moment of the posterior probability density $p(x_t|y_{1:t})$ obtained from the Bayesian update process respectively. These estimates provide the best linear unbiased estimate of the state and its uncertainty, given the available measurements up to time t.

$$\hat{x}_t = \int x_k \bullet p\left(x_t \middle| y_{1:t}\right) dx_t \tag{6.16}$$

The estimated state $\hat{x}_t$ can be obtained by calculating the mean of the posterior probability density $p\left(x_t \middle| y_{1:t}\right)$, which represents the statemost likely given the measurements.

$$P_t = \int \left(x_t - \hat{x}_t\right)\left(x_t - \hat{x}_t\right)^T \bullet p\left(x_t \middle| y_{1:t}\right) dx_t \tag{6.17}$$

The covariance matrix P_t represents the uncertainty or spread of the estimated state $\hat{x}_t$ around its mean. It can be calculated using the second moment of the posterior probability density $p\left(x_t \middle| y_{1:t}\right)$. After obtaining the posterior probability density using Bayesian estimation, a Kalman filter or its variant can be employed further to refine the state estimate and its covariance matrix.

6.2.2 2D Point Mass Filter

The 2D point mass filter (PMF) is a simple filtering technique used to estimate the position and velocity of an object moving in a 2D space. It assumes that the object's motion can be adequately represented by a constant velocity model, meaning that it moves at a constant speed and direction i.e., between the two measurements obtained. The PMF equations involve predicting the position and velocity of the object based on its previous state and the elapsed time, as well as updating these estimates based on new measurements.

The filter iterates between prediction and measurement update steps to continuously estimate the object's state based on available sensor measurements. Predicting the tidal depth bias ahead of time can compensate for it; however, this creates a 2D TAN estimation challenge. This is because the tidal depth bias is intended to be zero.

Let's denote the grid points in the search area as $x_{t_{i,j}}$, where $i = 1,\ldots, M$ and $j = 1,\ldots, N$. These grid points represent discrete positions within the search area. The probability density $p\left(x_t \middle| y_{1:t}\right)$ is approximated at each grid point by a weight $p\left(x_{t_{i,j}} \middle| y_{1:t}\right)$. So, the equation representing the approximation of each grid point can be written as:

$$p\left(x_{t_{i,j}} \middle| y_{1:t}\right) = p\left(x_{t_{i,j}} y_{1:t-1}\right) \cdot p\left(y_t x_{t_{i,j}}, y_{1:t-1}\right), \tag{6.18}$$

where $p\left(x_{t_{i,j}} \middle| y_{1:t-1}\right)$ represents the prior probability density at grid point $x_{t_{i,j}}$ based on measurements up to time $t-1$, $p\left(y_t \middle| x_{t_{i,j}}, y_{1:t-1}\right)$ represents the likelihood of observing measurement y_t given the state $x_{t_{i,j}}$ and past measurements $y_{1:t-1}$.

The above equation expands as follows:

$$p\left(x_{t_{i,j}} \middle| y_{1:t}\right) = \alpha_t^{-1} p_{v_t}\left(y_t - h\left(x_{t_{i,j}}\right) + H \bullet c_b^n \bullet q_t\right) \bullet p\left(x_{t_{i,j}} y_{1:t-1}\right), \tag{6.19}$$

where the process noise density $p(w_t)$ and measurement noise density p, the likelihood of observing measurement y_t at grid points can be expressed as:

$$p\left(x_{ti,j} \middle| y_{1:t}\right) = \sum_{m=1}^{M} \sum_{n=1}^{N} \left[p_{w_t}\left(x_{ti,j} - u_{t,t-1} - x_{t-1}(m,n)\right) \bullet p\left(x_{t-1}(m,n) y_{1:t-1}\right) \right] \Delta^2, \quad (6.20)$$

where α_t denotes a normalized factor expressed as:

$$\alpha_t = \sum_{m=1}^{M} \sum_{n=1}^{N} p_{v_t}\left(y_t - h\left(x_{ti,j}\right) + H \bullet c_b^n \bullet q_t\right) \bullet p\left(\delta x_{ti,j} y_{1:t-1}\right) \Delta^2 \quad (6.21)$$

Next, the covariance matrix and estimated location are computed by

$$\hat{x}_t = \sum_{m=1}^{M} \sum_{n=1}^{N} x_{ti,j} \, p\left(x_{ti,j} y_{1:t}\right) \Delta^2 \quad (6.22)$$

$$\hat{P}_t = \sum_{m=1}^{M} \sum_{n=1}^{N} \left(x_{ti,j} - \hat{x}_t\right)\left(x_{ti,j} - \hat{x}_t\right)^T p\left(x_{ti,j} y_{1:t}\right) \Delta^2 \quad (6.23)$$

6.2.3 Marginalized Point Mass Filter in Three Dimensions

The marginalized PMF in three dimensions is an extension of the traditional PMF to handle systems with a higher dimensional state space, typically encountered in applications such as 3D localization or tracking. In a marginalized PMF, the focus is on estimating a subset of the state variables while marginalizing the remaining variables, leading to a lower-dimensional estimation problem. Estimating and compensating for the unpredictable tidal depth bias is necessary. However, because of its high processing complexity with 3D grids, the 3D PMF is not feasible for real-time TAN. Given that the bias associated with tidal depth, it has a linear substructure in both state space and measurement models. A three-dimensional minimized PMF is suggested to simplify the computations.

6.2.3.1 Marginalized Point Mass Filter

By marginalizing out irrelevant variables, such as linear variables from state space and measurement model and also estimating them by a filter, the marginalized PMF simplifies the estimation process and reduces computational complexity while still providing accurate estimates of the variables of interest, such as position in a 2D space. This approach is particularly useful when the system dynamics are known to be simpler in certain dimensions or when velocity information is not crucial for the application. In the state vector equation, the tidal depth bias variable is linear, where the horizontal position is non-linear.

Let $x_t^n = \left[\left[x_{t,N} - x_{t,E}\right]^T\right]$ represent the concatenated state vector, where $x_{t,N}$ and $x_{t,E}$ E are the North and East position coordinates respectively, and τ is the tidal depth bias variable. $p(x_t|y_{1:t})$ denotes the joint posterior probability density, which represents the probability of the state vector x_t given all measurements up to time t. This probability density can be decomposed into two terms:

$$p\left(\delta z_t, x_t^n|y_{1:t}\right) = p\left(\delta z_t x_t^n, y_{1:t}\right) \cdot p\left(x_t^n y_{1:t}\right) \tag{6.24}$$

The $p\left(\delta z_t|x_t^n, y_{1:t}\right)$ represents the probability density of the tidal depth bias conditioned on the horizontal position. This term is analytically tractable and can be estimated using the Kalman filter. $p\left(x_t^n|y_{1:t}\right)$ is the conditional probability density of the concatenated state vector given the measurements up to time t. However, this term is intractable and is still estimated using the original 2D PMF.

If x_t^n has been approximated by a set of grid points with weights, the conditional probability density of the tidal depth bias at a grid point $x_t^{i,j}$ can be represented as

$$p\left(\delta z_t, x_t^{i,j}|y_{1:t}\right) = N\left(\delta z_t^{i,j}, P_{\delta z_t^{i,j}}\right) \tag{6.25}$$

Here, $N(\ )$ denotes a Gaussian probability density function, $\delta z_t^{i,j}$ and $P_{\delta z_t^{i,j}}$ represent the conditional mean and covariance calculated using the Kalman filter.

$$\delta z_t^{i,j} = \delta z_{t|t-1}(i,j) + G_t(i,j)\left(y_t - h\left(x_t^n(i,j) + H \bullet c_b^n \bullet q_t\right)\right), \tag{6.26}$$

$$P_{\delta z_t^{i,j}} = P_{\delta z_t}(i,j) - G_t(i,j) I P_{\delta z_t}(i,j), \tag{6.27}$$

where $G_t(i,j)$ denotes the Kalman gain,

$$G_t(i,j) = P_{\delta z_t}(i,j) I^T \left(I P_{\delta z_t(i,j)} I^T + R_t\right)^{-1} \tag{6.28}$$

$\delta z_{t|t-1}(i,j), P_{\delta z_t}(i,j)$ are the one-step prediction and covariance where $Q_{\delta z_t}$ is the noise of tidal depth bias.

$$\delta z_{t|t-1}(i,j) = \delta z_{t-1}(i,j) \tag{6.29}$$

$$P_{\delta z_t}(i,j) = P_{\delta z_{t-1}}(i,j) + Q_{\delta z_t} \tag{6.30}$$

6.2.3.2 Design of Grid and Adaptation

The entire filtering process in a marginalized or Rao-Blackwellized PMF for TAN typically employs a fixed grid configuration. As the grid support or the area that the grid covers grows over time, it becomes more challenging to use a fixed grid. During the entire duration of the filtering process, if the resolution is constant, as the grid support grows, there will be a rise in the grid point numbers. On the other hand, if the grid support grows and the grid point number stays steady, the resolution decreases and with it, the probability density approximation's accuracy is lost. To tackle these problems, an adaptive grid design technique similar to that employed in 2D PMF is required. When the number of effective grid points is less than a minimum threshold, the grid is refined by adding new grid points; when the number of effective grid points is more than a maximum threshold, the grid is decimated by eliminating grid points. This is the process of adaptive grid design. Compared with fixed grid design, which keeps the grid resolution constant during the filtering process, adaptive grid design dynamically modifies the grid resolution to capture systemic changes better, improving probability density estimation accuracy and efficiency.

In an adaptive grid design approach, the grid density is adjusted dynamically based on specific criteria that involve monitoring the effectiveness of the current grid in representing the underlying system. For example, one standard measure is the number of effective grid points N_{eff} that indicates how well the grid is capturing the system's dynamics and uncertainties. If N_{eff} falls below a predefined minimum threshold N_0, it suggests that the current grid is not adequately representing the system, and additionally, grid points are needed for better resolution; these new grid points are strategically placed based on the existing grid structure to maintain coherence and efficiency. Conversely, if N_{eff} exceeds a predefined maximum threshold N_1, it indicates that the current grid is too dense, potentially wasting computational resources, and some grid points can be removed to simplify the representation.

Every grid in marginalized PMF is equivalent to a Kalman filter with state and covariance. As a result, the appropriate Kalman filter state and covariance must be changed in accordance with grid changes during adaptive grid steps. To make it easier to add or remove grid points for Kalman filter updates, an index-based technique similar to resampling in particle filters is presented. A new grid point is sampled from the nearest existing grid point during the insertion step, and its index is recorded. In the removal process, each other row and column are removed in order to acquire the index of the new grid point. The Kalman filter state and covariance corresponding to the added or deleted grid points are copied or eliminated, respectively, based on these indices.

The modification of the position measurement update involves adjusting the equations used to integrate new measurements into the state estimation process, ensuring accuracy and consistency in the filtering algorithm's performance, particularly in the context of adaptive grid design. The adaptive grid design necessitates modifications in the position measurement update eqn. (6.31) to accommodate the changes in grid structure.

$$p\left(x_{t_{i,j}}{}^{n} \mid y_{1:t}\right) = \alpha_t^{-1} p_{v_t^{\cdot}}\left(y_t - h\left(x_{t_{i,j}}{}^{n} + H \bullet c_b^n \bullet q_t\right) - \delta z_{t|t-1}(i,j) \bullet I\right) \bullet p\left(x_{t_{i,j}} y_{1:t-1}\right) \quad (6.31)$$

6.3 SIMULATION EXPERIMENT

To evaluate the TAN performance of the suggested approach, simulated experiments are conducted in an underwater reference map in this section. The depth measurements from Multi-beam Echosounder System (MBES) equipment were used to create the bathymetric data for the underwater reference map in Figure 6.1, where the data was sourced from [2].

The vehicle is considered to move in the simulation at a constant speed of 2 m/s, similar to that of a lawnmower. There simulation time is 600 seconds in total. Two distinct terrain regions are selected and designated as Area A and Area B to examine the algorithm's performance in various types of terrain. The simulated accurate trajectories in two terrain areas are displayed in Figures 6.2 and 6.3, by taking the simulation setup from the reference [1]. A constant velocity error of 0.1 m/s in both north and east directions is added to the actual velocity to approximate the INS drift error, and 50 m is the maximum initial position error that the INS can collect in both directions. The multi-beam sensor generates 127 beams for every ping, and a uniform sampling of just 11 beams is selected from each ping due to the considerable measurement noise of beams on the outer border and the limitation of map grid resolution. A Gaussian with mean 0 and covariance P_t is used to describe the measurement noise that results from combining the noise from depth pressure sensors and multi-beam sensor noise. The update rate of the filter is 5 seconds. INS and

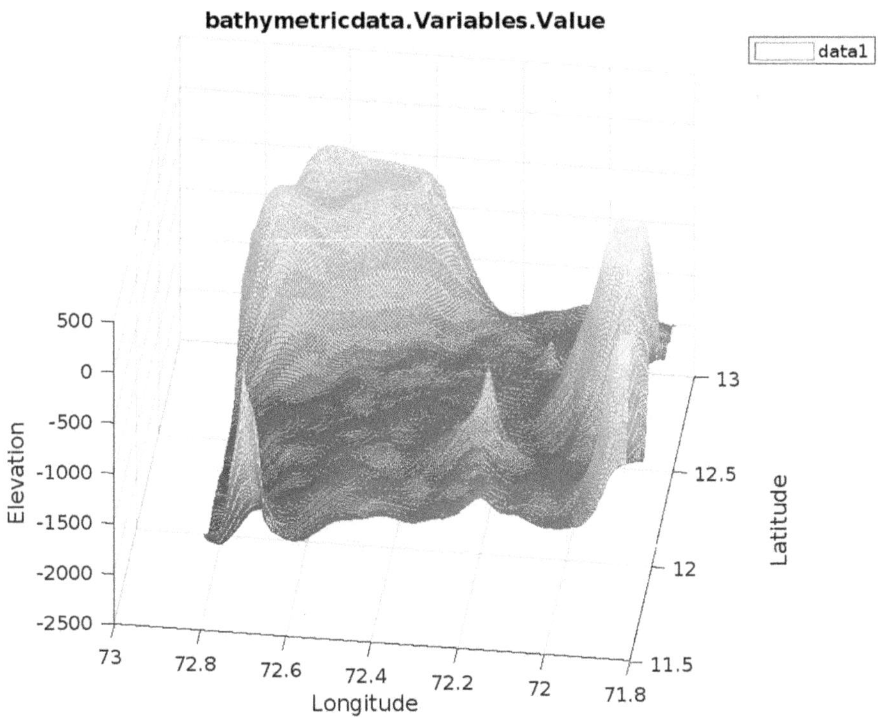

FIGURE 6.1 Underwater reference map.

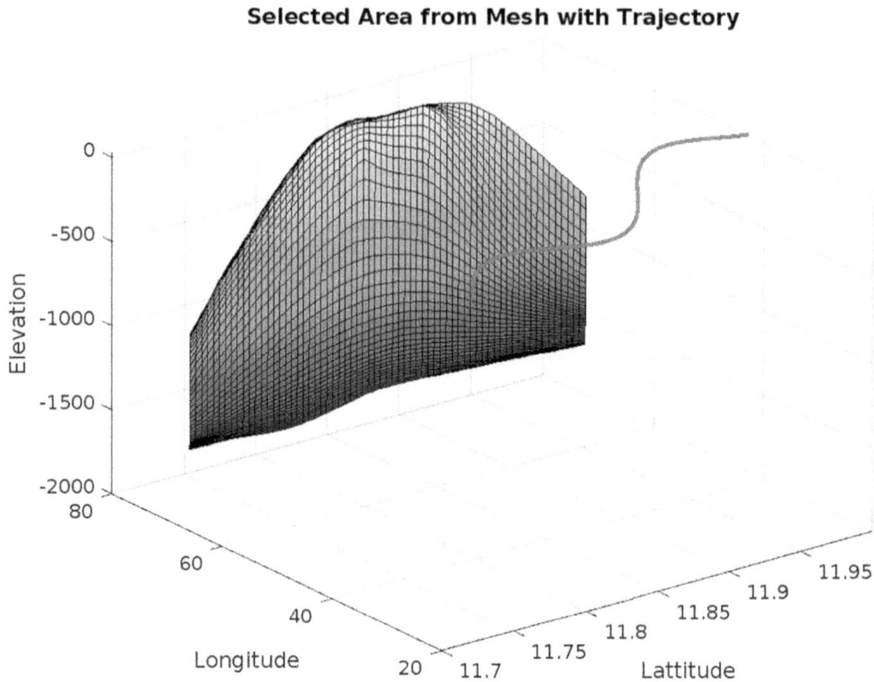

FIGURE 6.2 Simulated trajectory in area A.

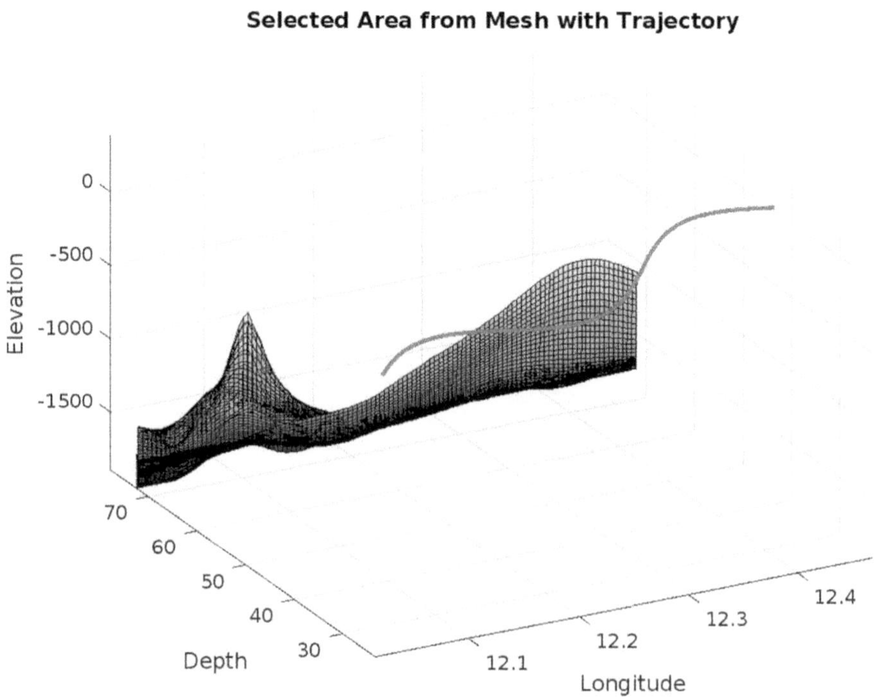

FIGURE 6.3 Simulated trajectory in area B.

MBES sample frequencies are 100 and 0.2 Hz, respectively. The horizontal position parameters in the 2D PMF and marginalized PMF are identical, except for the tidal depth bias parameter. It is assumed that the initial tidal depth bias follows a Gaussian distribution with a mean of zero. Fifty Monte Carlo runs are performed for each filter due to the randomness of the additional measurement noise. MATLAB is the environment in which simulations are implemented.

In order to examine and validate the marginalized PMF capacity to approximate the tidal depth bias, two distinct tidal depth biases are applied to the bathymetric measurement in isolation. Since the bias in tidal depth is negligible in real life, we set δz_t to 0 m and δz_t to 1 m. There are two regions where simulations are run, Area A and Area B, using Monte Carlo simulations. The TAN performance of the filters is assessed in terms of the root mean square error as the RMS error of the horizontal position and the tidal depth bias estimation.

Area A's results show differing tidal biases. Initially, two filters in region A are examined without any additional tidal depth bias ($\delta z_k = 0$ m). Figure 6.4 displays the tidal depth bias estimate from minimized PMF and the RMS errors of horizontal position from two filters for a single run. It is evident that after 50 seconds, the 2D PMF and the minimized PMF converge to astable estimate, with typical RMS

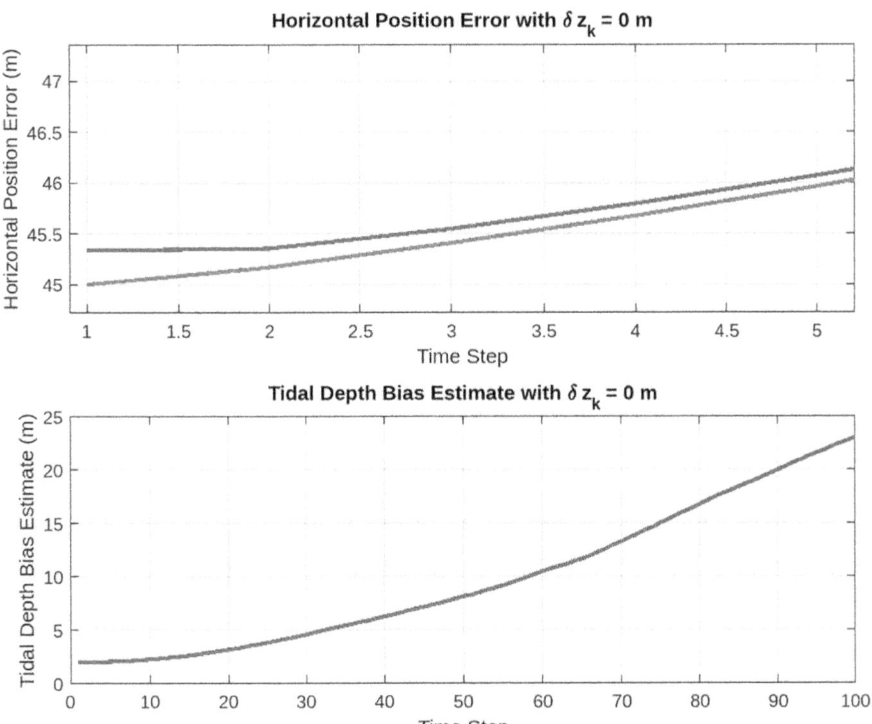

FIGURE 6.4 Horizontal position error from two filters and tidal depth bias estimate from mpmf for zero tidal depth bias for area A.

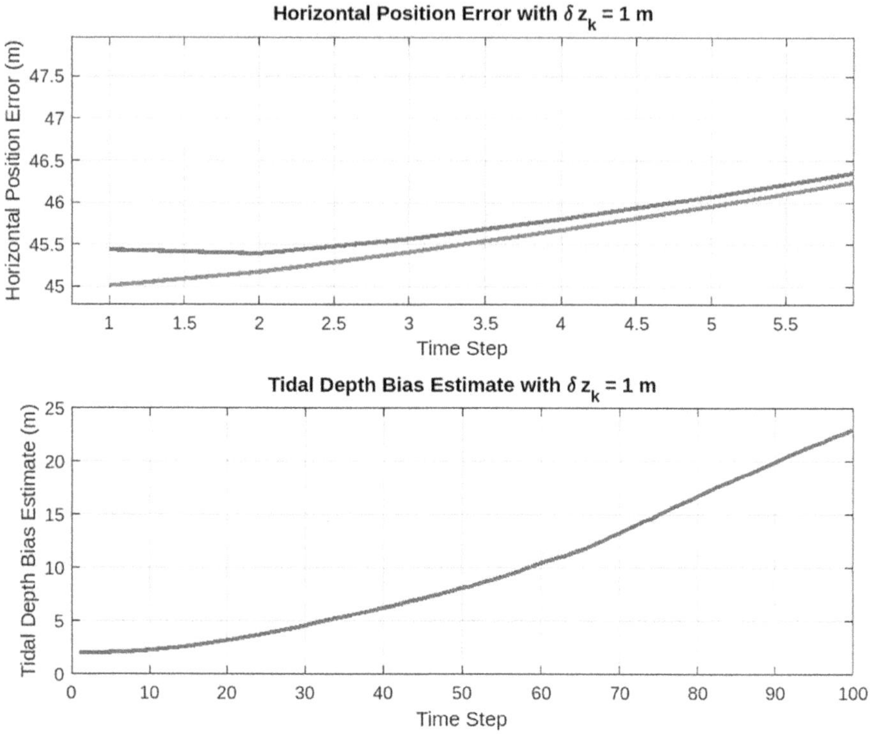

FIGURE 6.5 Horizontal position error from two filters and tidal depth bias estimate from mpmf for one tidal depth bias for area A.

errors of about 5 m. The minimized PMF exhibits a modest variation in the error curve compared to the 2D PMF. The marginalized PMF has a minor amount of tidal depth bias noise added to it, which is the cause. Furthermore, the estimation of tidal depth bias can be obtained near zero using the minimized PMF, as shown in Figure 6.4.

Next, two filters are tested in region A with the addition of tidal depth biases of 1 m ($\delta z_k = 1$ m). In this scenario, the marginalized PMF outperforms the 2D PMF, as illustrated in Figure 6.5. While the marginalized PMF is resistant to the tidal depth bias, the 2D PMF's performance steadily deteriorates as the bias increases. Furthermore, it is evident that even in the presence of 1 m tidal depth biases, the marginalized PMF can estimate tidal depth biases with accuracy.

Area B's results show differing tidal biases. Three distinct extra tidal depth biases are used in simulation experiments conducted in approximately area B to examine the filter's performance in various terrain types. Figure 6.6 displays the corresponding results for the two filters in region B. Figure 6.7 illustrates how they quickly converge after multiple measurement updates and how their final accuracy is around 5 m without additional tidal depth bias. This suggests that the multi-beam data, with their about 100 m coverage swath throughout the trajectory in this area, give sufficient

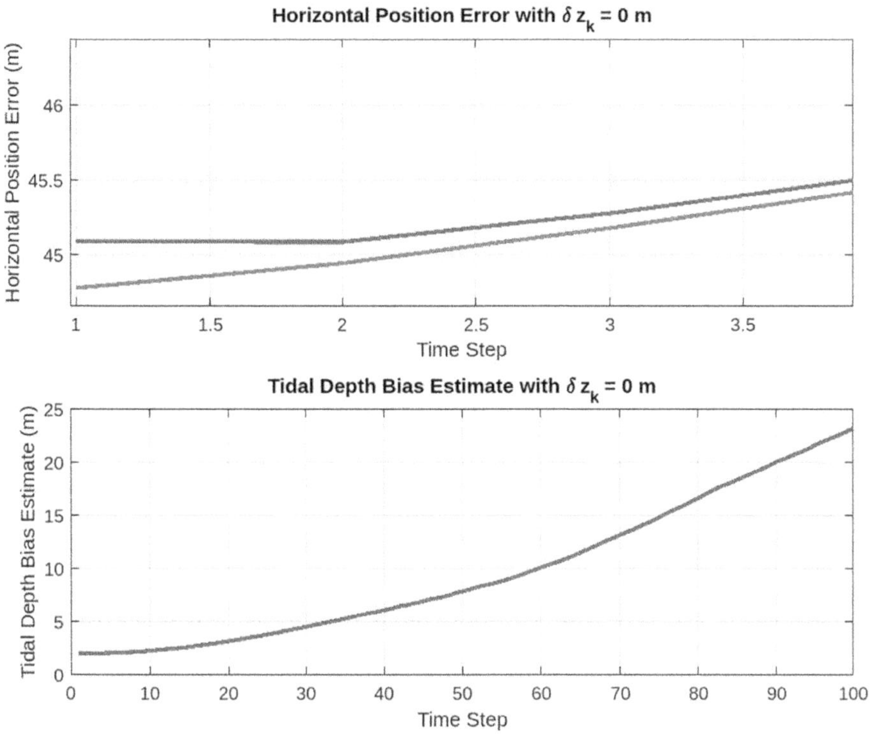

FIGURE 6.6 Horizontal position error from two filters and tidal depth bias estimate from mpmf for zero tidal depth bias for aera B.

topographical information. The 2D PMF diverges the maximum terminal horizontal position error with additional tidal depth bias.

It is noteworthy that, in contrast to the outcomes in the rugged terrain region A, the 2D PMF in the comparatively flat terrain area B is more susceptible to the tidal depth bias, even with a 1 m ($\delta z_k = 1$ m) modest tidal depth bias. Furthermore, compared to region A, it is evident that the variations in tidal depth estimate and horizontal position errors over time are far smoother. The interpolation error could be the source of this. There is a modest interpolation error in terrain region B and a substantial interpolation error in rough-terrain area A since the linear interpolation approach was applied in this instance.

For underwater terrain-aided navigation, we offer a 3D marginalization point mass filter (marginalized PMF) in this study, which simultaneously estimates and compensates for the tidal depth bias. Multi-beam sonar simulation tests have been carried out in various terrain types. Both the 2D PMF and minimized PMF function well and have good accuracy in two different types of terrain when there is no additional tidal depth bias. The 2D PMF is susceptible to the depth bias caused by tidal forces. Instead, the minimized PMF can retain a reliable navigation solution by estimating and compensating for the tidal depth bias.

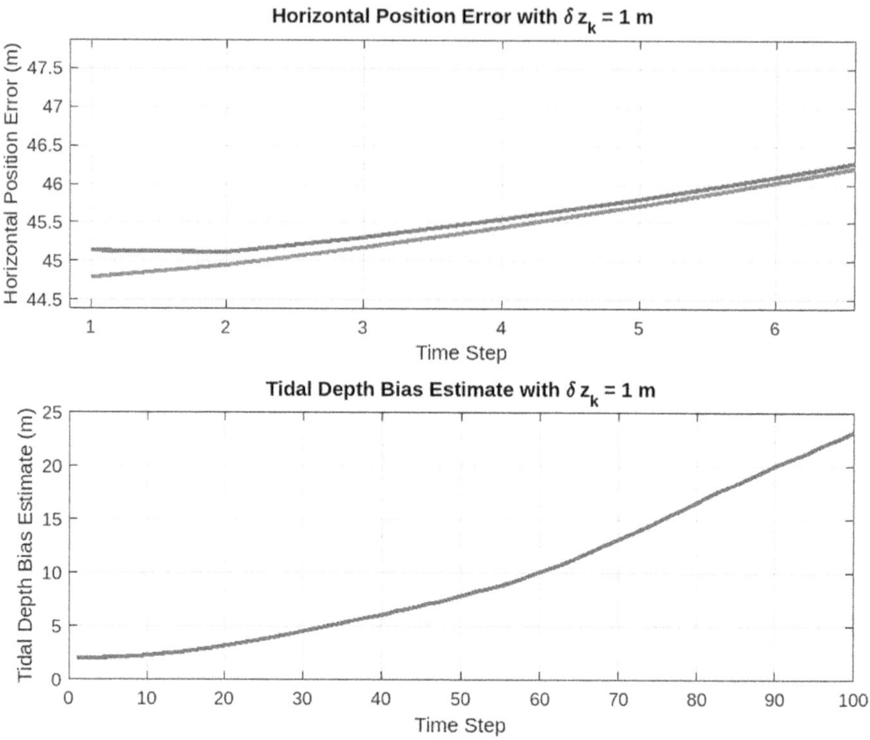

FIGURE 6.7 Horizontal position error from two filters and tidal depth bias estimate from mpmf for one tidal depth bias for area B.

6.4 ANALYSIS OF REAL-TIME TERRAIN-ASSISTED NAVIGATION

6.4.1 Navigation Algorithm

Navigation algorithms for AUVs are designed to enable these vehicles to navigate underwater environments autonomously, often with limited communication with the surface. These algorithms typically involve a combination of sensors, such as sonar, inertial measurement units (IMUs), depth sensors, and sometimes cameras, along with sophisticated computational methods for data processing and decision-making.

6.4.1.1 System Model

The system model provides a framework for understanding the AUV's behaviour, capabilities, and limitations, enabling engineers to design efficient algorithms and control strategies that leverage terrain data to enhance navigation accuracy, reliability, and autonomy in underwater environments. For our AUV range of motion, we are considering the following model.

$$X_{t+1} = X_t + P_t + N_t, \tag{6.32}$$

$$N_t = V_t' + V_t, \tag{6.33}$$

where the X_t is the position vector of an AUV, P_t is the change between the adjacent positioning, i.e. calculated by the INS, and N_t is white noise. Typically, the INS is paired with an underwater terrain-aided navigation (UTAN) system. After receiving rough location data from INS, the accurate position can be determined by performing the UTAN matching process close to the rough location. UTAN and INSs can be combined in two different ways: loosely coupled and closely coupled; the loosely coupled systems provide superior robustness and a more straightforward system structure, because they do not account for sensor error. Consequently, the UTAN and INS combination model is implemented using loosely linked mode. The measurement equation relates the measurement obtained from sensors to the variables being estimated.

Let's denote the measurement obtained from the INS as Z_t and the actual state of the system (which includes position, velocity, or other variables) as X_t. The system measurement equation can be written as follows:

$$Z_t = h(X_t) + n_t, \tag{6.34}$$

where $h(X_t)$ is the measurement function, which maps the true state at X_t to the expected measurements, n_t is the measurement noise at time t, which may include sensor noise, biases, or other sources of error. So, the measurement equation specifies how the sensor measurements relate to the underlying state of the system. It accounts for the fact that sensors and measurements are typically corrupted by noise, which introduces uncertainty into the estimation process.

Suppose we consider Z as the bottom depth measurement in a vector or a scalar based on the type of sensor used. The single beam echosounders (SBE) sensor is measured as a scalar, and the multi-beam echosounders (MBE) sensor is measured as a vector. The water depth at place X at time t is determined by $h(X_t)$ using a digital terrain map (DTM)—the n_t measures the sensor noise, i.e. Gaussian white noise.

To make the mathematical explanation of our algorithms simpler, we will represent our actual position as an offset, δx, from the INS's position estimate $\tilde{x}$

$$\delta x_t = x_t - \tilde{x}_t \tag{6.35}$$

$$\delta x_{t+1} = x_{t+1} - \tilde{x}_{t+1} = x_t + p_t + v_t' + v_t - \tilde{x}_t - p_t = \delta x_t + v_t' + v_t \tag{6.36}$$

6.4.1.2 Filter Model

The filter model processes sensor measurements and fuses them with predictions from the dynamic model to estimate the current state of the AUV. The filter iteratively performs two main steps: prediction and update. In the prediction using the dynamic model, the filter predicts the AUV's state forward in time from the previous estimate. It also provides an estimate of the uncertainty associated with the state prediction. In the updated step, the filter then incorporates new sensor measurements. It calculates a correction to the predicted state based on the difference between the predicted and actual sensor measurements. This correction is weighted by the uncertainties associated with both the predicted state and the measurements. The update step adjusts the state estimate to better align with the sensor data.

Here, we are limited to a two-dimensional state vector representing the vehicle's horizontal position due to the PMF's processing needs. All of our noises must be characterized as white because our state vector cannot contain any more error states, and we must take into account a simpler system than the one in (6.32)–(6.34). We utilize asterisks for filter model variables to help us distinguish between those in our system and the filter models. Applying the identical delta notation as in (6.35)–(6.36), our filter model reads.

$$\delta x_t^* = x^*_t - \tilde{x}_t, \tag{6.37}$$

$$\delta x^*_{t+1} = \delta x^*_t + v^*_t, \tag{6.38}$$

$$z_t = h^*\left(x^*_t\right) + n^*_t = h^*\left(\tilde{x}_t + \delta x^*_t\right) + n^*_t, \tag{6.39}$$

where $\delta x_t^* = \left(\delta x_{t,N}, \delta x_{t,E}\right)^T$ is the two-dimensional subvector of x_t. with the assumptions (by taking the reference from [3])

$$E\left\{v_t^* v_l^{*T}\right\} = Q_t^* \delta_{tl}, \tag{6.40}$$

$$E\left\{w_t^* w_l^{*T}\right\} = R_t^* \delta_{tl}, \tag{6.41}$$

where the noise sequences are uncorrelated from one time step to the next, and δ_{tl} stands for the Kronecker delta. In eqn (6.40), the white noise sequence v_t^* has replaced v_t and v_t^T. In addition, the densities of the noise sequences and the initial position offset, δx_0^*, must be specified.

$$p\left(\delta x_0^*\right) = N\left(0, P_0^*\right) \tag{6.42}$$

$$p\left(v_t^*\right) = N\left(0, Q_t^*\right) \tag{6.43}$$

$$p\left(n_t^*\right) = N\left(0, R_t^*\right) \tag{6.44}$$

According to eqn. (6.42), the predicted INS position $\tilde{x}_0$ is the centre of the Gaussian density of the initial position. We further assume that there is no correlation between the measurement noise, process noise, and beginning position. The depth at point x^*_t is indicated by the function $h^*\left(x^*_t\right)$, which is provided by the digital terrain map. We make use of gridded node terrain maps, and bilinear interpolation of the terrain database yields the depth values indicated by h^*. As a result, the noise sequence n^*_t in eqn. (6.44) models both sensor noise and map noise, including interpolation mistakes. The total sea depth (z_k) at the present AUV position is measured as the sum of the AUV altitude above the seafloor (provided by the bathymetric sensor) and the AUV depth (provided by a pressure sensor). Map errors, pressure sensor noise, and bathymetric sensor noise are thus components of the noise sequence n_t^*.

The researchers utilized a PMF, or histogram filter, for state propagation due to its known higher robustness over other non-parametric methods for low-dimensional problems. However, the increase in the number of states needed for the presented filter motivated a shift to a particle filter implementation.

6.4.1.3 The Recursive Bayesian Filter Equations

As discussed in Section 6.2, the recursive Bayesian equations extend Bayesian estimation by iteratively updating the probability distribution as new sensor measurements become available. This allows for real-time estimation of the vehicle state. Using the above filter model and Bayes formula, the measurement vector Z_t containing all measurements made up to time t

$$p\left(\delta x_t^* | Z_t\right) = \frac{p\left(z_t | \delta x_t^*, Z_{t-1}\right) p\left(\delta x_t^* | Z_{t-1}\right)}{p\left(z_t | Z_{t-1}\right)} \tag{6.45}$$

$$p\left(\delta x_t^* | Z_t\right) = \frac{pw_t^*\left(z_t - h^*\left(\tilde{x}_t + \delta x_t^*\right)\right) p\left(\delta x_t^* | Z_{t-1}\right)}{\int_{R^2} pw_t^*\left(z_t - h^*\left(\tilde{x}_t + \delta x_t^*\right)\right) p\left(\delta x_t^* | Z_{t-1}\right) d\delta x_t^*} \tag{6.46}$$

Let $\propto$ be the

$$\propto = \frac{1}{\int_{R^2} pw_t^*\left(z_t - h^*\left(\tilde{x}_t + \delta x_t^*\right)\right) p\left(\delta x_t^* | Z_{t-1}\right) d\delta x_t^*} \tag{6.47}$$

From the above equation, the measurement updated eqn. (6.46) is

$$p\left(\delta x_t^* | Z_t\right) = \propto^{-1} pw_t^*\left(z_t - h^*\left(\tilde{x}_t + \delta x_t^*\right)\right) p\left(\delta x_t^* | Z_{t-1}\right) \tag{6.48}$$

The minimum square error (MMSE) can be written as

$$\delta \hat{x}_t = \int_{R^2} \delta x_t^* p\left(\delta x_t^* | Z_t\right) d\delta x_t^* \tag{6.49}$$

The estimation covariance can be

$$\hat{P}_t = \int_{R^2} \left(\delta x_t^* - \delta \hat{x}_t^*\right)\left(\delta x_t^* - \delta \hat{x}_t^*\right)^T p\left(\delta x_t^* | Z_t\right) d\delta x_t^* \tag{6.50}$$

Our position distribution's time update comes from utilizing eqn. (6.38) and conditioning on the position offset from the previous time step.

$$p\left(\delta x_{t+1}^* | Z_t\right) = \int_{R^2} p\left(\delta x_{t+1}^*, \delta x_t^*\right) d\delta x_t^*$$

$$p\left(\delta x_{t+1}^{*} | Z_{t}\right) = \int_{R^2} p\left(\delta x_{t+1}^{*} | Z_{t}\right) p\left(\delta x_{t}^{*} | Z_{t}\right) d\delta x_{t}^{*}$$

$$p\left(\delta x_{t+1}^{*} | Z_{t}\right) = \int_{R^2} pv_{t}^{*}\left(\delta x_{t+1}^{*} - \delta x_{t}^{*}\right) p\left(\delta x_{t}^{*} | Z_{t}\right) d\delta x_{t}^{*} \tag{6.51}$$

It is now possible to derive the distribution of the position offsets for each time step by recursively using eqns. (6.46) and (6.51) given the distribution of the initial position, $p(x{*}0)$. However, we are unable to solve the integrals in the equations analytically. Thus, we must evaluate them numerically.

6.4.1.4 Point Mass and Particle Filter

The PMF is used in Bayesian estimation, particularly in scenarios where the state space is discrete or can be represented as a finite set of points. It's also known as the histogram filter or Discrete Bayes Filter. Instead of representing the state as a continuous distribution or as a set of particles, the PMF represents the state using a discrete set of points. Each point in this set is associated with a probability mass. The update process in the PMF involves updating the probabilities associated with each discrete state point based on the system model and measurements. This update is typically done using Bayes' rule, where the prior probabilities are multiplied by the likelihood of the measurements given the state (observation model) and normalized to ensure they sum to one.

The Particle filter, also known as the Sequential Monte Carlo (SMC) method, is a Bayesian filtering technique used for estimating the state of a system in the presence of noise and uncertainty. It's particularly useful for non-linear and non-Gaussian systems where traditional filters like the Kalman filter may need to perform better. In a Particle filter, the system's state is represented by a set of particles, where each particle is a hypothesis about the true state of the system. These particles are drawn from an initial distribution, often chosen to cover the possible range of states based on prior knowledge or assumptions. Each particle is then propagated through the system's dynamic model, incorporating the system's dynamics and control inputs. This step predicts where each particle will likely move in the next time step based on the system's behaviour.

The choice between PMF and Particle filter depends on factors such as the complexity of the system, the available computational resources, and the desired accuracy of the estimation; since we have dynamics and noise, we concluded that PMF will give more robust and accurate terrain navigation.

6.4.2 Underwater Terrain Navigation Sensors

Sensors play a crucial role in TAN for AUVs, which refers to the process of using environmental features that provide the key information needed to estimate the position and avoid obstacles in the underwater environment, such as the seafloor terrain, to aid in the navigation of underwater vehicles. Here is how they contribute:

6.4.2.1 Building the Map

Bathymetric Sensors: Sonar systems like multibeam sonar and sidescan sonar map the underwater terrain, creating a digital map of the seafloor. This map serves as a reference for the AUV to compare its sensor readings and estimate its location. These sonar systems emit sound waves and then listen for their echoes to create detailed maps of the underwater terrain. Sonar can be either active, where it emits sound pulses, or passive, where it listens for ambient sounds. MBE are advanced sonar systems that emit multiple sonar beams simultaneously in a fan-shaped pattern, which allows for rapid mapping of large areas with high resolution. The sidescan sonar systems are used to create detailed images of the seafloor by emitting sonar pluses to the sides of the sensors, allowing them to create detailed images of underwater terrain features such as wrecks, reefs, and geological formations.

Magnetometers: Measure the earth's magnetic field variations caused by underwater geological features, such as seamounts, ridges, and geological faults, and offer additional data points for map creation and refinement in areas lacking reliable bathymetric data. These variations can be used as additional navigational cues for detecting submerged objects like shipwrecks or pipelines.

6.4.2.2 Localization and Navigation

Doppler Velocity Logs (DVLs): These are critical sensors for underwater navigation, providing real-time velocity measurements essential for accurate positioning, navigation, and mapping of AUV's and ROV's in diverse underwater environments. These sensors utilize the Doppler effect to measure the vehicle's velocity relative to the surrounding water and measure the AUV's speed relative to the seafloor, providing importance for dead reckoning and correcting drift in INSs. The Doppler effect is a shift in the frequency of a wave (sound waves) due to the motion of the source or the observer. DVLs emit acoustic pulses downward and measure the Doppler shift of the return signals reflected off particles in the water. By analysing these Doppler shifts, the DVL can determine the vehicle velocity in three-dimensional measurements (along the x, y, and z axes) relative to the surrounding water. This information is crucial for accurately determining the vehicle's speed and direction of motion underwater because the vehicle's current position is estimated based on its known starting position and measured velocity over time. This helps to maintain accurate navigation even when GPS signals are unavailable or unreliable underwater. By measuring the vehicle velocity relative to the seabed, DVLs enable the vehicle to maintain a constant altitude above the seafloor or follow the contours of the terrain. Additionally, changes in velocity measured by DVL can indicate the presence of obstacles or changes in the underwater environment, allowing the vehicle to adjust its course to avoid collisions. These DVLs also generate high-resolution maps of the underwater environment, including features such as bathymetry, currents, and underwater structure.

Sonar: The multi-beam, side-scan, and sub-bottom sonars continuously scan the surrounding terrain, comparing features with the pre-existing map to localize the AUV's position within the mapped area. The sub-bottom profilers are used to image layers of sediment and rock beneath the seafloor. They emit high-energy sound pulses that penetrate the seafloor and bounce back to the sensors, revealing subsurface geological features. The sidescan sonar systems are used to create detailed images of the seafloor. They emit sonar pulses to the sides of the sensors, allowing them to create detailed images of underwater terrain features.

Inertial Navigation Systems (INS): Provides initial position and orientation data by measuring accelerations and rotations. These systems are crucial for maintaining accurate navigation in the absence of external references, such as when the autonomous underwater vehicle is submerged and unable to receive GPS signals but requires correction due to drift over time. TAN compensates this drift by providing external reference points. It's often used in conjunction with other sensors for precise navigation in underwater environments. INS utilizes accelerometers and gyroscopes to measure changes in velocity and orientation respectively. By integrating these measurements over time, INS estimates the AUV's position and orientation through dead reckoning navigation. This enables AUVs to maintain accurate navigation. Since it provides continuous navigation capabilities throughout the mission, it conducts long-span durations and large distances underwater. It is also integrated with other sensors, such as DVLs, depth sensors, and acoustic positioning systems, to enhance navigation accuracy. External sensors provide additional information to correct errors and drift in INS measurements, improving overall navigation performance. Some advanced AUVs employ redundant INS systems to ensure fault tolerance and robustness against sensor failures. Redundant systems can provide backup navigation capabilities in case of sensor malfunctions or inaccuracies.

6.4.2.3 Obstacle Avoidance and Path Planning

Sensors like sonar, LiDAR, and cameras are used to detect obstacles in the AUV's path, such as rocks, wrecks, or marine life. These sensors help ensure safe navigation and prevent collisions by detecting and avoiding obstacles.

Forward-Looking Sonar (FLS): Scans ahead of the AUV, identifying and mapping underwater obstacles or structures in real-time, thus providing early warning of potential hazards, allowing the navigation system to plan a safe path to adjust its course to avoid around them. It provides information about the distance, size, and shape of underwater obstacles or features in the AUV's path before they come into contact with the vehicle by analysing the time delay and intensity of the echoes. It is mainly used in search and rescue operations to locate underwater objects or features. The ability to scan a wide area ahead of the AUV helps expedite search efforts and improve the chances of locating targets. It can integrate with other navigation sensors, such as INSs or global navigation satellite systems, to enhance accuracy, navigation, and reliability.

Side-scan Sonar: This can also be used for obstacle detection, providing wider coverage than FLS but with lower resolution. By scanning the area to the sides of the vehicle, it detects underwater real-time information about obstacles, and by analysing the images generated by it, path planning algorithms can identify areas of open water and avoid regions with dense clusters of obstacles or hazardous terrain. This information is used to create high-resolution maps of the seafloor, identify underwater structures, and conduct surveys of marine habitats and ecosystems.

6.4.2.4 Sensor Fusion

By combining or integrating data from multiple sensors like DVL, Sonar, and magnetometers, TAN algorithms can create a more robust and accurate picture of the AUV's surroundings and reliable navigation capabilities, improving navigation performance, obstacle detection, and enabling autonomous operation in challenging underwater environments while optimizing efficiency and safety.

These sensors are mainly used in terrain navigation systems for autonomous underwater terrain navigation systems. There are more sensors available for AUVs, like underwater LiDAR systems that use laser pulses to measure the distance to objects and create high-resolution 3D underwater maps, which are similar to their above-water counterparts. Acoustic Doppler Current Profilers (ADCPs) measure water currents by analysing the Doppler shift of sound waves reflected off particles or bubbles in the water; this device can also provide valuable information about underwater currents, which is essential for navigation and mapping. CTD sensors provide information about the conductivity, temperature, depth, and salinity of underwater environments.

Overall, the specific types and configurations of sensors are essential for TAN to depend on the mission requirements, available resources, and environmental factors in AUVs by providing data for map creation and localization, correcting drift in INSs, identifying and avoiding obstacles in real-time, and enabling accurate and safe navigation in complex underwater environments.

6.4.3 The Hugin Real-Time Terrain Navigation System

The Hugin AUV represents a paradigm shift in real-time terrain navigation systems, offering unparalleled accuracy and reliability. The Kongsberg Maritime and the Norwegian Defence Research Establishment (FFI) collaborated to develop the Hugin AUVs. Here, we discuss how extensively the Hugin AUV is examined, exploring its functionality, features, applications, and prospects. Through detailed analysis, we uncover the transformative potential of this cutting-edge technology across various industries. Navigating complex terrains poses significant challenges across various domains in offshore surveying for the oil and gas industry, naval mine countermeasures (MCM) and rapid environment assessment (REA) operations, Marine research, and hydrography. The Hugin emerges as a ground-breaking solution, promising to revolutionize real-time terrain navigation. Hugin operations have taken place in all parts of the world, from shallow to deep water and arctic and tropical waters, which will help improve the vehicle's robustness and quality.

6.4.3.1 Understanding the HUGIN

At its core, it is a sophisticated terrain navigation system designed to provide real-time data and insights for navigating through challenging landscapes. The Hugin delivers precise navigation capabilities in dynamic environments by leveraging advanced sensors, machine learning algorithms, and geospatial data. Its adaptive nature enables it to analyse terrain features, anticipate obstacles, and chart optimal navigation routes with unparalleled accuracy and efficiency. The Hugin vehicle's notable accomplishments include their commercial fieldwork from 1997; over 150,000 km of commercial survey was conducted in 2007 with a toolkit of position-aiding strategies. DVL assisted the INS system in operating in autonomous, semi-autonomous, and supervised modes. A large payload suite and flexible payload system with pressure-tolerant power sources that can operate for up to 60 hours at four knots uses a secure launch and recovery mechanism for up to five sea states.

A real-time TAN system for the Hugin vehicles named TerrP was developed; this system is used as an external position-aiding method for the main INS. The Hugin AUV family constitutes three basic models, i.e. HUGIN 1000 (1000 m depth), HUGIN 1000 (3000 m depth), Hugin 3000, and HUGIN 4500. These Hugin AUVs are based on a shared, well-established technological platform. This mainly refers to the systems for propulsion, navigation, control, payload, and emergencies. The three main variations between the vehicles are payload sensor setup, size, battery technology, and endurance. To know more about the Hugin family, refer to [4].

6.4.3.2 Features and Capabilities

Hugin's design philosophy aims to achieve high operational reliability, maximum data quality, robustness, and operational reliability to complete AUV concepts for civilian and military applications. Modern payload sensors should be operated from a steady, low-noise platform near the sea bottom to get the maximum data quality. Integrating concurrently operating acoustic sensors is a critical capability of the Hugin group. Every sensor's output has an exact time stamp. The Hugin-assisted INS provides precise location and attitude data that is georeferenced with the payload data. Positioning precision is unmatched when using the NavLab suite for optimal navigation post-processing.

The keys to robustness and operational reliability are obtained by the rugged vehicle designs, well-proven launch and recovery systems for open rough seas, built-in redundancy, and extensive error handling and emergency systems. Because of the enormous operational experience gained with the Hugin vehicle, its robustness and dependability are constantly increasing. Low water drag is the goal of Hugin AUV's body shape, which combines minimal water resistance, excellent hydrodynamic stability, great manoeuvrability, and low acoustic noise levels. Because of the vehicle's incredibly strong body, the payload sensors and navigation system are always oriented in the same direction. Hugin contains steering pins that allow payload and navigation sensors to be mounted precisely and repeatedly. These features ensured seabed mapping accuracy.

The two distinct pressure-tolerant battery designs, Lithium polymer battery, and Aluminium oxygen semi-fuel cell, are the foundation for the HUGIN power sources. The intrinsic risk of explosions from battery threats is eliminated by pressure-tolerant batteries, which also remove the necessity for pressure containers. The pressure-tolerant designs increase battery energy and endurance, particularly in deeper depths. The control and communication of the Hugin AUVs are operated in three different modes: supervised mode, which provides unparalleled position accuracy with USBL position, updates to operator supervision and has the capacity to reprogram a mission; autonomous mode, which operates independently of a support ship; and semi-autonomous mode, which is a combination of autonomous and supervised mode.

A sophisticated real-time Assisted Inertial Navigation System (AINS) called the Hugin NavP system is fitted on Hugin. Using acceleration and angular rate information from an IMU, the INS determines the vehicle's position, velocity, and attitude. The vehicle navigation sensors are used in a theoretically ideal way by a Kalman filter to assist the INS. Compared to separate navigation sensors, the Kalman filter offers significantly better overall navigation performance since it is based on an error-state model. To reduce the position error drift in the navigation system, regular position updates are required even if the DVL is quite effective at minimizing the error drift.

GPS provides position measurements when the AUV is on the surface in autonomous mode. GPS surface fixes aren't always an option for AUV applications because of their covertness or inefficiency in deep-water operations. A terrain-referenced navigation system for submerged location updates is an optional addition to Hugin. Bathymetric measurements are compared with an apriori map by the terrain navigation system. The NavP Underwater Transponder Positioning (UTP), which works similarly to the long baseline system, works on a function called a range measurement to one or more underwater transponders and can also be used to get position updates while submerged. For NavP, no more than one underwater transponder is required to find a position error draft. Utilizing multiple transponder ranges maximizes accuracy and integrity because doing so increases the range of transponders.

Although the HUGIN navigation system's Kalman filter is the greatest real-time estimator, the best algorithm for post-processing is optimal smoothing, which also makes use of 'future' observations. NavLab has an implementation of optimal smoothing. When compared to the real-time solution, the smoothed estimate of location, attitude, and velocity offers several significant advantages. In general, mission plan pattern (trajectory), DVL precision, and IMU accuracy affect navigation accuracy. For example, a lawnmower pattern compensates for mistakes in heading and velocity very slowly. The position update accuracy is also crucial. HUGIN can be configured with the necessary position measurement, DVL, and IMU systems.

Hugin's payload system has a SUBSEA payload system and a TOPISDE payload system. The standard Hugin payload suite includes a multi-beam echosounder, Side-scan or synthetic aperture sonar, sub-bottom profiler, conductivity temperature density (CTD), turbidity sensor, and ADCP. With the help of the topside payload system, the operator can set up the payload sensors before the mission and manage and oversee them while the survey is being conducted. A specific payload operator station (POS) is included with the majority of payload sensors like graphic user interface (GUI) for instructing the payload, GUI components for showing events, errors, and payload plug-in command responses and the information about the payload state is displayed.

The launch and recovery system of Hugin is one of the main challenges during the AUV operation; it has a stinger-based system for ship stern installation. Hugin AUV is launched while the ship is travelling at a speed of two to three knots against the wind. The hydraulically actuated stinger is slanted down into the water, and a disconnect mechanism releases the vehicle. While recovering, the ship is situated between 50 and 100 m away from the AUV's surface. The L/R system winch is linked to the vehicle drop nose, and the recovery line is attached. The stinger is then raised and lowered while the vehicle is drawn upon it.

6.4.3.3 Applications and Implications

The versatility of the Hugin extends across a multitude of applications, with far-reaching implications in various industries like naval, research, offshore, and hydrography. An extensive research effort has made military uses for the HUGIN AUV technology possible. Specifically, FFI, Kongsberg Maritime, and The Royal Norwegian Navy (RNoN) have effectively executed a variety of military AUV tasks, such as route mapping, rapid environmental assessment (REA), and mine countermeasure (MCM), within the HUGIN Mine Reconnaissance Program (HUGIN MRS).

Hugin AUVs are equipped with a broad variety of sophisticated acoustic payload sensors, such as sidescan sonars (SSS), synthetic aperture sonars (SAS), sub-bottom profilers (SBP), MBE, and fishery research sonars, to satisfy the demands of diverse applications. The HUGIN AUVs are outfitted with a cutting-edge integrated INS that delivers superior location accuracy and high-quality, high-resolution data.

The Hugin enhances situational awareness and navigation capabilities in military operations, enabling troops to manoeuvre through complex combat environments with precision. The Hugin serves as a critical component in autonomous vehicle navigation systems, enabling safe and reliable operation in diverse environments, from urban streets to off-road terrain. Researchers leverage the Hugin for environmental monitoring missions, enabling precise data collection in remote or hazardous locations and facilitating better understanding and management of ecosystems.

6.4.3.4 Future Directions

The future of the Hugin holds immense promise, with potential advancements including:

Enhanced Sensory Capabilities: Integration of advanced sensors for improved environmental perception and obstacle detection, further enhancing navigation accuracy and reliability.

AI-Driven Navigation: Leveraging artificial intelligence for predictive navigation algorithms, enabling proactive route planning and optimization.

Miniaturization and Portability: Developing compact and portable versions of the Hugin, expanding its applicability across a wider range of platforms and environments.

Integration with IoT: Seamless integration with Internet of Things (IoT) devices for enhanced connectivity and data exchange, enabling more robust and interconnected navigation systems.

6.4.4 Results and Discussion

The TerrP system was tested in different sea environments to assess its performance under various conditions. These tests included areas with different underwater terrains, such as flat seabeds and valleys. Such as between the Norwegian coast and Bear Island and in the Oslofjord. The system was integrated into the HUGIN 1000 HUS vehicle, which was equipped with a range of sensors to aid in navigation. These included an IMU to track the AUV's orientation and motion, a DVL to measure the vehicle's speed relative to the seabed, a pressure sensor to gauge depth, and acoustic transponders for underwater positioning. The primary goal of the tests was to evaluate the TerrP system's ability to maintain precise navigation without relying on external positioning systems like GPS. Instead, it relied on terrain data from sensors like MBE or DVLs.

The TerrP system's performance was continuously monitored in real-time. In [2], they have plotted the real-time results; here are some key points regarding the real-time results:

Navigation Accuracy: The system demonstrated consistent navigation accuracy throughout the tests carried out. It successfully maintained the AUV's position within acceptable margins of error over extended periods, even when relying solely on terrain data for navigation updates.

TerrP Measurement Integration: The real-time results showed how the TerrP measurements were integrated into the navigation system. When a TerrP measurement was accepted, there was a noticeable improvement in position accuracy, particularly in areas where the navigation uncertainty was high.

Comparison with Reference Solution: The real-time navigation solution was compared with a reference solution, which excluded terrain navigation updates but included measurements from external systems like HiPAP (ultra-short baseline) for underwater positioning. The comparison highlighted the effectiveness of the TerrP system in maintaining navigation accuracy, with deviations typically within expected bounds.

Effect of Turning Off DVL Aiding: During the test, DVL aiding to the INS was temporarily turned off to trigger a higher navigation drift. This action demonstrated the system's ability to adapt and recover from degraded navigation conditions, and turning DVL aiding back on led to an immediate decrease in estimated uncertainty and improved navigation quality.

The robustness testing assessed the TerrP system's ability to recover from sizeable initial navigation uncertainties. Here's more detail on this aspect:

Recovery from Uncertainty: The tests, simulated scenarios where the AUV entered previously unmapped areas with significant initial navigation errors. And despite starting with large errors, the TerrP system gradually converged to the correct position over time, showcasing its ability to recover and reduce overall navigation uncertainty.

Impact of Initial Errors: The tests varied the magnitude of initial errors, ranging from 0 to 100m. Results showed in [3] potriait that while more significant initial errors prolonged the convergence time, the system was still able to achieve accurate positioning, albeit with a slightly longer delay.

Effect of Uncertainty on Convergence Time: The uncertainty in the system's estimated position, reflected by the covariance matrix in the probability density function (PDF), influenced the convergence time. Higher initial uncertainties resulted in longer convergence times, highlighting the system's sensitivity to initial error conditions.

The sensor and map analysis focused on comparing depth data obtained from the DVL with map data to identify discrepancies and improve data quality:

Comparison of Depth Data: The depth data from the DVL were compared with map data at corresponding positions in the reference solution. Discrepancies between the two datasets were identified, with some extreme deviations exceeding 15 m.

Potential Causes of Discrepancies: Factors contributing to the discrepancies included errors in the bottom tracking algorithm of the DVL, limitations in bottom tracking under high pitch angles or steep bottom slopes, and biases in in-depth measurements.

Compensation Techniques: Compensation techniques were proposed to mitigate these discrepancies. These techniques involved ray tracing with sound velocity profiles, tide compensation, pressure-to-depth conversion, and sensor alignment adjustments. These compensation efforts significantly improved data quality, illustrating the potential for enhancing navigation accuracy through data refinement.

Overall, the real-time results demonstrated the TerrP system's reliability in maintaining navigation accuracy, while robustness testing highlighted its resilience to initial uncertainties. Additionally, sensor and map analysis provided insights into improving data quality and enhancing navigation performance. The Hugin represents a monumental advancement in real-time terrain navigation technology, offering unparalleled precision, reliability, and versatility. As its applications continue to expand across diverse industries, the impact of this revolutionary system is poised to reshape the landscape of navigation. With ongoing advancements and strategic implementations, the Hugin holds the potential to unlock new possibilities and redefine the boundaries of navigation excellence in the years to come.

EXERCISES

1. Which measurement model is better and why?
2. What is Bayesian estimation?
3. What factors are involved in the PMF equations for predicting the position and velocity of an object?
4. How does a marginalized PMF differ from a traditional PMF?
5. When is the marginalized PMF approach particularly useful?
6. Can you explain the linear and non-linear variables in the state vector equation?
7. State the difference between fixed grid and adaptive grid design?
8. Which factor indicates how well the grid captures the system's dynamics?
9. Which technique is more accessible for adding or removing grid points for Kalman filter updates?
10. How does the system model provide a framework to understand the AUV's behaviour, capabilities, and limitations?
11. How many ways can underwater terrain-aided and inertial navigation systems be combined?

REFERENCES

1. Peng, D., Zhou, T., Xu, C., Zhang, W., and Shen, J. Marginalized point mass filter with estimating tidal depth bias for underwater terrain-aided navigation. *J. Sens.*, vol.2019, Article ID 7340130, 2019.
2. https://www.earthdata.nasa.gov
3. Anonsen, K. B. and Hagen, O. K. An analysis of real-time terrain aided navigation results from a HUGIN AUV. *Oceans 2010 MTS/IEEE*, Seattle, WA, USA, 2010, IEEE, pp. 1–9.
4. Marthiniussen, R., Vestgard, K., Klepaker, R. A. and Storkersen, N. HUGIN-AUV concept and operational experiences to date. *Proceedings of the Oceans '04 MTS/IEEE Techno-Ocean '04 (IEEE Cat. No. 04CH37600),* Kobe, Japan, 9–12 November 2004, Volume 2, pp. 846–850.

7 Object Detection and Tracking from UAV with Thermal Camera

7.1 INTRODUCTION

The recent surge in the commercial availability of small unmanned aerial vehicles (UAVs) has opened up a wide array of applications spanning inspections of structures, surveillance, and search and rescue operations. UAVs are often lauded for their remote sensing capabilities, positioning them as versatile platforms for various tasks [1–3]. Cameras, in particular, have emerged as indispensable remote sensing tools for UAVs. With the advent of smaller, lighter, and more affordable thermal cameras, their integration into UAV systems has become increasingly feasible.

Equipping low-cost UAVs with thermal cameras offers distinct advantages, allowing for unique perspectives and enhanced visibility of the environment compared to traditional visual spectrum cameras. Moreover, advancements in processing platforms have made it possible to conduct real-time onboard analysis of thermal image data. This shift toward onboard processing reduces reliance on continuous communication links with ground stations, thereby extending the operational range of UAVs.

In scenarios where multiple moving objects need to be tracked in an Earth-fixed coordinate frame using a single UAV and camera, manual tracking quickly becomes impractical as the number of tracked objects increases. Automation is the logical solution, necessitating UAV payloads capable of autonomously detecting, tracking, and recognizing multiple objects in real time during the flight. Such capabilities are vital for revisiting and verifying the movements of tracked objects while ensuring sufficient camera time for each object to accurately estimate their movement.

Multitarget tracking, or the process of monitoring multiple moving objects simultaneously, poses a complex challenge with various approaches found in the literature. While this problem extends beyond UAVs to other remote sensing platforms, exploring solutions specific to thermal cameras in UAVs remains essential. Unlike radar-based multitarget tracking, thermal cameras offer unique capabilities and constraints that necessitate tailored solutions. In the following sections, we delve into the existing literature on (multi)target detection and tracking in the context of UAVs equipped with thermal cameras.

7.2 SYSTEM OVERVIEW

This section presents a framework tailored for object tracking utilizing UAVs equipped with thermal cameras, offering example implementations of each module

DOI: 10.1201/9781003429869-7

specifically tailored for detecting and tracking objects on the ocean surface within a maritime setting. It's important to note that while these modules are designed with ocean surface tracking in mind, the framework's structure allows for easy adjustment or re-implementations to adapt the system for diverse applications in varied environments.

Object tracking in this context refers to the UAV's function of monitoring the position and velocity of multiple objects within an Earth-fixed coordinate frame throughout its flight. This process goes beyond merely tracking known object positions; it also involves identifying, recognizing, and tracking potential objects of interest whose positions are yet unknown. To achieve this, the system relies on an onboard sensor, typically a thermal camera mounted on a pan-tilt gimbal, in conjunction with a path controller.

Various methods can be employed to estimate the positions and velocities of objects of interest. One autonomous approach involves utilizing an onboard computer to analyze images captured by the onboard camera, thereby automatically detecting, recognizing, and tracking objects over a sequence of images. In this method, no prior knowledge of the location or number of objects is assumed. However, users typically provide input to a machine vision module regarding certain features of the objects of interest, such as size or shape [4].

Alternatively, there are scenarios where the locations of the objects to be tracked are known, yet additional verification or surveillance is desired. In such cases, the object detection, tracking, and recognition module can be initialized with a predetermined list of object positions and velocities.

7.3 OBJECT DETECTION AND RECOGNITION

The object detection, tracking, and recognition module employs onboard camera images to autonomously perform image segmentation. This involves using machine vision techniques to classify pixels into foreground (object) or background (non-object) regions. For automatic detection of objects of interest, the module utilizes the machine vision algorithm developed by Leira, Fossen et al. [5], which has been extended to include a feature vector. This feature vector enhances the system's ability to classify and recognize objects, facilitating the process of data association. Let us understand this in more detail.

Image Frame to Camera Frame: Photographic cameras are instruments capable of capturing images from the three-dimensional environment onto two-dimensional surfaces known as image planes. Their design is straightforward, typically comprising a lightproof enclosure with a small aperture. Photons reflected from objects in the three-dimensional space pass through this aperture, forming an image on the opposite side of the dark box, where an optical sensor can be positioned. The system's mathematical representation involves establishing a coordinate system at the camera aperture, termed the camera frame (C), where the object's position is described by coordinates i_c, j_c, and k_c [6]. Another reference frame, the image frame (i), is positioned on the image plane or focal plane, with image positions represented by coordinates (a_i,b_i). A pinhole camera model is considered, which simplifies the mathematical operations. A geometrically equivalent setup is achieved by mirroring the image or projection plane across the camera frame. This straightforward procedure ensures that calculations involve only positive values along the axes, streamlining the process.

The relationship between the coordinates a_i in the image frame (i) and i_c in the camera frame (C) is derived, utilizing the focal distance $f\left(f_i, f_j\right)$. The image frame has two axes m, n, and c_i, c_j are the coordinates of the principal point of the image. For calculation, rotation and translation of the matrices operation are necessary. Therefore, it takes the form.

$$k_c\left[m\ n\ 1\right]=\left[f\ 0\ c_i\ 0\ f\ c_j\ 0\ 0\ 1\right]\left[i_c\ j_c\ k_c\right] \tag{7.1}$$

The image frame I has coordinates m and n, which are conveniently used in pixel units. The camera frame (C) object coordinates are usually expressed in meters. Therefore, the focal distance and principle point should obviously be given in pixels. For the conversion of pixels, which is done because the focal distance is given in millimeters, we apply the following equation.

$$f\left[\text{px}\right]=f\left[\text{mm}\right]\frac{\text{image size }\left[\text{px}\right]}{\text{sensor size}\left[\text{mm}\right]} \tag{7.2}$$

To calculate the pixel, f is calculated based on the focal length in millimeter (similar to the above equation), the image dimensions, which are the image width and image height in pixels, and the sensor dimensions, which are the sensor width and sensor height in millimeters. The k in eqn. (7.1) is the intrinsic parameters' camera matrix, also known as calibration matrix, and can be written as:

$$k_c\ P_i = KP_c \tag{7.3}$$

$$K=\left[f\ 0\ c_i\ 0\ f\ c_j\ 0\ 0\ 0\right] \tag{7.4}$$

With respect to eqns. (7.1) and (7.3):

$$k_c\left[m\ n\ 1\right]=K\left[i_c\ j_c\ k_c\right] \tag{7.5}$$

As k_c is unknown, to resolve the equation P_c' vector can be written as:

$$P_c'=\left[\frac{1}{k_c}i_c\ \frac{1}{k_c}j_c\ \frac{1}{k_c}k_c\right]=\left[j_c'\ i_c'\ 1\right]=K^{-1}\left[m\ n\ 1\right] \tag{7.6}$$

Camera Frame to Gimbal Frame: Then comes the coordinates transformation from the camera frame (C) to the gimbal frame (G). The rotation matrix from the camera frame to the gimbal frame, $R_{C\to G}$, can be obtained throughout the projection of the gimbal frame in the camera frame, being given by:

$$R_{C\to G} = [0\ 0\ 1\ 1\ 0\ 0\ 0\ 1\ 0] \quad (7.7)$$

The object's position in the gimbal frame prime can be obtained by:

$$|i'_G\ j'_G\ k'_G| = [0\ 0\ 1\ 1\ 0\ 0\ 0\ 1\ 0][i'_c\ j'_c\ k'_c] \quad (7.8)$$

In Figure 7.1, we can see the difference in the position of frames; then, a translation vector will be inserted from the camera to the gimbal frame. The gimbal frame prime having object position is expressed as:

$$P'_G = R_{C\to G}P'_C + T_{C\to G} \quad (7.9)$$

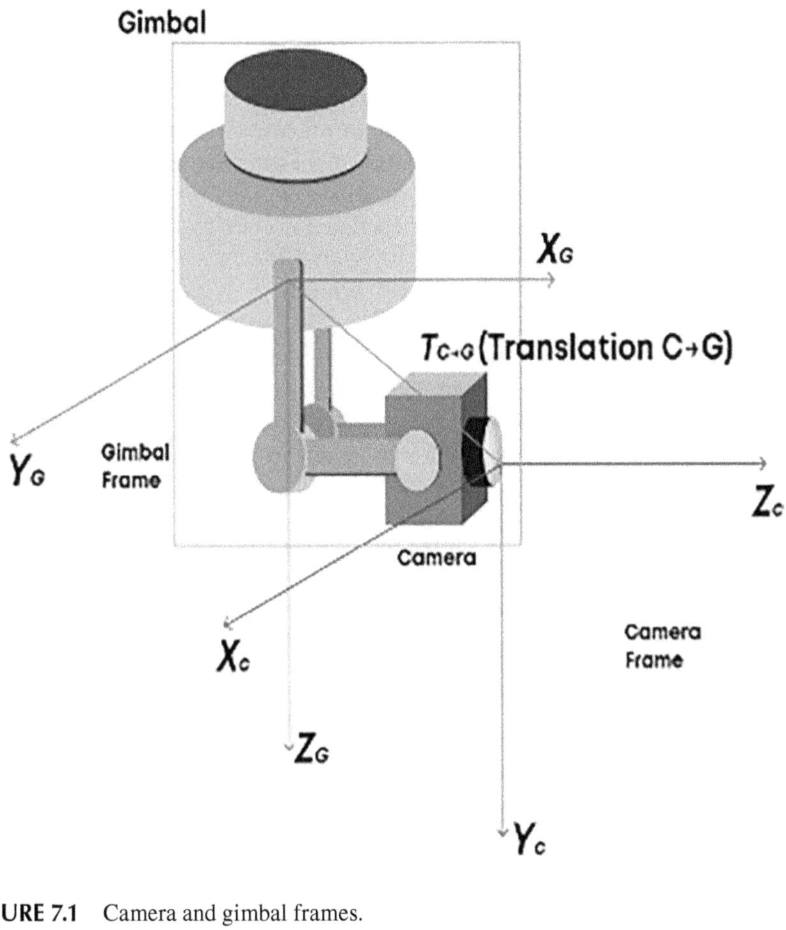

FIGURE 7.1 Camera and gimbal frames.

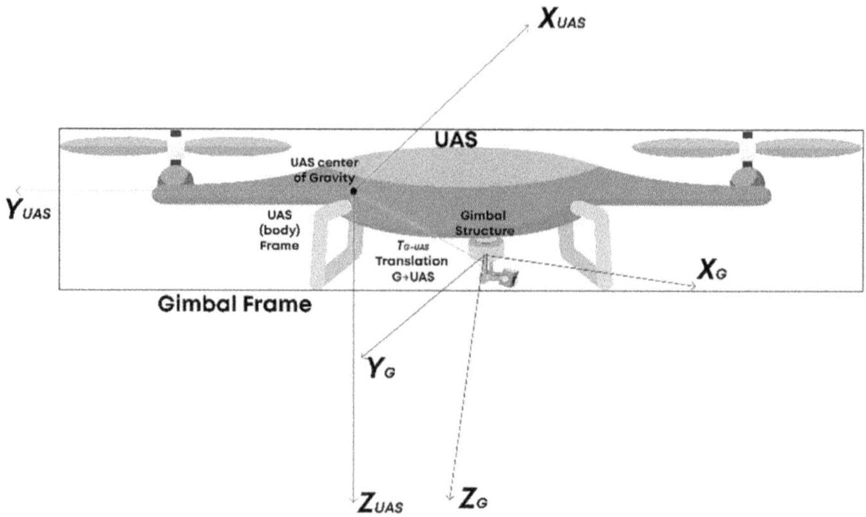

FIGURE 7.2 UAS and gimbal frames.

Gimbal Frame to Unmanned Aerial System (UAS) Frame: Next, we proceed with the transformation of the object's coordinates from the gimbal frame to the UAS frame, which is centered on the UAS center of gravity, as illustrated in Figure 7.2. This transformation necessitates the utilization of the direction cosine matrix (DCM) from the gimbal frame to the UAS frame, denoted as $R_{G\rightarrow \text{UAS}}$, which is determined by the Euler angles yaw, pitch, and roll. The operation is given by:

$$P'_{\text{UAS}} = R_{C\rightarrow G}\ P'_{G} + T_{C\rightarrow \text{UAS}} \tag{7.10}$$

Yaw, pitch, and roll are three rotational movements used to describe the orientation of an object, such as an aircraft or a drone, in three-dimensional space:

1. **Yaw**: Yaw refers to the rotation around the vertical axis of the object. Imagine a vertical axis passing through the object's center, and yaw involves turning the object left or right around this axis. For example, if you were to rotate your head left or right while keeping it level, you're performing a yaw movement.
2. **Pitch**: Pitch refers to the rotation around the lateral axis of the object. This axis runs horizontally from side to side. Pitch involves tilting the object up or down, like nodding your head. In aviation, pitching up means pointing the nose of the aircraft upward, while pitching down means pointing it downward.
3. **Roll**: Roll refers to the rotation around the longitudinal axis of the object. This axis runs from the front to the back of the object. Roll involves tilting the object from side to side, like when an aircraft leans to one side. Rolling to the left means tilting the left wing down and the right wing up, while rolling to the right means the opposite.

Together, these three movements allow us to precisely describe the orientation of an object in three-dimensional space. In the context of drones or unmanned aerial systems (UASs), yaw, pitch, and roll are crucial for controlling the flight and positioning of the drone. They are often adjusted using remote control inputs or programmed commands to achieve desired flight trajectories and orientations.

UAS Frame to NED Frame: The subsequent step involves converting the object's coordinates from the UAS frame to the NED frame (north-east-down frame). Figure 7.3 illustrates this process, taking into account a translation vector $T_{\text{UAS}\to\text{NED}}$. This vector represents the position of the origin of the UAS frame relative to the NED frame, accounting for any disparities between the origins of the reference systems. In order to execute the coordinates transformation, it is necessary to utilize the DCM between the frames, denoted as $R_{\text{UAS}\to\text{NED}}$. Consequently, accurate knowledge of the Euler angles yaw, pitch, and roll is imperative. The position is given by:

$$P'_{\text{NED}} = R_{\text{UAS}\to\text{NED}}\ P'_{\text{UAS}} + T_{\text{UAS}\to\text{NED}} \tag{7.11}$$

NED Frame to East–North-Up (ENU) Frame: In numerous scenarios, a requirement might arise to utilize the east–north-up (ENU) reference system. This reference frame aligns with the Universal Transverse Mercator (UTM) coordinate system, directly correlated with the World Geodetic System (WGS84) utilized by global positioning systems (GPSs). The WGS84 system is well-known for its utilization of latitude, longitude, and height coordinates. Figure 7.4 illustrates the connection between the NED and ENU frames. The rotation matrix from the NED frame to the ENU frame, $R_{\text{NED}\to\text{ENU}}$, can be obtained:

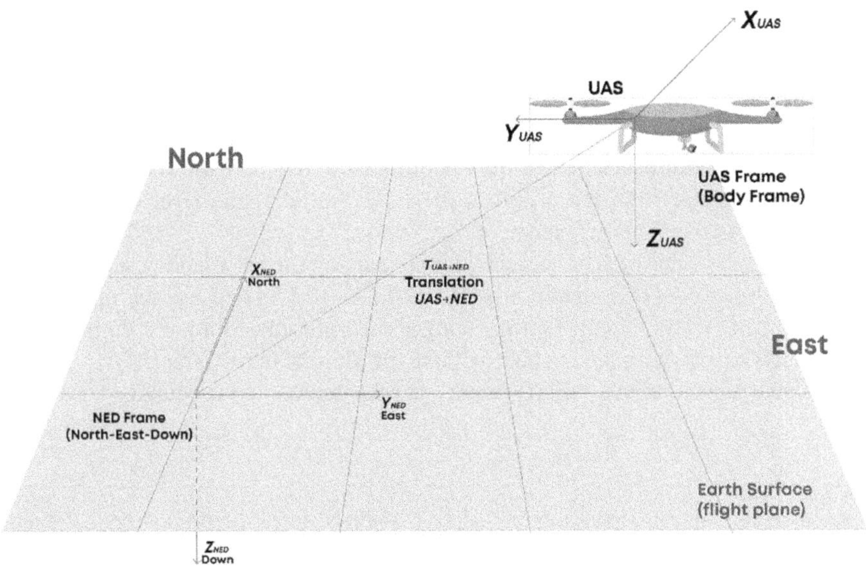

FIGURE 7.3 NED and UAS frames.

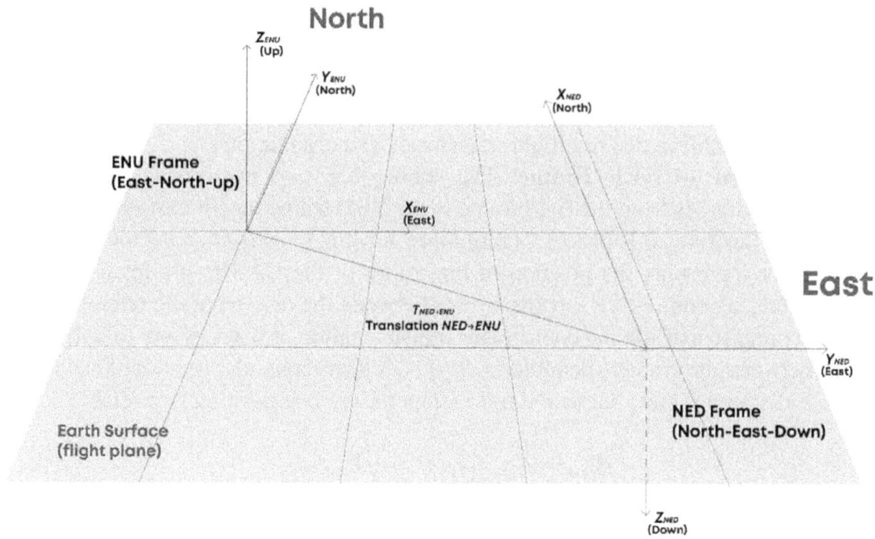

FIGURE 7.4 ENU and NED frames.

$$R_{\text{NED}\to\text{ENU}} = \begin{bmatrix} 0 & 1 & 0 & 1 & 0 & 0 & 0 & 0 & -1 \end{bmatrix} \tag{7.12}$$

The ENU frame, having the primed position of the object, is given by:

$$P'_{\text{ENU}} = R_{\text{NED}\to\text{ENU}} P'_{\text{NED}} + T_{\text{NED}\to\text{ENU}} \tag{7.13}$$

The equation below represents a matrix equation with a solution in the z coordinates. This is because the k_{ENU} coordinate, or the target object's z coordinate in the ENU frame, is predetermined: it corresponds to the object's altitude, given that the origin of the ENU frame is established at sea level. In instances where the ENU frame is positioned at a different elevation, the z coordinate will denote the vertical distance, akin to the height, from the origin of the ENU frame to the level of the object. Figure 7.5 illustrates this arrangement.

Knowing the value of k_{ENU} is crucial for conducting the calculations, and it's quite reasonable to expect this condition to be fulfilled. This is because the operational region of the UAS is typically known, along with its altitude. Moreover, digital elevation models can be leveraged to approximate the altitude of the target.

The object position in the ENU frame can be obtained by the equation:

$$P_{\text{ENU}} = k_C P'_{\text{ENU}} - k_C T_{C\to\text{ENU}} + T_{C\to\text{ENU}} \tag{7.14}$$

According to Hanning, the equation below is referred to as the pinhole model equation. In this equation, L represents the camera matrix utilized to describe a projective mapping from world coordinates ($i_{\text{ENU}}, j_{\text{ENU}}, k_{\text{ENU}}$) to pixel coordinates ($m$, n).

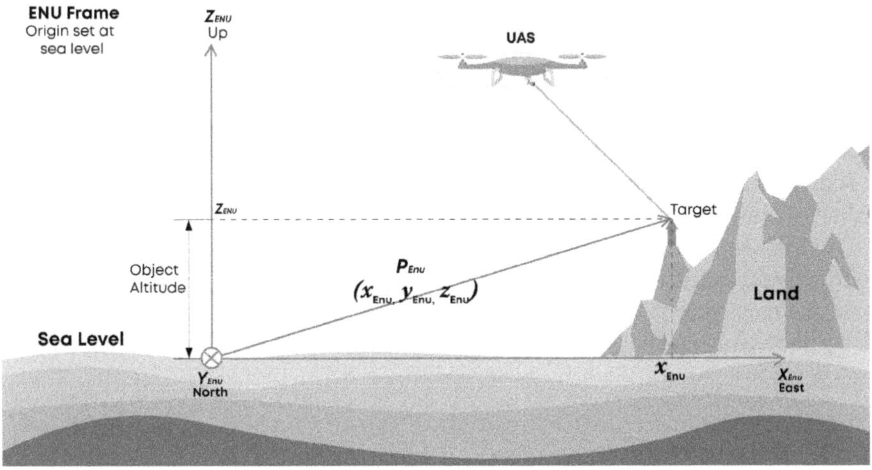

FIGURE 7.5 Object height in ENU frame.

$$k_c\left[m\ n\ 1\right] = K\left[R_{\text{ENU}\to C}\ T_{\text{ENU}\to C}\right]\left[i_{\text{ENU}}\ j_{\text{ENU}}\ k_{\text{ENU}}\right] = L\left[i_{\text{ENU}}\ j_{\text{ENU}}\ k_{\text{ENU}}\ 1\right] \quad (7.15)$$

In pursuit of the objective of this study, which is to provide a comprehensive and detailed explanation of the entire direct georeferencing process, it is essential to briefly elucidate the significance of this expression. The critical aspect of this equation lies in the utilization of homogeneous coordinates derived from projective geometry rather than Euclidean geometry coordinates. This representation is indispensable to facilitate calculations in matrix form, employing a single matrix (the camera matrix L) for mapping a three-dimensional image from the real world to a two-dimensional camera image and vice versa. Within the equation, K represents the intrinsic parameter matrix, $R_{\text{ENU}\to C}$ denotes the 3×3 matrix embodying the composite of all rotations from the camera frame to the ENU frame, and $T_{\text{ENU}\to C}$ signifies the 3×1 matrix, which constitutes the summation of all rotated translations from the ENU frame to the camera frame.

The extrinsic matrix, according to Hanning:

$$\left[R_{\text{ENU}\to C}\ T_{\text{ENU}\to C}\right] = \left[R_{\text{ENU}\to C}\ T_{\text{ENU}\to C}\ 0_{1\times3}\ 1\right]$$

$$= \left[r_{11}\ r_{12}\ r_{13}\ t_{11}\ r_{21}\ r_{22}\ r_{23}\ t_{21}\ r_{31}\ r_{32}\ r_{33}\ t_{31}\ 0\ 0\ 0\ 1\right] \quad (7.16)$$

The camera matric L is:

$$L_{3\times4} = \left[K\ 0_{3\times1}\right]\left[R_{\text{ENU}\to C}\ T_{\text{ENU}\to C}\ 0_{3\times1}\ 1\right] = \left[f\ 0\ c_i\ 0\ 0\ f\ c_j\ 0\ 0\ 0\ 1\ 0\right]$$

$$\left[r_{11}\ r_{12}\ r_{13}\ t_{11}\ r_{21}\ r_{22}\ r_{23}\ t_{21}\ r_{31}\ r_{32}\ r_{33}\ t_{31}\ 0\ 0\ 0\ 1\right] \quad (7.17)$$

7.4 OBJECT TRACKING

The object tracking module is responsible for estimating and maintaining the position and velocity of detected objects. This is achieved through the use of Kalman filters, which provide estimates and predictions for each object's position and velocity. When a detected object is not likely to belong to any of the objects currently being tracked, the tracking module assumes it's a new object and creates a new instance of a Kalman filter associated with it. Subsequent detections of the tracked object serve as measurements in the Kalman filter to refine its estimated position and velocity.

Additionally, the tracking module must be capable of associating new detections with existing tracked objects. This is accomplished by matching object detections with the most probable among the tracking histories, referred to as tracking gaits. A tracking gait encompasses the complete historical state of an object, including its past positions and velocities.

7.4.1 KALMAN FILTER

The Kalman filter employed in the detection, recognition, and tracking module is derived from Ali and Terada's work [7]. It utilizes a motion model premised on the assumption that the UAV is tracking objects positioned on a flat surface, such as the sea surface, and these objects move with a constant velocity.

$$X_{k+1}^{\text{obj}} = x_k^{\text{obj}} + \Delta t V_{y,k}^{\text{obj}} + \frac{1}{2}\Delta t^2 W_k \tag{7.18}$$

$$y_{k+1}^{\text{obj}} = y_k^{\text{obj}} + \Delta t V_{y,k}^{\text{obj}} + \frac{1}{2}\Delta t^2 Z_k \tag{7.19}$$

$$V_{x,k+1}^{\text{obj}} = V_{x,k}^{\text{obj}} + \Delta t W_k \tag{7.20}$$

$$V_{y,k+1}^{\text{obj}} = V_{y,k}^{\text{obj}} + \Delta t Z_k \tag{7.21}$$

An overview of the tracking process begins with the machine vision (MV) module receiving an infrared image $I(i,j)$ from the infrared camera and detecting the centroids (P_m, P_n) of objects of interest within the image. These detected data are then combined with navigation data obtained from the autopilot and gimbal, including camera attitude (R_{NED}^{C}) and position (P_C), to georeference the pixel location of the objects. Once the object's measured location in the NED frame is determined, it is associated with a specific Kalman filter track based on both the measured position and the object's features, such as size and infrared radiation. Subsequently, the Kalman filter track linked with the measurement is utilized to update the Kalman filter estimates. In this process, IR refers to infrared, and NED stands for north-east-down. In the given equation, $x_k^{obj}, y_k^{obj}, V_{x,k}^{obj}, V_{y,k}^{obj}$ represents the position and linear velocities of an object in north-east (NE) coordinates, where x denotes north and y denotes east, at time step k and Δt signifies the time elapsed from time step k to $k+1$. w_k and z_k denote Gaussian white noise terms, accounting for changes in the velocity of an object. This results in the following model in state-space form:

$$x_{k+1}^{obj} = Ax_k^{obj} + Ew_k \tag{7.22}$$

$$y_k^{obj} = Cx_k^{obj} + v_k \tag{7.23}$$

The matrices A, E, and C equal to:

$$A = [1\ 0\ 0\ 1\Delta t\ 0\ 0\Delta t\ 0\ 0\ 0\ 0\ 1\ 0\ 0\ 1] \tag{7.24}$$

$$E = \left[\frac{1}{2}\Delta t^2\ 0\ 0\ \frac{1}{2}\Delta t^2\Delta t\ 0\ 0\Delta t\right] \tag{7.25}$$

$$C = [1\ 0\ 0\ 0\ 0\ 1\ 0\ 0\ ] \tag{7.26}$$

The following are the centroid of an object of the pixel coordinates, and assuming that the object is located at sea level, the object position in the NE frame can be calculated as:

$$\frac{1}{\lambda_k}\left[X_k^{obj}\ Y_k^{obj}\ 1\right] = G_{xyk}^{-1}K^{-1}\left[p_{m_k}^{obj}\ p_{n_k}^{obj}\ 1\right] \tag{7.27}$$

The efficiency of the Kalman filter depends on the magnitude of the measurement of noise v_k and the process noise w_k, which should in fact be tuned to fit the sensor accuracy and the motion model uncertainty. Once a new object is detected, a Kalman filter is initialized with the object's measured position as the filter's position states $\hat{x}_k^{obj}$, $\hat{y}_k^{obj}$. The object's estimated velocity is initialized to 0 because the object's velocity cannot be known by only one measurement. The initial position and velocity of the object are uncertain; therefore, the Kalman filter must be tuned in such a way.

7.4.2 Data Association

When it comes to tracking objects, an essential aspect is the accurate execution of data association and recognition. In the current system, a global nearest neighbor (GNN) approach akin to the one described in Konstantinova et al. [8] is employed for data association. This entails utilizing the following distance metric to measure the distance between measurement I and tracking gait J:

$$D_{I,J} = (1-\gamma)\tilde{y}_{I,J}^T S_J^{-1} \tilde{y}_{I,J} + \gamma \tilde{x}_{I,J}^T \Gamma \tilde{x}_{I,} \tag{7.28}$$

where:

$$\tilde{y}_{I,J} = \left[y^{\text{obj},I} - \hat{y}^{\text{obj},J} \right] \tag{7.29}$$

$$\tilde{x}_{I,J} = \left[x_I - \widetilde{x_J} \right] \tag{7.30}$$

To establish an association between a measured object position and the most probable tracking gaits, a matrix D representing the distances from all n measurements to all m tracking gaits is computed. Thus, the matrix D assumes the following structure:

$$D = \left[D_{1,1} \;\cdots\; D_{1,m} \;\vdots\; \ddots \;\vdots\; D_{n,1} \;\cdots\; D_{n,m} \right] \tag{7.31}$$

After calculating D and removing superfluous rows and columns, the data association problem is a matter of finding the combination of distances, $D_{I,J}$, that yield the global minimum distance. This implies that the combination that is selected assigns exactly one measurement to exactly one tracking gait in a way such that the total distance between all measurements and their assigned tracking gaits is the shortest achievable distance.

7.5 RESULTS

Simulation is carried out for four flight scenarios. Each flight scenario has different considerations. However, in all the cases, UAV is hovering at an altitude of 120 m (about 393.7 ft) above the mean sea level, 300 m (about 984.25 ft) north, and zero meters toward east direction. The speed of the boats is about 14 m/s. As discussed in Section 7.4, the GNN approach is used for data association where distance metric between measurement and tracker gait is involved.

7.5.1 Flight 1

The first flight is of the scenario where a single boat is speeding toward the northeast direction and its path crosses the hovering UAV. To evaluate the performance of the tracker, ground truth is to be utilized. It is necessary to check the performance of the tracker against the ground truth. Figure 7.6 shows the track data plotted against the detections obtained.

Tracking metrics such as root mean square error and number of missed targets are shown in Figures 7.7 and 7.8 from which we can say that the tracker is performing well, but its performance should also be satisfactory in a noisy, clutter environment.

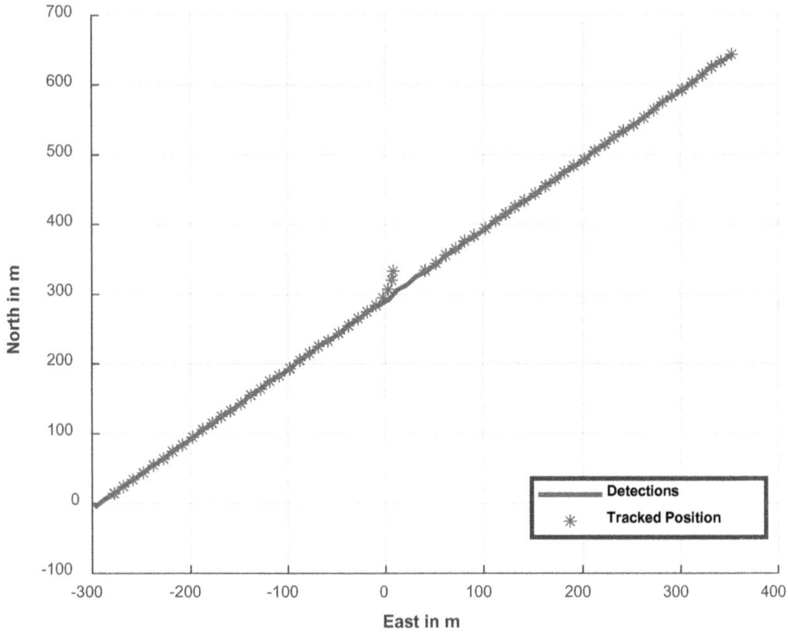

FIGURE 7.6 Single boat tracking output against detections (without noise) for first flight.

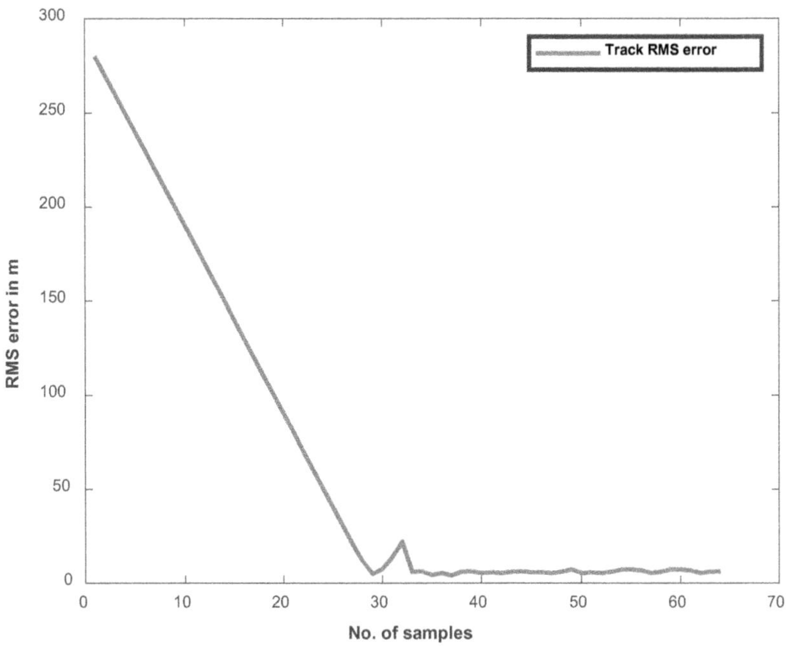

FIGURE 7.7 Root mean square error of the tracks for first flight.

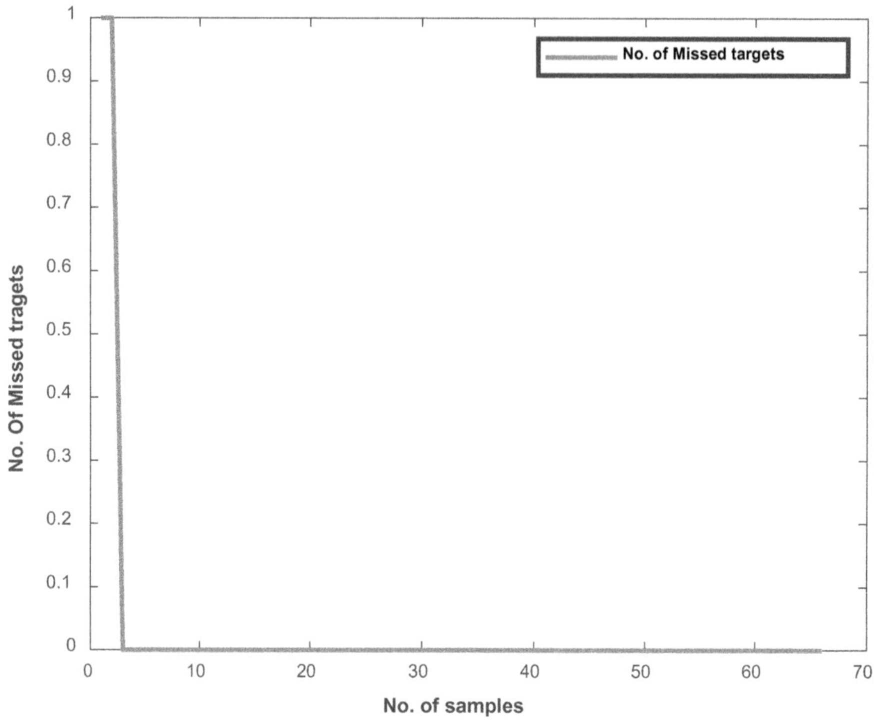

FIGURE 7.8 Number of missed targets for first flight.

7.5.2 Flight 2

The second flight is of the scenario where a single boat is speeding toward the northeast direction and its path crosses the hovering UAV. White Gaussian noise is added at the measurement level. Figure 7.9 shows the track data plotted against the noisy detections obtained.

The performance of the tracker can be said to be satisfactory from Figures 7.10 and 7.11.

7.5.3 Flight 3

In this third flight scenario, UAV is hovering, and this scenario has two boats which are crossing each other right below the UAV as shown in Figure 7.12. This scenario does not have either noisy measurement or clutter. One target is moving toward northeast, and another target is moving northwest at the same speed of about 14m/s. This scenario was to test how the tracker does multi-object tracking. The performance of the tracker can be said to be satisfactory from Figures 7.13 and 7.14.

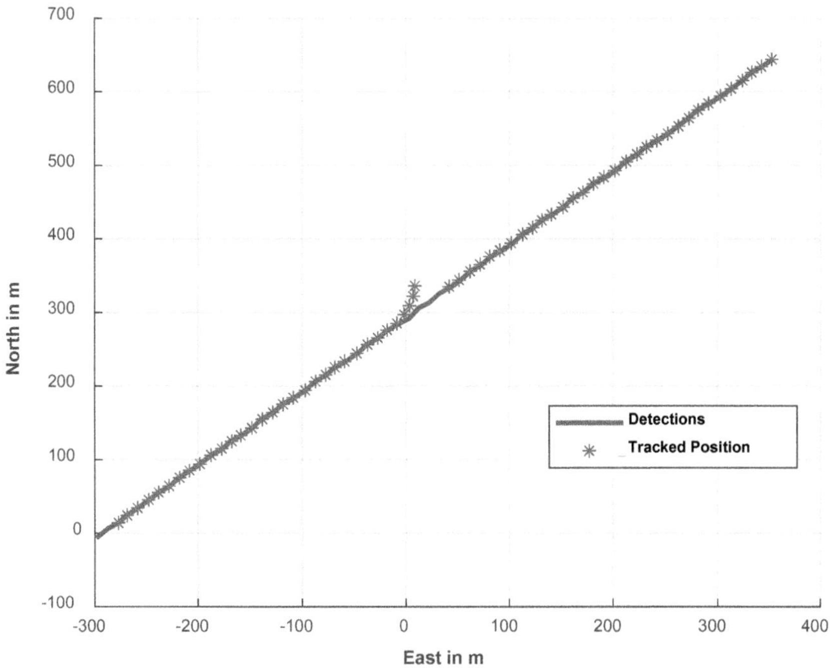

FIGURE 7.9 Single boat tracking output against noisy detections for second flight.

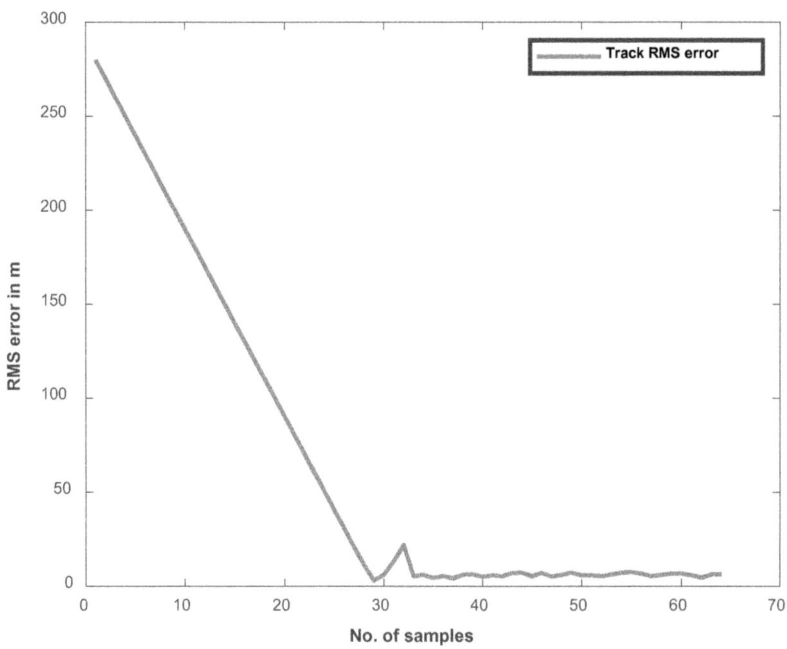

FIGURE 7.10 Root mean square error of the tracks for second flight.

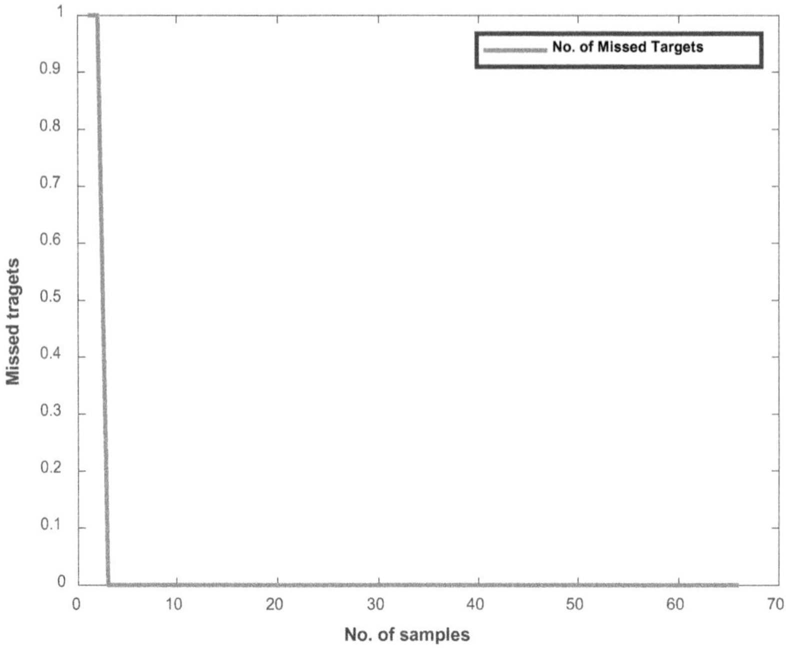

FIGURE 7.11 Number of missed targets for second flight.

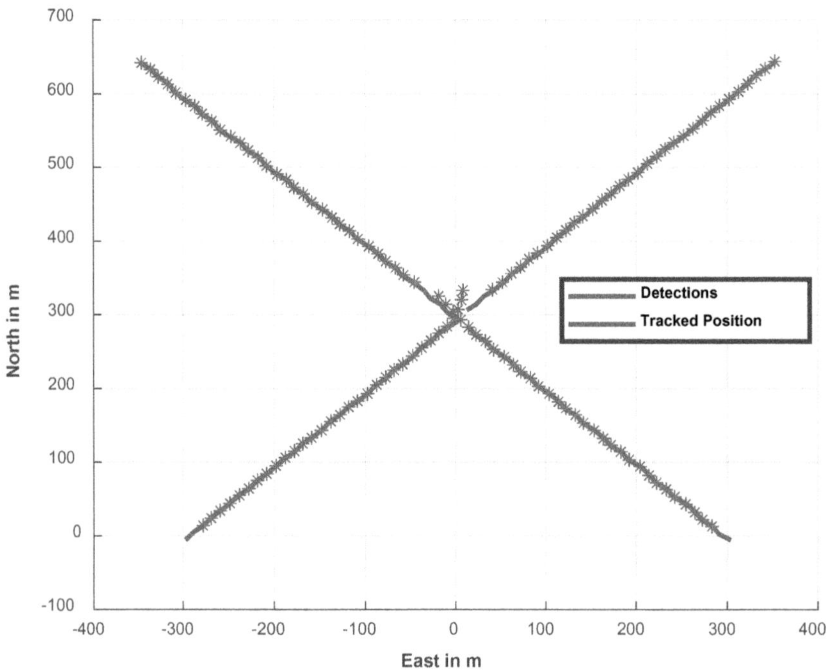

FIGURE 7.12 Two boat tracking output against detections (without noise) for third flight.

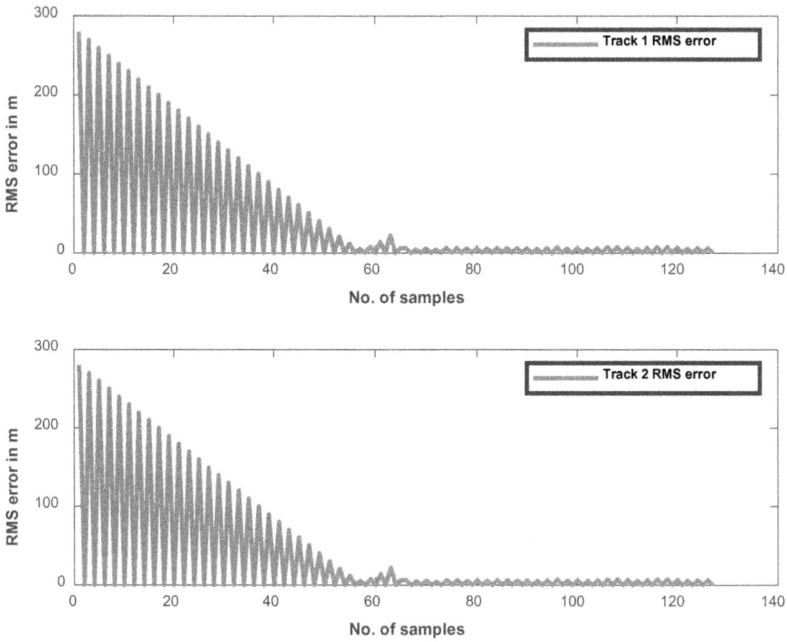

FIGURE 7.13 Root mean square error of the tracks for third flight.

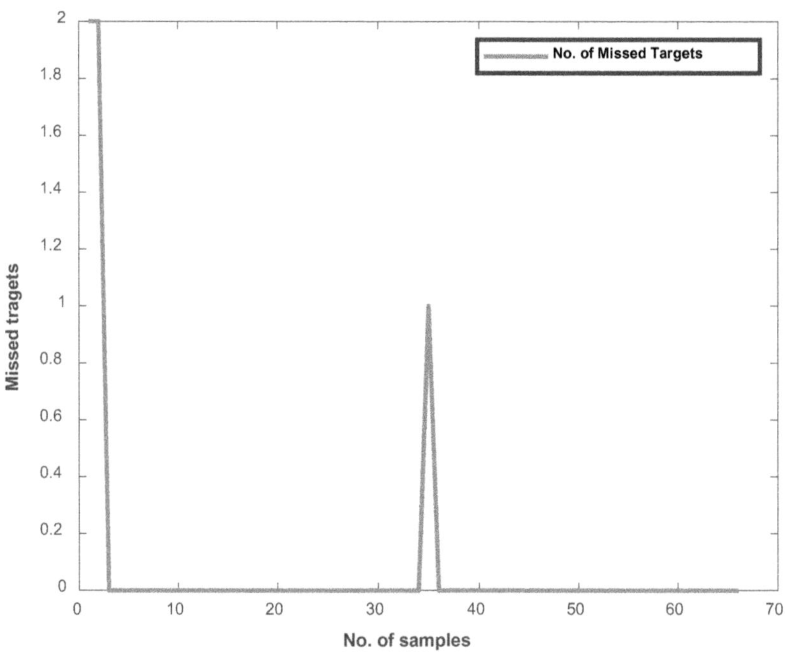

FIGURE 7.14 Number of missed targets for third flight.

7.5.4 Flight 4

The fourth flight is of the scenario, where two boats which are crossing each other right below the UAV. White Gaussian noise is added at the measurement level. Figure 7.15 shows the track data plotted against the noisy detections of the boats. The performance of the tracker can be said to be satisfactory from Figures 7.16 and 7.17.

7.6 CONCLUSION

The detection algorithm uses a blend of dilation/erosion techniques for a thermal camera to detect objects on ocean surface and filter out objects that are of no interest. The developed object detection, recognition, tracking algorithm is to be used for ocean surface object tracking, enabling UAVs to do multi-object tracking in real life. The tracking algorithm utilizes a Kalman filter with constant velocity motion model. To associate a detected object whether it is already being tracked or not, Kuhn–Munkres algorithm is used. The thermal signature is assumed to be uniformly distributed, which is not true, and hence the deduction that the tracking algorithm should also be able to track in such kind of environment. Initially, after getting few

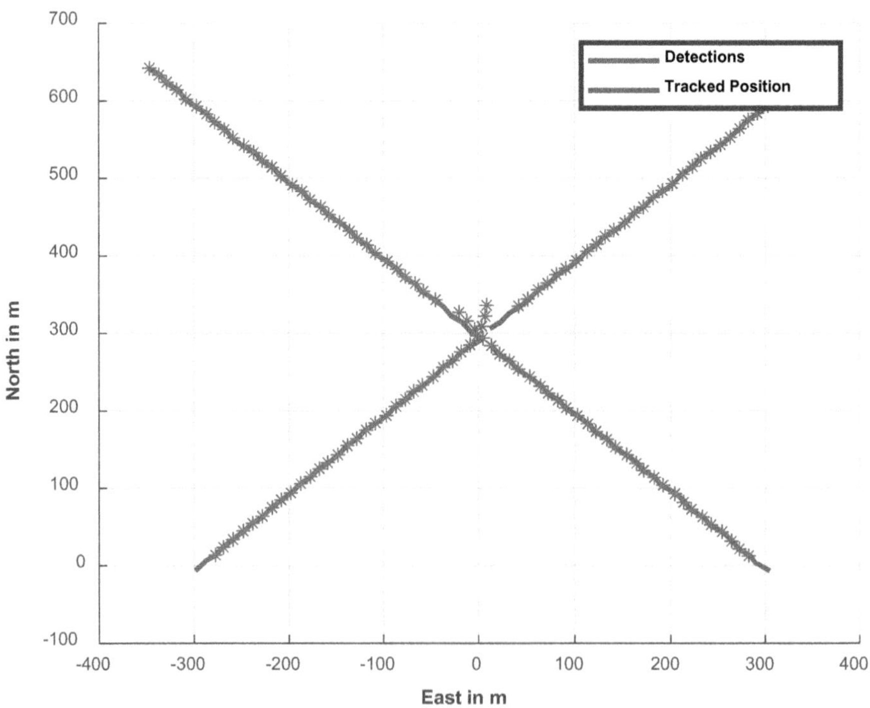

FIGURE 7.15 Two boat tracking output against noisy detections for fourth flight.

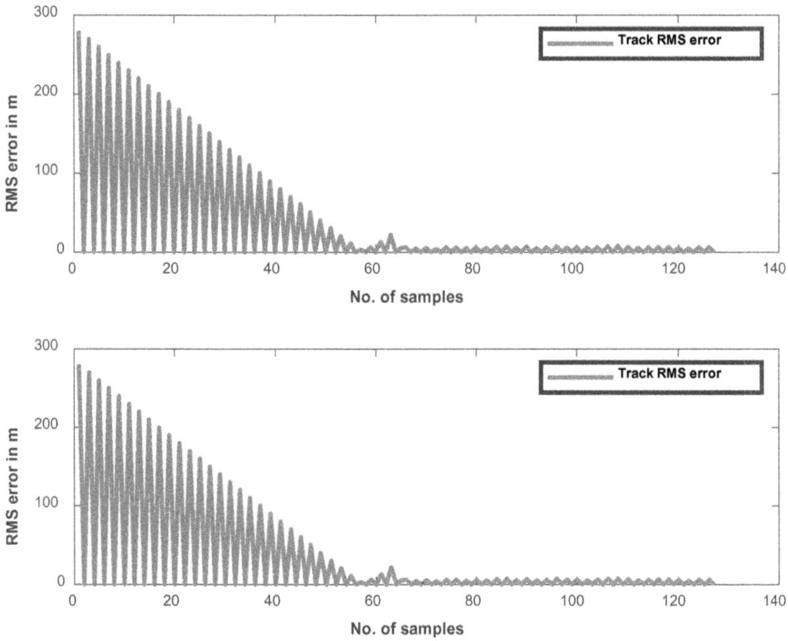

FIGURE 7.16 Root mean square error of the tracks for fourth flight.

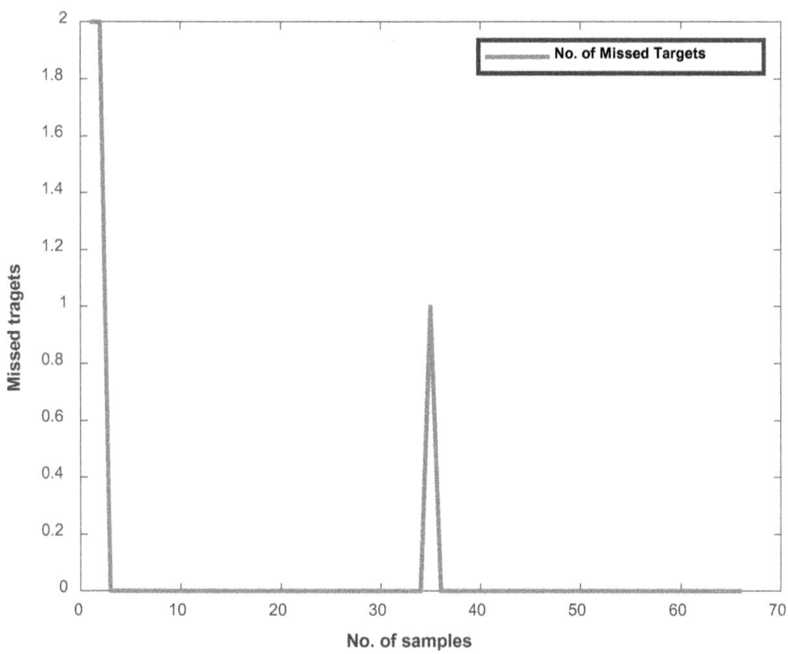

FIGURE 7.17 Number of missed targets for fourth flight.

measurements only the tracking algorithm starts giving starts giving outputs and it can be seen that they are converging with the truth values. From the tracking algorithm developed, multiple objects can also be tracked.

EXERCISES

1. What are the primary applications of UAVs equipped with thermal cameras?
2. Why are thermal cameras advantageous for UAVs compared to traditional visual spectrum cameras?
3. How has the commercial availability of small UAVs impacted their applications?
4. What advancements have made the integration of thermal cameras into UAVs more feasible?
5. What is the benefit of onboard processing of thermal image data in UAVs?
6. Why is automation necessary for tracking multiple moving objects with UAVs?
7. What is multitarget tracking in the context of UAVs?
8. What challenges are associated with multitarget tracking using UAVs?
9. How do advancements in processing platforms benefit UAV operations?
10. What are some approaches to solving the multitarget tracking problem in UAVs?

REFERENCES

1. Nonami, K. Prospect and recent research & development for civil use autonomous unmanned aircraft as UAV and MAV. *J. Sys. Des. Dyn.*, 1(2), 120–128, 2007.
2. Pai, P., Pratyusha, P.L. and Naidu, V.P.S. Geo-Localization of Target in 3-Dimensional Coordinate Using SRTM Data. *J. Inst. Eng. India Ser. B*, 100, 53–60, 2019. https://doi.org/10.1007/s40031-018-0359-z
3. Pai, P., and Naidu, V.P.S. Target geo-localization based on camera vision simulation of UAV. *J Opt*, 46, 425–435, 2017. https://doi.org/10.1007/s12596-017-0395-0
4. Leira, F.S., Helgesen, H.H., Johansen, T.A., Fossen, T.I. Object detection, recognition, and tracking from UAVs using a thermal camera. *J Field Robotics*, 38: 242–267, 2021. https://doi.org/10.1002/rob.21985.
5. Leira, F., Fossen, T., and Johansen, T. Automatic detection, classification and tracking of objects in the ocean surface from UAVs using a thermal camera. In *IEEE Aerospace Conference Proceedings*, 2015, IEEE Computer Society.
6. Correia, C.A.M., Andrade, F.A.A., Sivertsen, A., Guedes, I.P.,Pinto, M.F., Manhães, A.G., and Haddad, D.B. Comprehensive Direct Georeferencing of Aerial Images for Unmanned Aerial Systems Applications. *Italian National Conference on Sensors*, 2022. https://doi.org/10.3390/s22020604.
7. Ali, A., and Terada, K. A general framework for multi-human tracking using Kalman filter and fast mean shift algorithms. *J. Univers. Comput. Sci.*, 16(6), 921–937, 2010
8. Konstantinova, P., Udvarev, A., and Semerdjiev, T. A study of a target tracking algorithm using global nearest neighbor approach. In *Proceedings of the 4th International Conference on Computer Systems and Technologies: E-Learning, CompSysTech '03*, pp. 290–295, 2003, New York, NY: ACM.

8 Feature Tracking/ Mapping Using a Vision and GPS/INS System on a UAV Platform

8.1 INTRODUCTION

The multi-target tracking problem can be said to involve tracking all features of interest within a given area accurately, thereby creating a comprehensive picture of the objects present. This task is essential in surveillance systems that rely on one or more sensors to interpret the environment. Common onboard sensor systems like radar, infrared, vision, and laser provide measurements from these features of interest. Extensive research has been conducted in the field of multi-feature tracking. The work carried out has been focused on various critical aspects, such as developing efficient algorithms that can handle real-time tracking while minimizing computational resources (Computational Efficiency), solving the challenge of associating measurements with the correct targets, especially in scenarios with noisy or ambiguous data (Data Association), improving the accuracy of tracking models to ensure reliable result (Model Accuracy), exploring approaches that use multiple models to adapt to different tracking scenarios (Multiple-Model Techniques), and investigating methods that consider multiple hypotheses to handle uncertainties (Multiple Hypothesis Techniques), which are effective ways to represent spatial information for tracking purposes. Notable contributions include the work of Blackman [1] and Bar-Shalom and Blair [2].

In practical scenarios, the performance of tracking algorithms is influenced not only by their inherent design but also by the accuracy of the onboard navigation system. This impact arises during the feature registration process. Even a minor attitude error in the navigation system can lead to significant deviations in feature location, especially when combined with range information. Consequently, many commercial remote mapping and tracking systems incorporate high-grade navigation systems to mitigate the effects of vehicle attitude errors.

However, when utilizing an Uninhabited Air Vehicle (UAV) as the tracking platform, additional challenges emerge due to the constraints posed by compact, low-power navigation systems. Unfortunately, commercially available UAV navigation systems are either limited in functionality or prohibitively expensive for most academic research endeavours.

In this chapter, a cost-effective airborne feature-tracking system that combines a low-cost inertial sensor and a Global Positioning System (GPS) unit for UAV

DOI: 10.1201/9781003429869-8

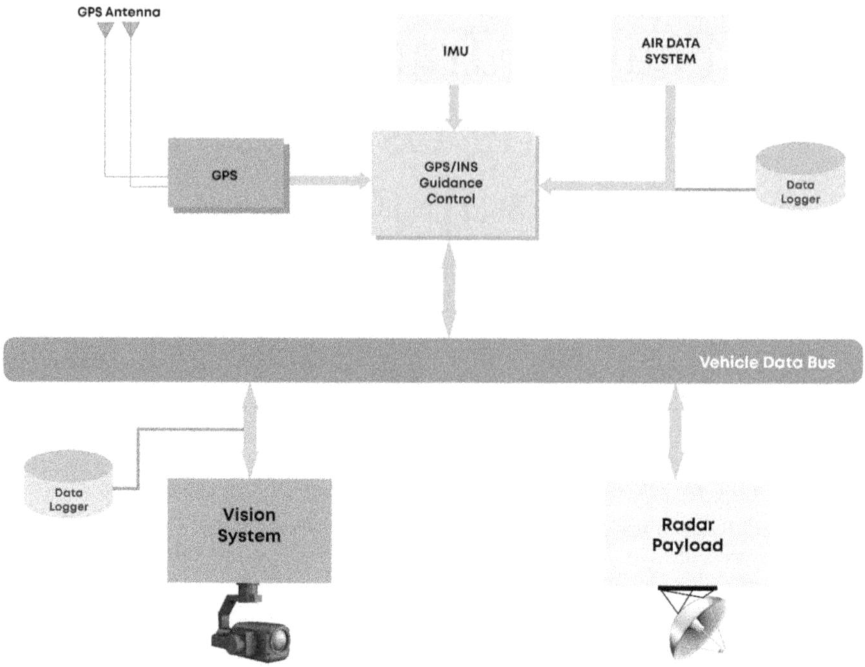

FIGURE 8.1 Architecture of feature tracking and navigation system.

navigation is introduced. Additionally, the system utilizes a low-cost passive vision camera for feature observation.

The system architecture, as depicted in Figure 8.1, is designed for Brumby, a UAV platform developed at the University of Sydney. To ensure modularity and scalability, the sensor node follows a decentralized design. This design allows for easy connection and disconnection. The navigation solution is computed by the flight control computer, which handles GPS/INS fusion and provides guidance and control for autonomous flight.

The sensor payloads connected to the vehicle bus utilize this navigation solution for several critical functions:

- Feature Tracking: The accurate navigation data aids in tracking features of interest.
- Radar Gimbal Control: It ensures precise control over the radar gimbal system.
- Time Synchronization: The navigation solution contributes to maintaining synchronized time across the system.

Here's a concise overview of the upcoming sections where Section 8.2 will delve into the vision system, covering both the sensor and the tracking algorithm. One can expect insights into how system observes and processes visual information. Section 8.3 explores the integration of the GPS and the Inertial Navigation System (INS).

Understanding how these systems work together is crucial for accurate navigation. Section 8.4 will dive into specifics about the flight vehicle itself and the components of the onboard system. Expect details about the UAV's design, hardware, and overall architecture are covered. Section 8.5 will showcase the simulation results, where tracking performance, feature observations, and any relevant metrics are covered. Finally, with Section 8.6 we'll wrap up with conclusions drawn from the study.

8.2 VISION SYSTEM

8.2.1 Passive Vision Camera

The system employs a low-cost, lightweight, monochrome CCS-SONY-HR camera manufactured by Sony. This camera has a resolution of 600 horizontal lines and operates on a 12-V power supply. Its composite video output provides images at either 50 or 25 Hz (when interlaced). Before flight, artificial landmarks (made of plastic sheets) were strategically placed on the ground. These landmarks serve as reference points for the vision system, aiding in feature identification. Given the limited processing resources, the vision system employs a simple yet fast point-based feature extraction algorithm. All pixels above a certain threshold are converted into line segments for further analysis. The range gate is responsible for associating data with the identified line segments. By analysing the centre of mass of the pixels within these segments, the system determines their collective properties. The mass, aspect ratio, and density of the pixel cluster are crucial for identifying features. These characteristics help distinguish different objects or landmarks observed by the vision system. Once the feature is identified, the system calculates its bearing (horizontal direction) and elevation (vertical angle) relative to the sensor's position. Although the vision sensor does not directly provide range information, an estimated value can be generated. This estimation relies on the known size of the features observed. In summary, the vision system combines various measurements and calculations to identify and locate features in the environment.

8.2.2 Feature Tracking Algorithm

The integration of the sensor and platform locations from the GPS/INS system via the vehicle bus plays a crucial role in the feature-tracking process. The GPS/INS system provides precise location data for both the sensor and the UAV platform. This information is essential for converting relative observations into a global Cartesian frame. In the sensor pre-processing stage, all relative observations are transformed into a global reference frame. This conversion ensures that tracking occurs consistently across different viewpoints and orientations. The filter observation model simplifies to a linear model after this conversion. This linear model facilitates efficient tracking computations. In global coordinates, the x and y positions and velocities are modelled using an integrated Ornstein–Uhlenbeck process [5]. This process characterizes velocity as Brownian motion, which can be bounded by adjusting the model parameter γ. The z position is modelled as a simple Brownian process. This is expressed in the following way:

$$x(k+1) = F(k)x(k) + G(k)w(k) \tag{8.1}$$

$$z(k) = H(k)x(k) + v(k) \tag{8.2}$$

The state vector is as follows

$$x(k) = \begin{bmatrix} x(k) & \dot{x}(k) & y(k) & \dot{y}(k) & z(k) \end{bmatrix}^T \tag{8.3}$$

The state transition matrix [4] for this system is:

$$F(k) = \begin{bmatrix} 1 & \Delta t & 0 & 0 & 0 \\ 0 & F_v & 0 & 0 & 0 \\ 0 & 0 & 1 & \Delta t & 0 \\ 0 & 0 & 0 & F_v & 0 \\ 0 & 0 & 0 & 0 & 1 \end{bmatrix} \tag{8.4}$$

$$Q(k) = \begin{bmatrix} \sigma_{vx}^2 & 0 & 0 \\ 0 & \sigma_{vy}^2 & 0 \\ 0 & 0 & \sigma_z^2 \end{bmatrix}$$

where, $F_v = e^{-\Delta t \gamma}$ using Brownian motion parameter.

To simplify the filter observation model, the sensor observations in range, bearing, and elevation are converted into Cartesian coordinates $\begin{bmatrix} x & y & z \end{bmatrix}^T$ in a global reference frame during the sensor pre-processing stage.

The observations are in the form of:

$$z(k) = \begin{bmatrix} x & y & z \end{bmatrix}^T \tag{8.5}$$

Hence, the observation matrix and noise strength matrix are:

$$H(k) = \begin{bmatrix} 1 & 0 & 0 & 0 & 0 \\ 0 & 0 & 1 & 0 & 0 \\ 0 & 0 & 0 & 0 & 1 \end{bmatrix}, R(k) = \begin{bmatrix} \sigma_x^2 & \sigma_{xy} & \sigma_{xz} \\ \sigma_{yx} & \sigma_y^2 & \sigma_{yz} \\ \sigma_{zx} & \sigma_{zy} & \sigma_z^2 \end{bmatrix} \tag{8.6}$$

The noise strength matrix is computed by using the Jacobians of the polar to Cartesian transformation function; hence, it contains cross-correlation terms. The tracking filter utilizes a feature model and an observation model derived from the vision system. These models help estimate the position and velocity of ground features. To link observations with specific features, the system performs data association. The goal is to associate each observation with the correct feature it corresponds to. Within the tracking filter, the innovation gate method is employed for data association. This method ensures that only relevant observations are associated with features. It helps filter out noisy or irrelevant data.

8.3 NAVIGATION SYSTEM

8.3.1 Inertial Navigation

The primary purpose of the inertial navigation algorithm is to predict high-dynamic vehicle motions using data from the Inertial Measurement Unit (IMU). The specific implementation chosen is a quaternion-based strap-down INS algorithm. The algorithm operates in the earth-fixed tangent frame. The approach is based on the work by Kim [3].

$$\begin{bmatrix} p^n(k) \\ v^n(k) \\ q^n(k) \end{bmatrix} = \begin{bmatrix} p^n(k-1) + v^n(k-1)\Delta t \\ v^n(k-1) + \Big[\big(q^n(k-1) \otimes f^b(k)\big) \\ \otimes \big(q^n\big)^*(k-1) + g^n\Big]\Delta t \\ q^n(k-1) \otimes \Delta q^n(k-1) \end{bmatrix} \tag{8.7}$$

where,

$p^n(k), v^n(k), q^n(k)$ represent position, velocity, and quaternion, respectively, at discrete time k. Δt is the time for the position and velocity update interval,

$\big(q^n\big)^*(k)$ is a quaternion conjugate for the vector transformation,

$\otimes$ represents a quaternion multiplication, and

$\Delta q^n(k)$ is a delta quaternion computed from gyroscope readings during the attitude update interval.

8.3.2 GPS/INS Integration

In the complementary GPS/INS architecture, the fusion filter estimates the errors in INS by observing vehicle states. Hence, the state is defined as the error in the position, velocity, and attitude expressed in the navigation frame.

$$\delta x(k) = \begin{bmatrix} \delta p^n(k) & \delta v^n(k) & \delta\psi\psi^n(k) \end{bmatrix}^T \tag{8.8}$$

The system dynamic and observation equations in discrete time are as follows:

$$\delta x(k+1) = F(k)\delta x(k) + G(k)w(k) \tag{8.9}$$

$$\delta z(k) = H(k)\delta x(k) + v(k) \tag{8.10}$$

where, the system transition matrix, the system input noise matrix, and noise strength are given by [4]:

$$F(k)=\begin{bmatrix} I & \Delta tI & 0 \\ 0 & I & \Delta tf^n \\ 0 & 0 & I \end{bmatrix} G(k)=\begin{bmatrix} 0 & 0 \\ \sqrt{\Delta t}C_b^n & 0 \\ 0 & -\sqrt{\Delta t}C_b^n \end{bmatrix} \quad (8.11)$$

where,

C_b^n is the direction cosine matrix formed from the quaternion, and

$\sigma_{\delta f}$ and $\sigma_{\delta\omega}$ are the noise strengths of acceleration and rotation rate, respectively.

Position and velocity are the observations of GPS, so the observation matrix is:

$$H(k)=\begin{bmatrix} I & 0 & 0 \\ 0 & I & 0 \end{bmatrix}, R(k)=\begin{bmatrix} \sigma_p^2 & 0 \\ 0 & \sigma_v^2 \end{bmatrix} \quad (8.12)$$

To estimate the errors in INS, subtract the position and velocity of the GNSS from the INS data, respectively, which results in the input $\delta z(k)$ given to the fusion filter. This loop continuously computes a navigation solution based on inertial measurements. It relies on data from the Inertial Measurement Unit (IMU) to estimate the vehicle's position, velocity, and attitude. Whenever GPS information becomes available, the filter assesses and estimates inertial errors. By comparing the INS solution with GPS data, it refines the navigation accuracy.

8.4 PHYSICAL SYSTEM

The ANSER (Autonomous Navigation and Sensing Experiment Research) project involved multiple Brumby Mk III UAVs. Let's explore the key features of the Brumby Mk III: The Brumby Mk III can achieve speeds of up to 180 km/h. It boasts an impressive roll rate of up to 300°/s. The Brumby Mk III features a modular construction, allowing easy replacement or upgrades of individual components. Its delta fixed-wing design eliminates the need for a tail, resulting in a compact structure relative to the wing area. The minimal component count simplifies maintenance and enhances reliability. The Brumby UAV follows a pusher configuration, where the engine is located at the rear. This design choice creates a clear space in the nose for accommodating radar and other sensors. Sensors positioned in the front section remain unaffected by exhaust contamination. The Mk III features a larger and transparent nose cone. A rear fairing seamlessly blends the nose cone with the fuselage. The embedded PC104 platform serves as the flight control system. It performs critical functions such as navigation, guidance, and control. Data from various sensors like IMU, GPS receivers, and two tilt sensors is fused within this platform. Provided by Inertial Science Inc., the IMU is lightweight and compact and has a small form factor, making it ideal for UAV applications. Stacked on the flight control computer are two CMS Allstar GPS receivers with antennae installed on each of the wings.

The vision camera is strategically positioned next to the IMU. This arrangement minimizes the lever-arm offset between the camera and other components. The camera captures visual information for feature extraction and tracking. The vision camera is connected to a secondary PC104 vision computer. This dedicated computing node handles feature extraction and tracking tasks. All computing nodes communicate via the Ethernet bus. This ensures seamless data exchange and coordination. In summary, the Brumby Mk III's hardware configuration optimizes sensor placement, leverages lightweight components, and facilitates efficient communication for reliable feature tracking and navigation.

8.5 RESULTS

The simulation is carried out in MATLAB, and scenario environment is created in Simulink. UAV model is done using Simulink block and camera model is utilized. GPS and IMU models are attached to the UAV. The scenario is simulated and the data is logged. Then, using this data, GPS/INS fusion as mentioned in Section 8.3.2 is carried out and later the feature tracking is done as mentioned in Section 8.2.2. As stationery targets are being tracked, the tracking model is tuned in such a way that velocity decays to zero. Figures 8.2 and 8.3 show the true flight path, GNS/INS

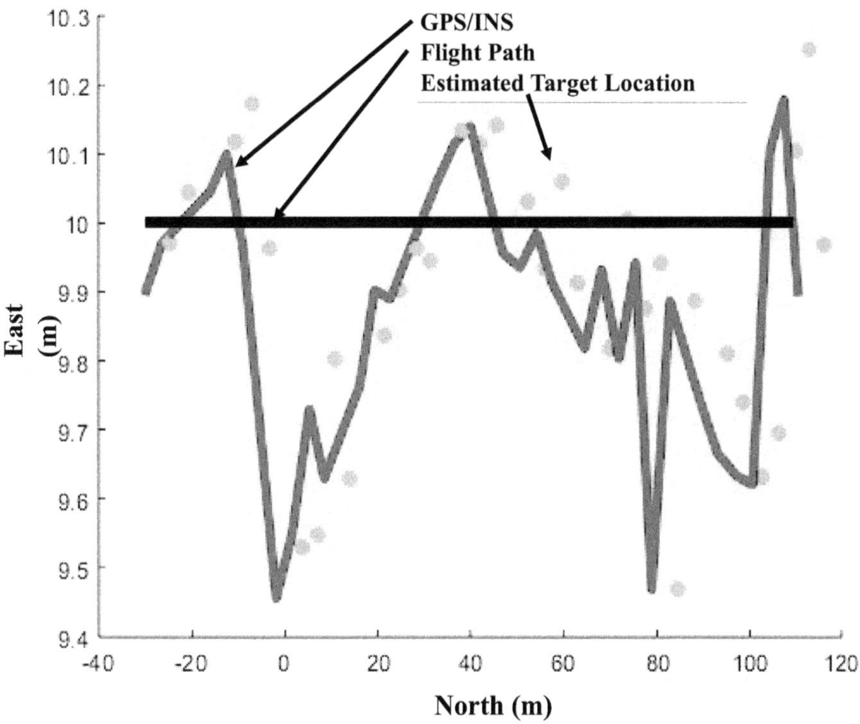

FIGURE 8.2 Estimated target and GPS/INS data plotted for a flight path with respect to horizontal view.

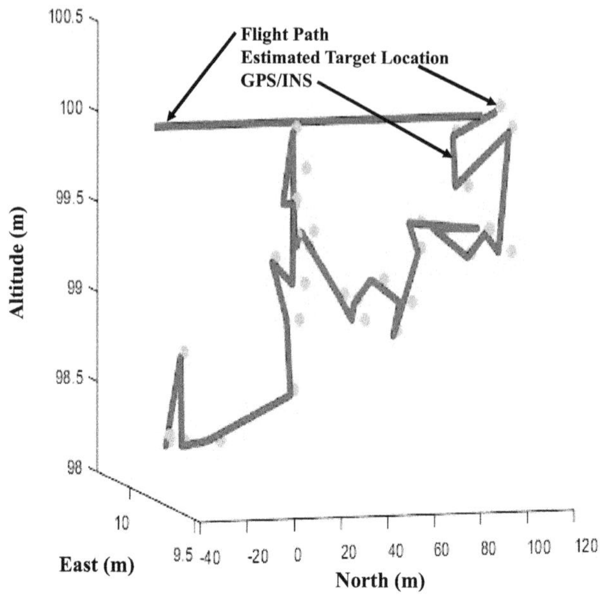

FIGURE 8.3 3D view of estimated target and GPS/INS data plotted for a flight path.

integrated data, and the estimated target positions. In both the figures, error can be said to be around at most one meter. From results, it can be inferred that the GPS/INS navigation system developed for airborne feature-tracking purposes is effective.

EXERCISES

1. What is the purpose of the vision node in the system?
2. How does the accuracy of the onboard navigation system affect the feature tracking performance?
3. What is the significance of the multi-target tracking problem?
4. What is the limitation of using a UAV as the platform for feature tracking?
5. How does the vision system estimate the range of the features, since the camera does not provide range directly?
6. What is the purpose of the sensor pre-processing stage in the feature tracking algorithm, and what conversion is performed?
7. What is the observation model used in the feature tracking algorithm, and how are the sensor observations converted?
8. What is the purpose of the noise strength matrix in the observation model, and how is it computed?
9. What are the sources of error in the feature-tracking system, and how do they affect the results?
10. What is the method used for point-based feature extraction in the vision system, and what are the steps involved?

REFERENCES

1. Blackman, S. *Multiple-Target Tracking with Radar Application.* Artech House, Norwood, OH, 1986.
2. Bar-Shalom, Y. and Blair, W. *Multitarget-Multisensor Tracking: Applications and Advances*, Vol. 3. Artech House, Norwood, OH, 2000.
3. Kim, J. *Autonomous Navigation for Airborne Applications.* PhD thesis, Australian Centre for Field Robotics, Sydney: The University of Sydney, 2004.
4. Ridley, M., Nettleton, E., Sukkarieh, S., and Durrant-Whyte, H. Tracking in decentralised air-ground sensing networks. *International Conference on Information Fusion*, July, 2002, Annapolis, MD.
5. Stone, D. L., Barlow, C. A., and Corwin, T. L. *Bayesian Multiple Target Tracking.* Artech House, Norwood, OH, USA, 1999.

9 Satellite Orbit Determination

9.1 INTRODUCTION

In this chapter, we present various aspects of satellite orbit determination. Since the orbit determination process (ODP) involves system models, measurements, and estimation techniques, we discuss these aspects to illuminate their merits and demerits. The main aspect is to take a step in the direction of evolving an orbit determination method (ODM) that has attributes of good numeric stability, efficiency, and accuracy.

9.2 ORBIT DETERMINATION PROBLEM AND PROCESS

Orbit determination (OD) is defined as a process of obtaining the values of those parameters that completely specify the motion of an orbiting body, e.g., a satellite moving through space, based on a set of observations of the orbiting body. A high-precision OD process takes measurement noise into account and determines an orbit that obtains the best fit to these measurement data subject to the system dynamics of the orbital motion of a satellite. Thus, the OD is an estimation/filtering problem.

Accurate OD and ephemeris generation are required for: (i) determination of (satellite) launch windows, (ii)) precise control of a satellite, (iii) anticipating events such as eclipse, and (iv) determination of the position of a distressed platform or an aircraft for search and rescue operations, the latter being a SARSAT problem (search and rescue satellite-aided tracking); this system operates mainly in/over the polar region countries, and expeditions, e.g., Canada. The OD process (ODP) involves the satellite's trajectory models, measurement models, measurement data (obtained from radars/tracking systems), and estimation techniques/algorithms. The system models include the system description required for the dynamics of orbital motion of a satellite, measurement models, gravitational perturbation models, and Earth's rotation [1–3]; thus the major constituents of an ODP are system models, measurements, and estimation techniques.

9.2.1 Systems Models—Orbital Trajectory Dynamics

By system models, we mean the system description required for representing the dynamics of orbital motion of a satellite, measurement models, effect of Earth's rotation, and perturbation models.

The parameters that completely specify the motion of a satellite are called orbital parameters. Several alternative sets of such parameters are available for

DOI: 10.1201/9781003429869-9

the description of the trajectory dynamics of a satellite. A single set for trajectory formulation would not be very efficient for all mission requirements. One set might have some advantages over the other set, in terms of computational efficiency and achievable accuracy.

9.2.1.1 Classical Orbital Parameters

This set employs an Earth-centered inertial frame and six elements to provide a complete description of a satellite's (elliptical) orbit; specifying the size, shape, and orientation of the orbit: semi-major axis, semi-minor axis, eccentricity, inclination, argument of perifocus, argument of ascending node, and true anomaly (C6), the latter parameter describes the instantaneous position of the satellite. This angle is measured in the orbital plane from perigee in the direction of motion to the current position of the satellite. These parameters provide direct and easy visualization of the orbital path but have a few limitations: for synchronous orbits the elements, arguments of perifocus, and true anomaly are not defined.

9.2.1.2 Inertial Coordinates

Since this set is relatively well defined, it is widely used in actual computations of the orbits; the inertial frame components of the position and velocity vectors completely determine the orbit. The orbital trajectory dynamics for the inertial coordinates (IC6) set have the state-space representation as

$$d/dt\begin{bmatrix} x \\ y \\ x \\ v_x \\ v_y \\ v_z \end{bmatrix} = \begin{bmatrix} v_x \\ v_y \\ v_z \\ -\mu/r^3 x + a_x \\ -\mu/r^3 y + a_y \\ -\mu/r^3 x + a_z \end{bmatrix} \tag{9.1}$$

Here, x, y, and z are the IC components of r, the (slant) range vector; v_x, v_y, v_z are the IC components of the satellite's velocity vector, v; μ is the gravitational parameter of the Earth; and a_x, a_y, a_z are the IC components of the perturbing accelerations. The IC set directly obtains the position and velocity of the satellite and hence, it is very useful compared to the classical set, C6. However, since all the six parameters of IC6 vary rapidly, the integration of the equations of motion requires a small step size.

9.2.1.3 Unified State Model

In the unified state model (USM), the orbital trajectory and attitude dynamics of a satellite are defined in a unified manner [3]. The USM is based on the orbit description in the velocity space as against the conventional position space. The

C6 and IC6 are the descriptions of the orbital mechanics in the position space. The circular, elliptic, parabolic, and hyperbolic orbits (shapes in the position space) transform into circular hodographs in the velocity space. It is interesting to note that the velocity space map or hodograph remains geometrically invariant as the orbital energy level changes. As a result of this, the orbital states resulting from perturbing forces will not experience singularities (undefinable parameters). These velocity space parameters are regularized and hence offer a greater advantage in computations. However, the retrograde and deep mission-type (rectilinear) orbits cannot be represented by this set. In this set, the state variables (U7) are the momenta and the coordinate variables are the Euler parameters. The velocity state parameters (C, R) are the functions of the radial and angular momenta:

$$C = \mu/h_s$$
$$R = \sqrt{v_r^2 + ((h_s/r) - (\mu/h_s))^2} \tag{9.2}$$

Here, v_r is the radial velocity and h_s is the specific angular momentum. The coordinate variables, the four Euler parameters, describe the rotation from the inertial frame to the velocity space orbital frame. Thus, the orbital dynamics for this seven-parameter set is

$$d/dt\begin{bmatrix} e_{o1} \\ e_{o2} \\ e_{o3} \\ e_{o4} \end{bmatrix} = \frac{1}{2}\begin{bmatrix} 0 & \omega_3 & 0 & \omega_1 \\ -\omega_3 & 0 & \omega_1 & 0 \\ 0 & -\omega_1 & 0 & \omega_3 \\ -\omega_1 & 0 & \omega_3 & 0 \end{bmatrix}\begin{bmatrix} e_{o1} \\ e_{o2} \\ e_{o3} \\ e_{o4} \end{bmatrix} \tag{9.3}$$

and

$$d/dt\begin{bmatrix} C \\ R_{f1} \\ R_{f2} \end{bmatrix} = \begin{bmatrix} 0 & -p & 0 \\ \cos\lambda & -(1+p)\sin\lambda & \dfrac{-vR_{f2}}{V_{e2}} \\ \sin\lambda & (1+p)\cos\lambda & \dfrac{vR_{f1}}{v_{e2}} \end{bmatrix}\begin{bmatrix} a_{e1} \\ a_{e2} \\ a_{e3} \end{bmatrix} \tag{9.4}$$

Here, $R = eC$, where e is the eccentricity of the orbit; $R_f(.)$ represents the components of vector R (R = magnitude (R) and $R_{f3} = 0$), and $e_o(.)$ denotes the four Euler parameters (EP) describing the rotation from the inertial frame to the velocity orbital frame. Also, $e_{o1}^2 + e_{o2}^2 + e_{o3}^2 + e_{o4}^2 = 1$, to maintain the regularization, and these parameters are the quaternions (EP). The auxiliary equations are associated with other variables in eqs. (9.3) and (9.4), which are given as [1–3]

$$\sin\lambda = \frac{e_{03}e_{04}}{e_{34}};\ \ \cos\lambda = (e_{04}^2 - e_{03}^2)/e_{34} \tag{9.5}$$

$$V_{e1} = R_{f1}\cos\lambda + R_{f2}\sin\lambda$$

$$V_{e2} = R_{f2}\cos\lambda - R_{f1}\sin\lambda + C \tag{9.6}$$

$$p = \frac{C}{V_{e2}};\ \omega_1 = a_{e_3}/V_{e2};\ \omega_3 = CV_{e2}^2/\mu$$
$$v = (e_{01}e_{03} - e_{02}e_{04})/e_{34};\ e_{34} = e_{03}^2 + e_{04}^2 \tag{9.7}$$

The local sidereal time or the instantaneous longitude of the site is given by

$$\theta = \theta_1 + \theta_2 T_C + \theta_3 T_C^2 + (t)\frac{d\theta}{dt} + \lambda_T$$

$$\theta_1 = 99^{\circ}.6909833;\ \theta_2 = 36{,}000^{\circ}.7689$$

$$\theta_3 = 0^{\circ}.000038708 \tag{9.8}$$

$$T_C = (J.D. - 2{,}415{,}020)/36{,}525 \text{ centuries}$$

$$J.D. = \text{Julian Date}$$

$$\frac{d\theta}{dt} = 0.25068477 \text{ deg./min.}$$

The matrix E_T is given as

$$E_T = \begin{bmatrix} -\sin\theta & \cos\theta & 0 \\ -\cos\theta\sin\phi_T & -\sin\theta\sin\phi_T & \cos\phi_T \\ \cos\theta\cos\phi_T & \sin\theta\cos\phi_T & \sin\phi_T \end{bmatrix} \tag{9.9}$$

The operational equations for determining the perturbing accelerations due to atmospheric drag and wind forces and the zonal and tesseral harmonics are given in ref. [3].

At times, it would be advantageous to combine two parameter sets (defining orbital mechanics) in order to derive certain merits from both. One such combination is the IC6 set for the (say Kalman) filter and U7 (USM) for the integration of the dynamic equations; the former directly represents the position and velocity of the satellite, and the U7 set does not require a small integration step size and is a regularization set (because the four orbits are circular in this representation). Hence, say in EKF, the U7 is used in the propagation part of the filter, and IC6 is used in the measurement update part of the filter; however, back-and-forth conversion from one to the other coordinate system is required.

9.2.2 Measurement Equations

These equations and associated models are defined by the type of measurements to be used in the orbit determination process. These include the specification of the coordinates of the observation sites, Earth (rotation/curvature) model, and certain coordinate transformation equations. The measurements may be obtainable from the ground-based tracking station or from the sensors onboard the satellite.

9.2.2.1 Types of Measurements

The choice of the type of measurements that should be utilized in the ODP depends on the problem at hand. For autonomy of navigation, the onboard measurements obtainable from the star trackers or the landmark trackers can be used. The information from these sensors can be transmitted to the ground station and processed by the OD software. Alternatively, a microcomputer onboard the satellite can process this information utilizing a computationally efficient, numerically stable, and very accurate filtering algorithm. Subsequently, the estimated orbital positions and velocity information can be utilized for orbit control and ephemeris generation onboard the satellite, and can also be transmitted to the ground station for mission data processing.

The conventional measurement data type is a set of azimuth (from North in the horizontal plane), elevation (upward/downward from the horizontal plane as a reference), range, and range rate measurements obtainable from a ground station of known coordinates. These angle measurements are with respect to a topocentric coordinate system located at the observation site. Often the position space observations can be transformed into the velocity space maps. These transformations and maps can be used for developing the orbit observation matrix to be used with the USM in an estimation algorithm. This velocity space representation has been found to be very appropriate for the ODP, and it could be very suitable for the onboard OD. Occasionally, various types of measurements can be suitably combined from different sites in order to achieve desired accuracies in position and velocity determination.

9.2.2.2 Measurement Models

For the measurement model, we use the position space description since it is so often used. The coordinates of the observation site are measured with respect to the reference ellipsoid approximating the Earth's shape, considering the cross-section of the Earth in the plane of the meridian of the site. The coordinates of the Earth-based observation site are

$$R_c = \begin{bmatrix} X_c \cos\theta \\ X_c \sin\theta \\ Y_c \end{bmatrix} \tag{9.10}$$

Here, $X_c = [\frac{a_e}{\sqrt{(1-e_e^2\sin^2\phi_T)}} + H_T]\cos\phi_T$; $Y_c = [\frac{a_e(1-e_e^2)}{\sqrt{(1-e_e^2\sin^2\phi_T)}} + H_T]\sin\phi_T$

a_e is the semi-major axis of the ellipsoid of Earth; e_e the eccentricity of the Earth; H_T the height/altitude of the observation site; ϕ_T the geodetic latitude of the site; λ_T the longitude of the site; θ the local sidereal time; and R_c the site position vector.

The auxiliary equations for this measurement model are given in [2]. The inertial frame components of the vector ρ from the site to the satellite are given by $\rho = r - R_c$. Then the topocentric reference frame is defined with its origin at the site. The primary plane is defined as a plane parallel to the tangent plane of the ellipsoid of the Earth. The x-axis is in the direction parallel to the equatorial plane. The z-axis is in the upward direction. These three directions are denoted as $T1$, $T2$, and $T3$. The transformation matrix from the inertial to the $\{T1\ T2\ T3\}$ frame is defined as E_T [2]. Thus, the components of ρ in the $\{T1\ T2\ T3\}$ frame are given by

$$\rho_T = E_T \rho = [\rho_{T1}, \rho_{T2}, \rho_{T3}] \tag{9.11}$$

Consequently, the observables, azimuth, elevation, and range, are given by

$$A_z = \tan^{-1} \frac{\rho_{T1}}{\rho_{T2}} \tag{9.12}$$

$$E_l = \tan^{-1} \frac{\rho_{T3}}{\sqrt{\rho_{T1}^2 + \rho_{T2}^2}} \tag{9.13}$$

$$\rho = \sqrt{\rho_x^2 + \rho_y^2 + \rho_z^2}\text{; with inertial frame components} \tag{9.14}$$

9.2.3 Hardware and Software

9.2.3.1 Vehicle-Mounted Sensor System (VMSS)

The equipment complex for an autonomous satellite navigation system consists mainly of sensors for observing navigation quantities and a space-borne computer (SBC) to process prestored and sensor data. The space-sextant system (SSS) uses the angle measurements between the moon and second-magnitude (or brighter) stars in a Kalman navigation filter (KNF) to estimate vehicle position and velocity. The SSS is a lightweight, low-power, high-performance, onboard navigation system. It has a very high level of autonomy. It provides an accuracy of 0.241 km within 24 hours of commencement of navigation and achieves about 1.8 km (~ I nmi) accuracy within 10 minutes. The space sextant can be used also as a sensor to allow spacecraft altitude determination to an accuracy of 1 arcsec. It requires data storage for lunar ephemeris and terrain height compensation if ultimate accuracy is required.

The passive ranging interferometer system (PRAIS) also uses a KNF. But, it measures angles to existing radiometric landmarks and signal time of arrival from some of these landmarks with special characteristics. It is a completely strapped down sensor which only requires small antennas to be mounted on the Earth-pointing face of the spacecraft. The PRAIS navigation system uses simultaneous measurements to three landmark radars.

Pseudorange (**PRA**) and interferometer landmark tracker (**ILT**) measurements are employed for each landmark. The ILT measurements quickly reduce navigation errors, while the PRA measurements contribute to accurate steady-state performance. The PRAIS provides potential accuracy on most orbits up to about 0.015 km position error. It has no moving parts.

9.2.3.2 Electronics and Space-Borne Computers (SBC)

Autonomous systems require electronics and computers for the acquisition and preprocessing of measurements and implementation of Kalman filter-type algorithms. In most such systems, the electronics package is an integral part of the entire system. The cost, weight, and power to perform a given function have decreased very significantly. As missions become more complex and as spacecraft operate for the deep-space mission, the spacecraft must be capable of much greater autonomy.

9.2.3.3 Software

The ODP/OBOD requires subroutines (software/SW) for coordinate transformation and numerical integration of orbital dynamics. If the IC6 set is used in the filtering algorithm, and U6 or U7 (USM) is used for the integration, then two-way coordinate transformation is required at every sampling instant of the measurement data processing. If a coordinate set other than the IC6 set is used, then transformation to the IC6 set is required in order to obtain the position and velocity of the spacecraft. Numerical integration of the dynamic equations is required for computing the time propagation of the states of the satellite; this is generally accomplished by using the Runge-Kutta/RK4 method, other alternative is a predictor-corrector method. Other SW is related to the orbit estimators and filtering algorithms.

9.3 ORBIT ESTIMATORS

The satellite OD is considered as a state estimation problem, given the noisy measurements. We briefly describe some filtering algorithms that are widely used for the purpose. The extended KF equations are given in Chapter 3 for target tracking, and the similar equations can be used for the estimation of orbital states of a satellite.

9.3.1 Generalized Least Squares Differential Correction Technique (GLSDCT)

In this section, the nonlinear least squares parameter estimation method is described. The method is based on a differential correction technique. This algorithm can be used to estimate the initial conditions of states and the parameters of a nonlinear dynamic model. It is a batch iterative procedure and can be regarded as complementary to other nonlinear parameter estimation procedures like the output error method. One can use this technique to obtain the start-up values of the aerodynamic parameters for other methods. To describe the method used to estimate the parameters of a given model, let us assume a nonlinear system as

$$\dot{x} = f(x,t,C) \tag{9.15}$$

$$y = h(x,C,K) + v \tag{9.16}$$

Here, x is $n \times 1$ state vector, y is $m \times 1$ measurement vector, and v is a random white Gaussian noise process with covariance matrix R. The functions f and h are vector-valued nonlinear functions, generally assumed to be known. The unknown parameters in the state and measurement equations are represented by vectors C and K. Let x_0 be a vector of initial conditions at t_0. Then the problem is to estimate the parameter vector

$$\hat{\beta} = \left[x_0{}^T C^T K^T\right]^T \tag{9.17}$$

It must be noted that the vector C appears in both state and measurement equations. Such situations often arise for aircraft parameter estimation. The iterative differential correction algorithm is applied to obtain the estimates from the noisy measured signals as

$$\hat{\beta}^{(i+1)} = \hat{\beta}^{(i)} + \left[\left(F^T W F\right)^{-1} F^T W \Delta y\right]^{(i)} \tag{9.18}$$

where

$$F = \left[\partial y/\partial x_0 \middle| \partial y/\partial C \middle| \partial y/\partial K\right] \tag{9.19}$$

We use "∂" to denote partial differentiation here. W is a suitable weighting matrix and Δy is a matrix of residuals of observables, $\Delta y = z(t_k) - y(t_k)$, where $k = 1,2,...,N$. The first submatrix in F is given as

$$\partial y(t_k)/\partial x(t_0) = \left[\partial h(x(t_k))/\partial x(t_k)\right]\left[\partial x(t_k)/\partial x(t_0)\right] \tag{9.20}$$

With

$$d/dt\left[\partial x(t)/\partial x(t_0)\right] = \left[\partial f(t,x(t))/\partial x(t)\right]\left[\partial x(t)/x(t_0)\right] \tag{9.21}$$

the transition matrix differential eqn. (9.16) can be solved with the identity matrix as an initial condition. The second submatrix in F is

$$\partial y/\partial C = \left(\partial h/\partial x\right)\left(\partial x/\partial C\right) + \partial h/\partial C \tag{9.22}$$

where $\left(\partial x(t)/\partial C\right)$ is the solution of

$$d/dt\left[\partial x/\partial C\right] = \partial f/\partial C + \left(\partial f/\partial x\right)\left(\partial x/\partial C\right) \tag{9.23}$$

The last submatrix in F is obtained as

$$\partial y/\partial K = \partial h/\partial K \tag{9.24}$$

The state integration is performed by the RK4 method. Convergence to the optimal solution/parameters would help in finding the global minimum of the cost function. It is a well-known fact that the quality of the measurement data significantly influences the accuracy of the parameter estimates. The technique can be employed to assess quickly the quality of the measurements (aircraft maneuvers), polarities of signals, and to estimate bias and scale factor errors in the measurements.

9.3.2 Factorization Method and Smoother

There is as such no problem with the KF, per se, but its implementation on a finite word length computing machine could pose a problem. These effects are circumvented or greatly reduced by implementing the KF in its factorized form. These algorithms do not process covariance matrix P in its original form but process its square root. The main reason for the improvement is that the range of numbers handled is $10^{+/-0.5N}$, and hence, the factorization filter can be implemented on a 32-bit machine, if the original KF algorithms were implemented on a 64-bit machine. Such factorization implicitly preserves the symmetry and ensures the non-negativity of the covariance matrix P. There are several such algorithms available in literature; however, one such widely used is the UD factorization filtering algorithm; here, U and D are matrix factors of the covariance matrix P of the KF, where U is a unit upper triangular matrix (with 1's on diagonal elements) and D is a diagonal matrix.

9.3.2.1 UD Factorization Filter

The UD factorization filter has the merits [1]: (i) It is numerically reliable, accurate, and stable, (ii) it is a square root type algorithm that does not involve square rooting operations, (iii) the algorithm is most efficiently and simply mechanized by processing vector measurements (observables), one component at a time, and iv) for linear (linearized nonlinear) systems, the UD filter, UDF (EUDF) is algebraically equivalent to the KF (EKF). The major advantage of UDF comes from the fact that the square root type algorithm processes square roots of the covariance matrices and hence, they essentially use half the word length normally required by the conventional KFs. In the UDF, the covariance update formulae of the conventional KF and the estimation recursion are reformulated, so that the covariance matrix P does not appear explicitly. Specifically, we use recursions for U and D factors of covariance matrix $P = UDU^T$. Computing and updating with triangular matrices involve fewer arithmetic operations and thus greatly reduce the problem of round-off errors, which might (otherwise) cause ill-conditioning and subsequent divergence of the algorithm, especially if the filter is implemented on a finite word length machine. This is more so for real-time implementation in onboard computers where the word length could be small, e.g., 16 or 32 bit.

9.3.2.1.1 Time Propagation

We have the covariance propagation from the KF recursion

$$\tilde{P}(k+1|k) = \varphi\hat{P}(k)\varphi^T + GQG^T \tag{9.25}$$

Given $\hat{P} = \hat{U}\hat{D}\hat{U}^T$ (a priori factors are assumed known) and Q as the process noise covariance matrix, the time propagated factors $\tilde{U}$ and $\tilde{D}$ are obtained through the modified Gram-Schmidt orthogonalization process [1] as follows.

We define $V = \left[\varphi\hat{U}|G\right]$ and $\bar{D} = \text{diag}[\hat{D},Q]$, and $V^T = [v_1, v_2, ..., v_n]$, then P is reformulated as $\tilde{P} = \tilde{V}\tilde{D}\tilde{V}^T$. The U and D factors of $\tilde{V}\tilde{D}\tilde{V}^T$ can be computed by the following recursion: for $j = 1,\ldots,n$, the following equations are recursively evaluated:

$$\tilde{D}_j = < v_j, v_j >_{\bar{D}} \tag{9.26}$$

$$\tilde{U}_{ij} = (1/\tilde{D}_j) < v_i, v_j >_{\bar{D}} \quad i = 1,\ldots,j-1 \tag{9.27}$$

$$v_i = v_i - \tilde{U}_{ij} v_j \tag{9.28}$$

In eqn. (9.27), $< v_i, v_j >_{\bar{D}} = v_i^T \bar{D} v_j$ is the weighted inner product between v_i and v_j. Therefore, the time propagation algorithm directly and efficiently produces the required U and D factors, taking the effect of previous U and D factors and the process noise. Thus, it also preserves the symmetry of the (original) P matrix.

9.3.2.1.2 Measurements Data Update

The measurement update process in KF combines a priori estimates $\tilde{x}$ and error covariance $\tilde{P}$ (obtained from the time propagation) with a scalar observation $z = cx + v$ to construct an updated estimate and covariance as

$$\begin{aligned} K &= \tilde{P}c^T/s \\ \hat{x} &= \tilde{x} + K(z - c\tilde{x}) \\ s &= c\hat{P}c^T + R \\ \hat{P} &= \hat{P} - Kc\tilde{P} \end{aligned} \tag{9.29}$$

Here, $\tilde{P} = \tilde{U}\tilde{D}\tilde{U}^T$; c is the (scalar) measurement matrix, R is the measurement noise covariance, and z is the vector of noisy measurements. Kalman gain K and updated covariance factors $\hat{U}$ and $\hat{D}$ can be obtained from the following equations [1]:

$$g = \tilde{U}^T c^T; g^T = (g_1, \ldots, g_n); w = \tilde{D}g;$$
$$\hat{d}_1 = \tilde{d}_1 R/s_1, s_1 = R + w_1 g_1 \tag{9.30}$$

For $j = 2, \ldots, n$ the following equations are evaluated:

$$s_j = s_{j-1} + w_j g_j; \hat{d}_j = \tilde{d}_j s_{j-1}/s_j$$
$$\hat{u}_j = \tilde{u}_j + \lambda_j K_j, \quad \lambda_j = -g_j/s_{j-1} \tag{9.31}$$
$$K_{j+1} = K_j + w_j \tilde{u}_j; \tilde{U} = [\tilde{u}_1, \ldots, \tilde{u}_n]; K = K_{n+1}/s_n.$$

Here, $\tilde{d}$ is the predicted diagonal element, and $\hat{d}_j$ is the updated diagonal element of the D matrix. The time propagation and measurement update for the state vector (x) are just similar to KF. We also note that the measurement update/data processing can be done sequentially, meaning that each observable can be processed in turn, and the state estimate updated, this avoids the (whole) matrix inversion in the Kalman gain formulation.

9.3.2.2 RTS Smoother

A computationally efficient fixed interval smoother was first presented by Rauch, Tung, and Striebel [1] and is called the RTS smoother. In order to get the smoothed estimates, RTS smoother does not need computation of backward (in time) estimates or covariances. It uses only the forward pass KF estimates and covariances which are stored during the forward pass of the KF. For the smoothed covariance matrix, applying the matrix inversion lemma, we get

$$\hat{P}_s(j) = \hat{P}_f(j) - \hat{P}_f(j)(\hat{P}_f(j) + \tilde{P}_b(j))^{-1}\hat{P}_f(j) \tag{9.32}$$

The backward covariance is given by

$$\tilde{P}_b(j) = \varnothing^{-1}\left[\hat{P}_b(j+1) + Q\right]\varnothing^{-T} \tag{9.33}$$

Substituting (9.33) for $\tilde{P}_b(j)$ in the expression for $\hat{P}_s(j)$, eqn. (9.32), and simplifying we get

$$\hat{P}_s(j) = \hat{P}_f(j) + K_s(j)(\hat{P}_s(j+1) - \tilde{P}_f(j+1))K_s(j)^T \tag{9.34}$$

In eqn. (9.34), the smoother gain is given by

$$K_s(j) = \hat{P}_f(j)\varnothing^T \widetilde{Y_f}(j+1) \tag{9.35}$$

Thus, it is seen that the smoother covariance depends only on the forward filter covariance, which saves a lot of computational effort. For a smoother state update, we can obtain

$$\hat{x}_s(j) = \hat{x}_f(j) + K_s(j)\left(\hat{x}_s(j+1) - \tilde{x}_f(j+1)\right) \tag{9.36}$$

Note that this expression does not need the backward-in-time estimates. A summary of the RTS is given as follows:

System model $x(k) = \varphi x(k-1) + \mathrm{Bu}(k-1) + \mathrm{Gw}(k-1)$

$$z(k) = Hx(k) + \mathrm{Du}(k) + v(k);\ \text{with}\ w(k) \sim (0,Q)\ \text{and}\ v(k) \sim (0,R) \tag{9.37}$$

Forward filter initialization $\hat{x}_f(0) = E\left(x(0)\right)$

$$\hat{P}_f(0) = E\left[(x(0) - \hat{x}_f(0))(x(0) - \hat{x}_f(0))^T\right] \tag{9.38}$$

For the standard forward KF for $k = 1,2,\ldots,N$, the following recursions are evaluated:

$$\begin{aligned}
\tilde{x}_f(k) &= \varphi\, \hat{x}_f(k-1) \\
\tilde{P}_f(k) &= \varphi \hat{P}_f(k-1)\varphi^T + GQG^T \\
K_f(k) &= \tilde{P}_f(k)H^T(H\tilde{P}_f(k)H^T + R)^{-1} = \hat{P}_f(k)H^T R^{-1} \\
\hat{x}_f(k) &= \tilde{x}_f(k) + K_f(k)\ (z(k) - H\tilde{x}_f(k)) \\
\hat{P}_f(k) &= (I - K_f(k)H)\tilde{P}_f(k)
\end{aligned} \tag{9.39}$$

Smoother initialization:

$$\hat{x}(N) = \hat{x}_f(N),\ \hat{P}_s(N) = \hat{P}_f(N) \tag{9.40}$$

Smoother equations:

For $k = N-1,\ldots,1,0$, the following recursions are evaluated:

$$\begin{aligned}
\tilde{Y}_f(k+1) &= \left(\tilde{P}_f(k+1)\right)^{-1} \\
K_s(k) &= \hat{P}_f(k)\varnothing^T \widetilde{Y_f}(k+1) \\
\hat{P}_s(k|N) &= \hat{P}_f(k) + K_s(k)(\hat{P}_s(k+1|N) - \tilde{P}_f(k+1))K_s^{\ T}(k)
\end{aligned} \tag{9.41}$$

The smoothed state is obtained by eqn. (9.36). It is to be noted that this smoother does not use the measurements in the reverse/backward pass. Also, the smoothed equations use only the state/covariance estimates generated by the forward KF in the forward pass. So, the process is to use the Kalman filter starting from the initial $\hat{x}_0$ and P_0, and complete one forward pass through all data points sequentially, during which process, all the filtered estimates are stored. The smoother equations are used in the backward pass starting from the final values of the state/covariance estimates (saved from the KF) and arriving at the initial point which results in smoothed state/covariance estimates.

9.4 COMPUTATIONAL ASPECTS

The orbit determination module should include the features of data editing, which may be built into the OD program, or accomplished prior to the beginning of the filtering, this is for the post-flight OD. One can use the measurement residuals for online editing; the data is ignored if its residual is larger than some specified value, or if its residual deviates from the mean residual for that measurement type by more than some number of standard deviations, usually 3. If measurements are available more frequently than they can be processed, then prefiltering or data compression is required.

If a KF-type algorithm is to be used as a batch processor, then the observations are processed at the end of each satellite pass. If orbital estimates are required in real time, say for online corrective action, then a reasonably fast processor would be required. The entire computation for the estimation of states per measurement data should be done well within one sampling interval. If observations are available at a very fast rate and if all of them are to be considered, then the choice becomes very crucial. Computational requirements for implementation of the discrete Kalman filter are simpler than the continuous KF.

9.5 SEARCH AND RESCUE SATELLITE ORBIT ESTIMATION

This technology has been very important in areas where there is a risk to lives, e.g., coastal areas and polar regions, due to extreme weather conditions (heavy winds and snow). In Canada, due to vast and remote expanses and very bad weather conditions, the search and rescue satellite-aided tracking (SARSAT) technology has attained tremendous importance. The delay in the rescue operations should be avoided or be kept to a minimum in order to improve the efficiency of the SAR program, as well as to save the lives and property of the people living and/or operating in such conditions. The efficiency of the program depends, among other factors, on the accuracy of the position of the SARSAT satellite, since this satellite is used for receiving the distress signal from the platform or vehicle under distress.

We consider satellite OD as a nonlinear filtering problem and study the application of EKF, EUDF, and extended UD-RTS smoother (EUDRS). Here, we describe the OD process and present the results of numerical simulation of orbit estimation using EKF, extended UD factorization filter (EUDF), and EUDRS (it is UD-RTS smoother) and some sensitivity studies. In the case of EUDF and EUDRS, one uses the EKF approach of linearization and then the UD filter and UD-RTS smoother are used.

9.5.1 System Overview

The SARSAT system utilizes a near-Earth polar orbit satellite to support the operation of SAR forces, and the people involved in the rescue operation. The platform or the vehicle under an emergency condition activates a device called the ELT, an emergency locator transmitter, which in turn emits the ELT signals. These signals are received by the SARSAT satellite's instruments and then relayed in real time to a local user terminal (LUT); the satellite would receive such signals at least once during its 2 hours orbital time period over the polar region. At the LUT, the signal is detected and the frequency of this retransmitted signal is determined. The location of the ELT is then determined. The computation of the location of the position of the distressed vehicle uses information regarding the measured Doppler frequency and the satellite's orbital position. Based on the final information of the location of the ELT, a rescue mission is coordinated by the ground station/rescue center. Here, we consider the problem of estimation of orbital states of a (SARSAT) satellite. The errors in the position estimation will contribute to the error in the computation of the location of the ELT. The search times and location errors can be decreased if very accurate estimates of the position of the satellite and the Doppler frequency are obtained. The location estimates within 10 km were possible (by 1985) with the available ELTs. To keep such errors within acceptable bounds, a SARSAT satellite's orbital position errors must be always kept within 1 km during its pass.

9.5.2 System-Filter Configuration

A SARSAT system utilizes a near-Earth almost circular and polar orbit of 2 hours orbital period (one round about the Earth's polar region). The data are assumed to be available from a ground-based observation site of known coordinates. The observables used are azimuth, elevation, and range data chains that are available during a short pass of the satellite over the site. The four filter configurations were studied: (i) EKF with USM-U7 coordinate set, (ii) UDF with USM-U7 coordinate set, (iii) UDF with USM-U7 and IC6 coordinate sets, and (iv) UDRS with USM-U7 and IC6 coordinate sets. In the first two configurations, the seven-state U7 description of the USM is utilized for the measurement update and time propagation of the state estimates. In the case of the third and fourth configurations, the IC6 set is used for the implementation of the filter and U7 is used for the time propagation of the states.

9.5.3 Basic Simulation Studies

The various equations of these four system-filter configurations were coded in PASCAL (this structural programming language was popular at that time in 1985), and the implementation was done on a Cyber computer at McMaster University, Canada, in 1984–1985 (. Nowadays, such OD estimation programs can easily be coded in MATLAB for research purposes. The two aspects were studied: (i) the effect of the number of observables and (ii) the effect of a priori process noise statistics on the accuracy of the position and velocity estimates provided by these configurations. The SARSAT orbit for the study has the characteristics at epoch time: 1978-Day 246-0 hours., Universal time. The initial position and velocity were ~ 105 km and 0.05 km/s. The geopotential perturbations were included in all the system-filter configurations for the computer simulations and data generation. The other parameters were chosen as per Table 9.1.

The effect of using only angles (E_e, A_z) as observables and combining the range measurements with the angles can be observed from the plots of the rms position errors (averaged over ten simulation runs; ten realizations of the same dynamics with different seed numbers for random number generation) shown in Figures 9.1–9.4 for four system-filter configurations; it is seen that angle only gives poor

TABLE 9.1
Orbital Parameters

C6 Elements	Simulated Orbit	Estimator (Guess-Estimates Used)
a	7213 km.	7359.18 km.
e	0.01	0.078
i	98.9°	98.01°
Ω	269°	268.21°
ω	205°	338.07°
v	174°	40.95°
v	174°	40.95°

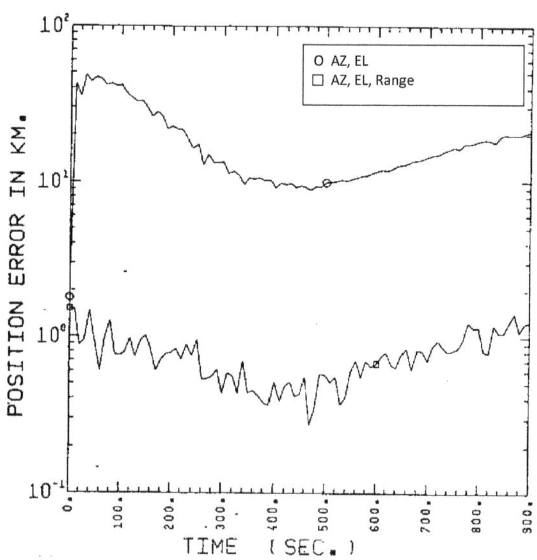

FIGURE 9.1 Position error—EKF (USM-U7): Effect of type of observables.

results compared to when these measurements are augmented with the range also. The improvement is by an order of magnitude in reduction of the errors; the combination set, U7 and IC6 yields better accuracy, hence subsequently three observables were used. In Figures 9.5 and 9.6, the position and velocity errors are compared for these four configurations. It is easily seen that the combination configurations yield better accuracy. Figures 9.7–9.14 show the effect of a priori process noise statistics on the position and velocity errors during the satellite's pass. In each case, the Q, the process noise covariance matrix was a diagonal matrix with entries related to velocity states (three in U7 and IC6 sets) as 10^{-8}. The entries related to position states (four in U7 and three in IC6 set) were varied by two orders of magnitude: PQ1 = $\{10^{-10}\}$, PQ2 = $\{10^{-8}\}$, and PQ3 $\{10^{-6}\}$. The values of the other statistics, initial values of P and UD covariance matrices and measurement noise covariance R, were selected

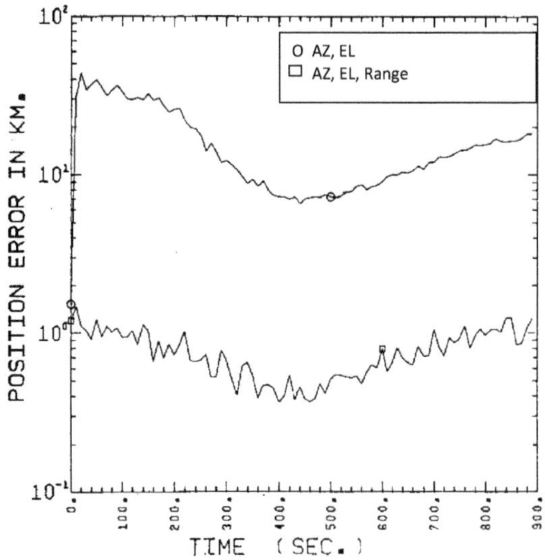

FIGURE 9.2 Position error—UDF (USM-U7): Effect of type of observables.

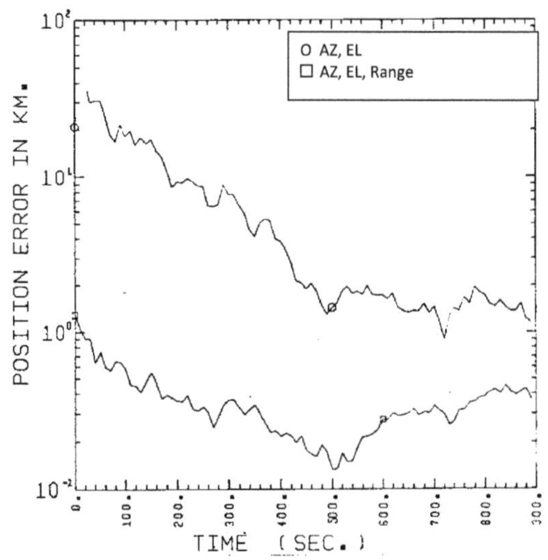

FIGURE 9.3 Position error—UDF (U7-IC6): Effect of type of observables.

after a few runs of these filters and then were kept unaltered for this study. It can be observed from these figures that the position and velocity errors are not affected by variation in Q. However, the velocity errors for EKF and UDF, in a single parameter set configuration, are affected by a marked difference. Interestingly, this is not the case for the combination set configurations, even for the position errors.

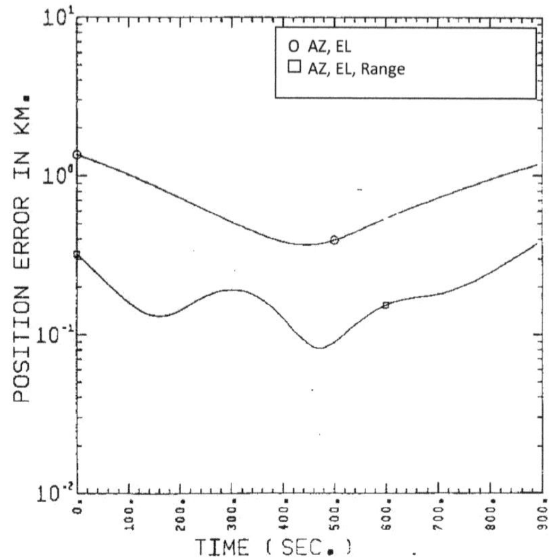

FIGURE 9.4 Position error—UDRS (U7-IC6): Effect of type of observables.

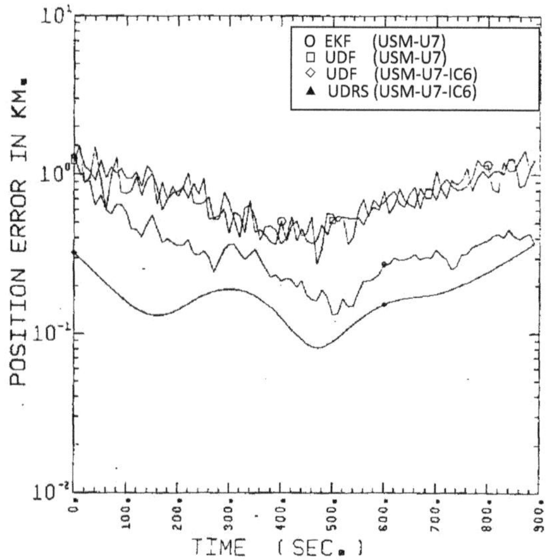

FIGURE 9.5 Four configurations compared—position error.

9.5.4 Study of Measurement Data Outliers

In many practical real-life systems, the measurements contain large (random) errors. Such errors might occur due to the failure of some sensors or the presence of large (random) disturbances. These errors are called outliers and adversely affect the

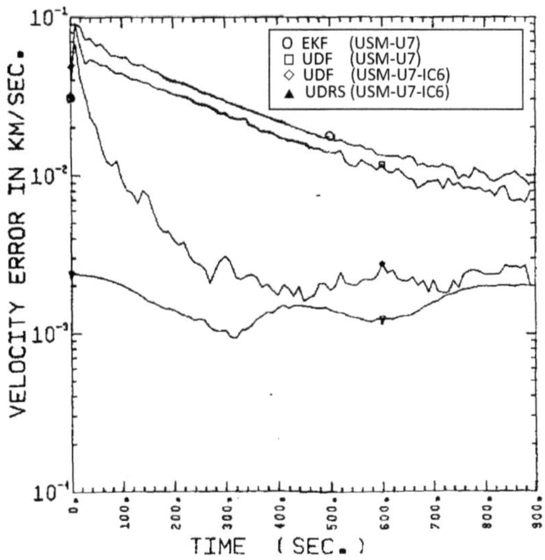

FIGURE 9.6 Four configurations compared—velocity error.

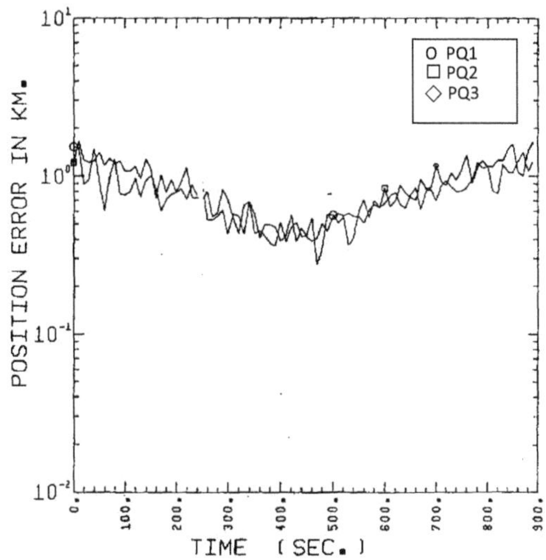

FIGURE 9.7 Position error—EKF (USM-U7): Effect of a priori Q-statistics.

performance of the filters. As a result, the accuracy of the position and velocity estimates will not meet the required specifications. Hence, it is necessary to use a procedure to minimize the adverse effect on the filter's performance. It is assumed that the measurement noise belongs to the (probability) distribution given as

$$P_\varepsilon = P_\varepsilon(v/0, R_1, R_2) = (1-\varepsilon)\mathbb{N}(v_1/0, R_1) + \varepsilon\mathbb{N}(v_2/0, R_2) \quad (9.42)$$

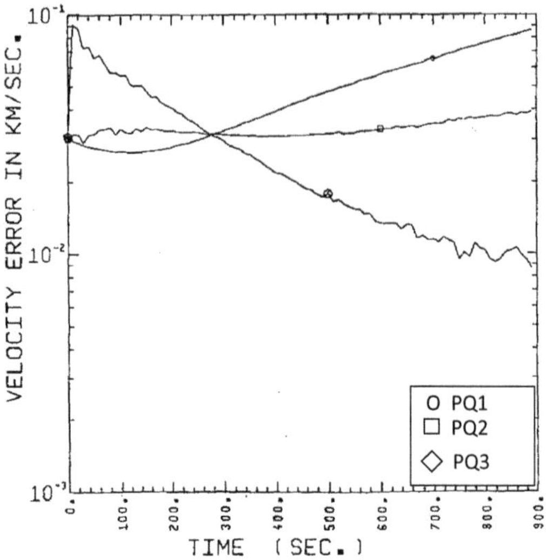

FIGURE 9.8 Velocity error—EKF (USM-U7): Effect of a priori Q-statistics.

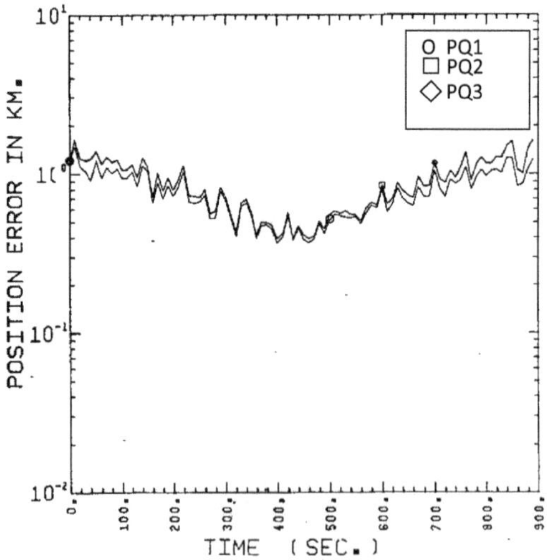

FIGURE 9.9 Position error—UDF (USM-U7): Effect of a priori Q-statistics.

Here, ℕ (./.,.) is the normal (Gaussian) density with zero mean and variance $R(.)$, and ε is the probability (numerical value) of the occurrence of a large error, the second term, $0 \le \varepsilon \le 1$. In expression (9.42), the occasional large errors are also assumed to be Gaussian but with large variance R_2. If the value of ε is known, then the scheme yields a robust estimation procedure. However, in the absence of the knowledge of ε, the following scheme can be used. Define the following:

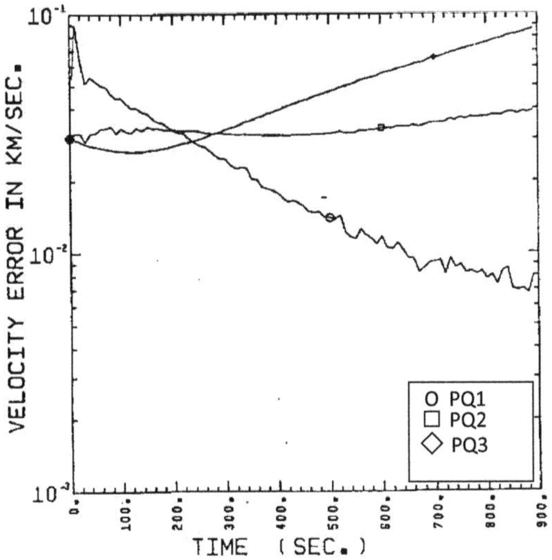

FIGURE 9.10 Velocity error—UDF (USM-U7): Effect of a priori Q-statistics.

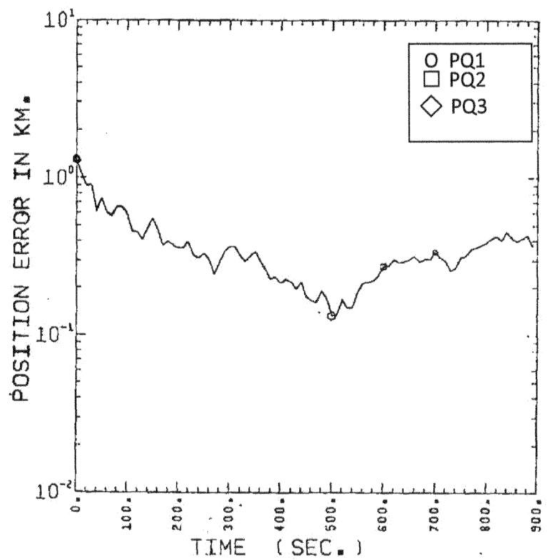

FIGURE 9.11 Position error—UDF (U7-IC6): Effect of a priori Q-statistics.

$$\begin{aligned} \delta(k) &= 0 \text{ for } e(k) \leq \text{Th} \\ &= 1 \text{ for } e(k) > \text{Th} \end{aligned} \tag{9.43}$$

Here, $e(k)$ is a scalar residual error and Th is some pre-assigned threshold/bound. With this, the modification of eqn. (9.42) is obtained as

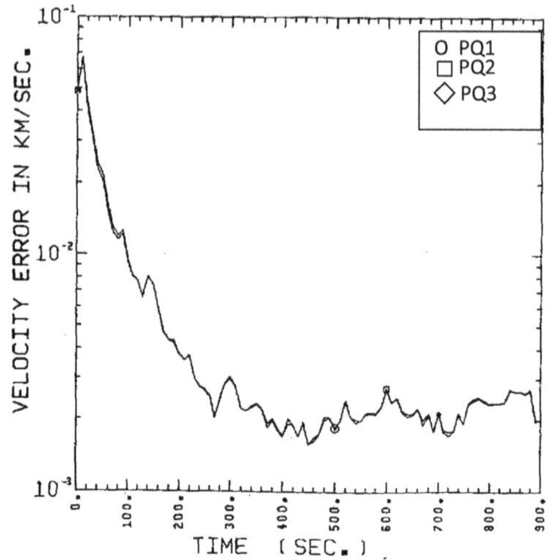

FIGURE 9.12 Velocity error—UDF (U7-IC6): Effect of a priori Q-statistics.

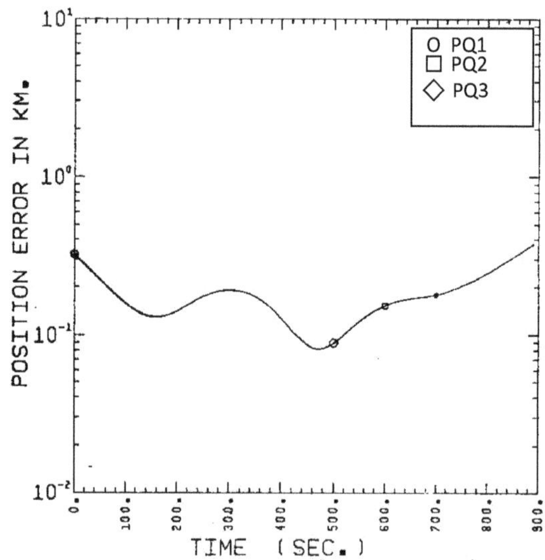

FIGURE 9.13 Position error—UDRS (U7-IC6): Effect of a priori Q-statistics

$$P_{\delta(k)} = P_{\delta(k)}\left(./ .,\sigma_1^2,\sigma_2^2\right) = \left(1-\delta(k)\right)\mathbb{N}\left(./ .,\sigma_1^2\right) + \delta(k)\mathbb{N}\left(./ .,\sigma_2^2\right) \quad (9.44)$$

This is the so-called mixed density function. This robustification scheme was used to handle measurement outliers in two of the system-filter configurations. The value of ε chosen was 0.2, implying that on average, 20 outliers were expected to occur in 100 measurement data (samples). The variance of the outlier errors was set to 400

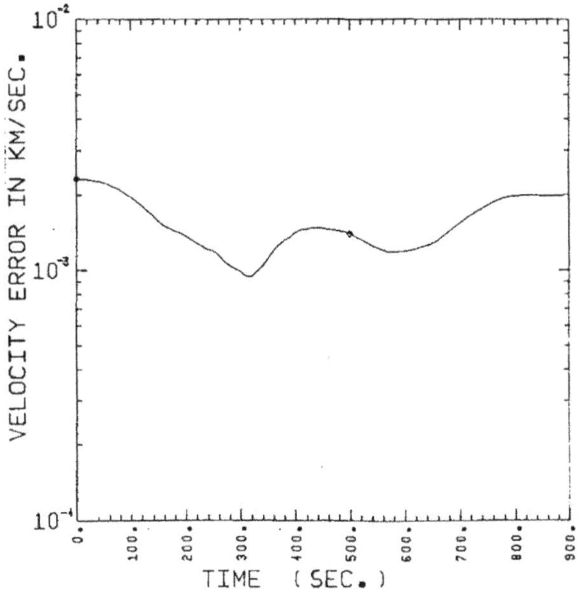

FIGURE 9.14 Velocity error—UDRS (U7-IC6): Effect of a priori Q-statistics.

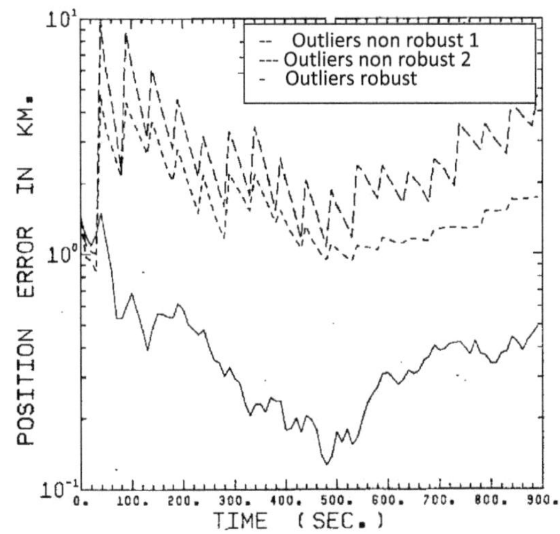

FIGURE 9.15 Position error—UDF (U7-IC6): Effect of outliers and robust scheme.

times the variance of the normal errors. The outliers were added to all the observables used for the estimation purpose. The results for the two combination set configurations are shown in Figures 9.15–9.18. When the outliers are present but the robustification scheme is not applied, the position and velocity errors in both the

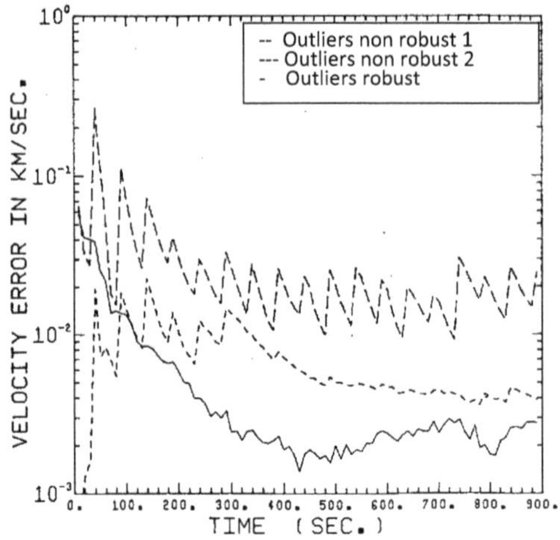

FIGURE 9.16 Velocity error—UDF (U7-IC6): Effect of outliers and robust scheme.

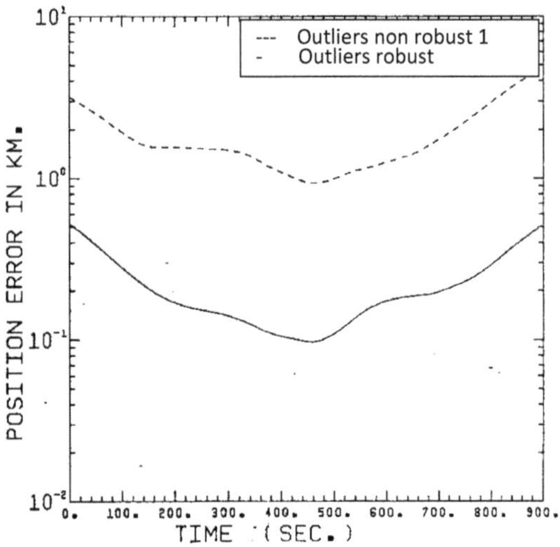

FIGURE 9.17 Position error—UDRS (U7-IC6): Effect of outliers and robust scheme.

cases are large. However, when the scheme is applied, the effect of outliers is either eliminated or drastically reduced. It is interesting to note that although the position and velocity errors exhibit smoothness in the case of UDRS (smoother) configuration (despite the presence of outliers with very large variance), they are quite large. Only when the robust scheme is utilized, that these errors are reduced considerably. Figures 9.19 and 9.20 show the range residuals for UDF and UDRS estimators with

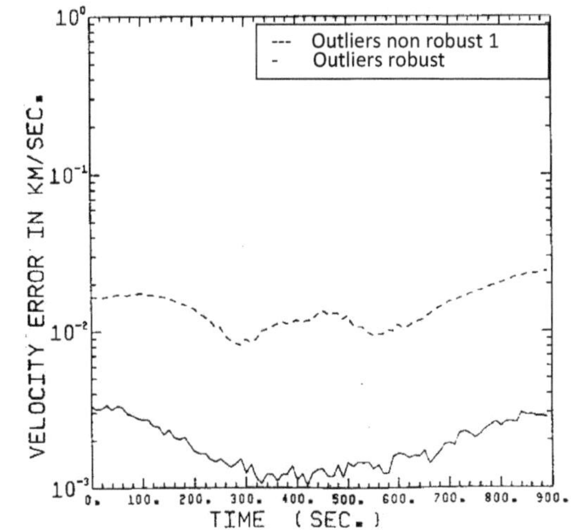

FIGURE 9.18 Velocity error—UDRS (U7-IC6): Effect of outliers and robust scheme.

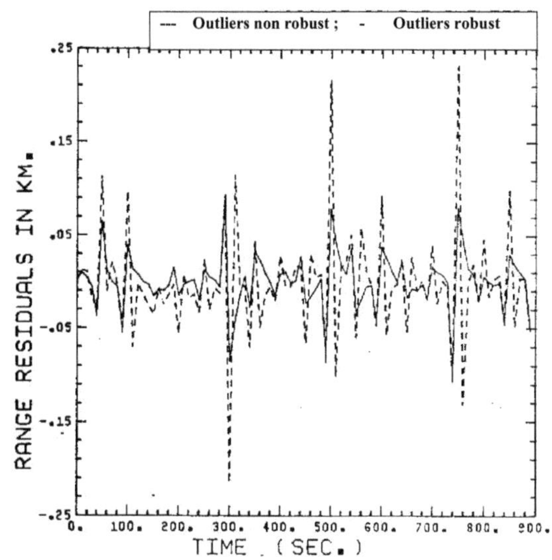

FIGURE 9.19 Effect of outliers and robust scheme on range residuals—UDF (U7-IC6).

outliers present in the data. The effects of the outliers and the robustness scheme are very clear. The results of this simulation exercise show that the EKF, UDF, and UDRS filtering/smoothing algorithms produce an acceptable performance of these OD programs.

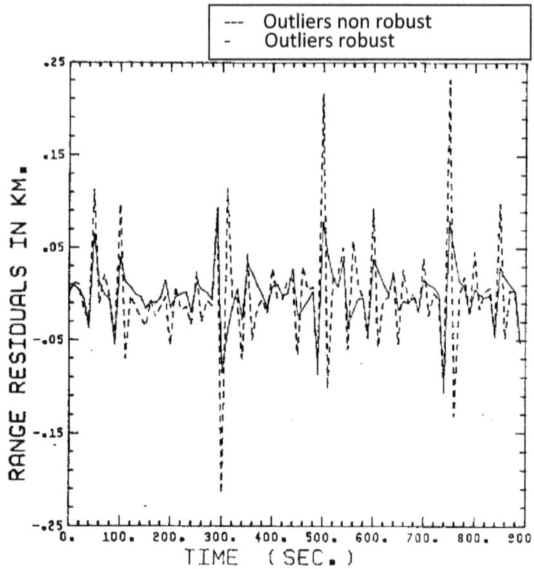

FIGURE 9.20 Effect of outliers and robust scheme on range residuals—UDRS (U7-IC6).

EXERCISES

9.1 What are the merits and demerits of the type of the three orbital coordinates sets?

9.2 What are the orbital disturbances?

9.3 Why the combination of two coordinates sets is used?

9.4 Why the Canadian SARSAT satellites have polar orbits?

9.5 What is the accuracy requirement of SARSAT orbit?

REFERENCES

1. Raol, J. R. *Stochastic State Estimation with Application to Satellite Orbit Determination.* PhD Thesis, Department of Electrical and Computer Engineering, McMaster University, Hamilton, Ont. Canada, 1986.
2. Raol, J. R. and Sinha, N. K. On the orbit determination problem. *IEEE Trans. Aerosp. Electron. Syst.*, 21(3), 274–291, 1985.
3. Altman, S. P. A unified state model of orbital trajectory and attitude dynamics. *Celest. Mech.*, 6(4), 425–446, 1972.

Appendix A
Modelling Based on Soft Computing

A.1 INTRODUCTION

In this appendix, we briefly discuss three main ingredients of soft computing: (i) artificial neural networks (ANNs), (ii) fuzzy logic (FL), and (iii) genetic algorithms (GAs) from the point of view of mathematical modelling [1]. Applications of these to modelling, parameter estimation, control, and sensor data fusion can be found in refs. [1–4]; following the illustrative examples discussed in these references, one can easily extend the applications of ANNs, FL, and GAs to the problems discussed in Chapters 2–9 of the present book.

A.2 ANN FOR MODELLING

An ANN possesses good ability to learn adaptively. One of the most successful ANNs is the so-called feedforward neural network (FFNN) which has found many applications in the pattern recognition, nonlinear curve fitting/mapping, flight data analysis, aircraft modelling, adaptive control, and system identification. An illustration and comparison of biological neuron (BNN) and artificial neuron are given in Table A.1.

The FFNN is an information processing system of a large number of simple processing elements (Figure A.1). These elements are called artificial neurons or nodes. These neurons are interconnected by links, which are represented by the so-called

TABLE A.1
Neural Systems: BNN and ANN

Biological Neuron (BNN, Human Brain)	Artificial Neuron (ANN)
Signals are received by dendrites and passed on to neuron receptive surfaces	Data enter through the input layer
Inputs are fed to the neurons through specialized contacts called synapse	Weights provide the connection between the nodes in the input and output layers
All logical functions of neurons are accomplished in soma	Nonlinear activation function operates upon the summation of the product of weights and inputs $f(\Sigma W x_i)$
Output signal is delivered by the axon nerve fibre	The output layer produces the network's predicted response

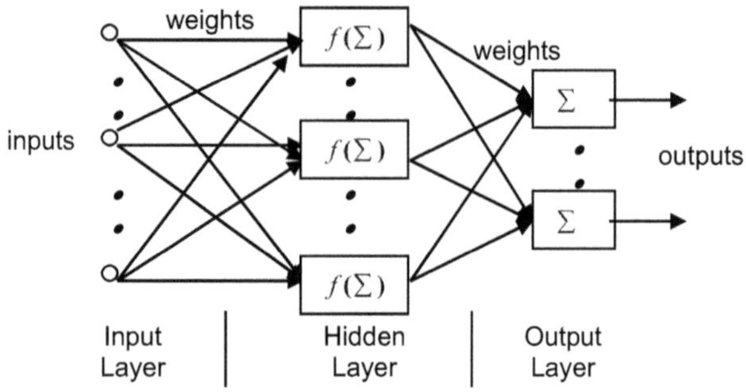

FIGURE A.1 FFNN structure.

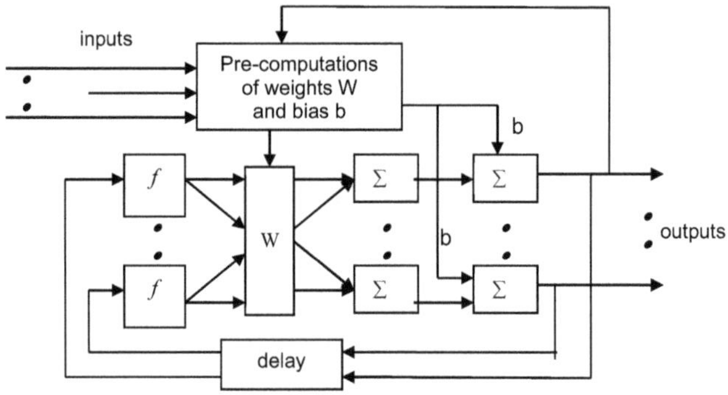

FIGURE A.2 Block schematic of RNN with one hidden layer.

weights, and they cooperate to perform parallel distributed computing in order to carry out a desired computational task (Figure A.2).

ANNs are used for I/O modelling because the basic neural network functions can adequately approximate the system behaviour in an overall sense. The FFNN can be thought of as nonlinear black box model structure, the parameters (weights/coefficients) of which can be estimated by conventional optimization methods. These are more suitable for system identification, time-series modelling & prediction, pattern recognition/classification, sensor failure detection, estimation of aerodynamic coefficients, and parameter estimation of a dynamic system. In this case, an FFNN is used for predicting the time histories of aerodynamic coefficients, and then, some regression method is used to estimate the aerodynamic parameters (the aerodynamic stability and control derivatives) from the predicted aerodynamic coefficients.

A.2.1 Feedforward Neural Networks

The FFNNs have non-cyclic and layered topology, very useful for nonlinear mapping between I/O of a system. The chosen network is first trained using the training set data (obtained from a real-life system), and then, it is used for prediction using a different input set, which belongs to the same class of the data; the validation set. The weights are determined using the so-called back-propagation/gradient-based procedure. Because of the layered disposition of weights of the FFNN, the estimation of the weights requires propagation of the error of the output layer in backward direction and hence the name back-propagation. The estimation algorithms are described using the matrix/vector notation for the sake of clarity and ease of implementation in PC-MATLAB.

The FFNN has the following variables: (i) u_0 = input to (input layer of) the network; (ii) n_i = number of input neurons = number of inputs u_0; (iii) n_h = number of neurons of the hidden layer; (iv) n_o = number of output neurons = number of outputs z; (v) $W_1 = n_h$ x n_i weight matrix between the input and hidden layers; (vi) $W_{10} = n_h$ x 1, bias weight vector; (vii) $W_2 = n_o$ x n_h weight matrix between the hidden and output layers; (viii) $W_{20} = n_o$ x 1, bias weight vector; and (ix) μ = learning rate or step size.

A.2.1.1 Back-Propagation Algorithm for Training

The algorithm is based on steepest descent (SD) optimization method. The forward pass signal computation is done using the following sets of equations, since u_0 is known and initial guesstimates of the weights are known.

$$y_1 = W_1 u_0 + W_{10} \tag{A.1}$$

$$u_1 = f(y_1) \tag{A.2}$$

Here,y_1 is a vector of intermediate values, and u_1 is the input to the hidden layer. The function $f(y_1)$ is a nonlinear sigmoidal activation function given by

$$f(y_i) = \frac{1 - e^{-\lambda y_i}}{1 + e^{-\lambda y_i}} \tag{A.3}$$

Next, the signal between the hidden and output layers is computed:

$$y_2 = W_2 u_1 + W_{20} \tag{A.4}$$

$$u_2 = f(y_2) \tag{A.5}$$

Here, u_2 is the signal at the output layer. The learning rule is derived next. From the optimal control theory, we have

$$\frac{dW}{dt} = -\mu(t)\frac{\partial E(W)}{\partial W} \tag{A.6}$$

With the output error defined as $e = z - u_2$, and a suitable quadratic cost function based on it, the expression for the gradient is obtained as

$$\frac{\partial E}{\partial W_2} = -f'(y_2)(z - u_2)u_1^T \tag{A.7}$$

Here, u_1 is the gradient of y_2 with respect to W_2. The derivative f' of the node activation function f is given by

$$f'(y_i) = \frac{2\lambda_i e^{-\lambda y_i}}{\left(1+e^{-\lambda y_i}\right)^2} \tag{A.8}$$

The modified error of the output layer can be expressed as

$$e_{2b} = f'(y_2)(z - u_2) \tag{A.9}$$

Thus, the recursive weight update rule for the output layer is given as

$$W_2(i+1) = W_2(i) + \mu e_{2b} u_1^T + \Omega[W_2(i) - W_2(i-1)] \tag{A.10}$$

The Ω is the momentum constant and is used to smooth out the weight changes and to accelerate the convergence of the algorithm. The back-propagation of the error and the weight update rule for W_1 are given as

$$e_{1b} = f'(y_1)W_2^T e_{2b} \tag{A.11}$$

$$W_1(i+1) = W_1(i) + \mu\, e_{1b} u_0^T + \Omega[W_1(i) - W_1(i-1)] \tag{A.12}$$

The data are presented to the network in a sequential manner; the same data are presented again but with initial weights as the outputs from the previous cycle. This process is stopped when the convergence is reached.

A.2.2 Parameter Estimation using FFNN

The parameter estimation procedure using FFNN has two steps: (i) the network is given the measured data and is trained to reproduce the predicted responses which are compared with the system responses in the sense of minimization of the output error, and (ii) these predicted responses are perturbed in turn for each parameter to be estimated and the changed predicted response is obtained.

Assume that $z = \beta x(+\text{noise})$, and the network is trained to produce clean z. The trained network is used to produce $z + \Delta z$ and $z - \Delta z$ when x is changed to $x + \Delta x$ and $x - \Delta x$. Then, β is obtained as $\beta = \frac{z^+ - z^-}{x^+ - x^-}$, and this method is called the Delta method. Since the variables are the signals, the parameter time histories are obtained, and hence, the estimates are obtained by averaging these respective parameter time histories.

A.2.3 Recurrent Neural Networks

The recurrent neural networks (RNNs) are dynamic neural networks, and hence amenable to explicit parameter estimation in state-space models.

A.2.3.1 Parameter Estimation with Hopfield RNN

Consider the dynamic system

$$\dot{x} = Ax + Bu \tag{A.11}$$

For parameter estimation, $\beta = \{A, B\}$ represents the parameter vector to be estimated, and n is the number of parameters to be estimated. Let the cost function be given as

$$E(\beta) = 1/2\sum_{k=1}^{N} e^T(k)\; e(k) = 1/2\sum_{k=1}^{N} (\dot{x} - Ax - Bu)^T (\dot{x} - Ax - Bu) \tag{A.12}$$

Here, *e(k)* is the equation error:

$$e = \dot{x} - Ax - Bu \tag{A.13}$$

From the optimization theory, we have

$$\frac{d\beta}{dt} = -\frac{\partial E(\beta)}{\partial \beta} = -\frac{1}{2}\frac{\partial\left\{\sum_{k=1}^{N} e^T(k)\; e(k)\right\}}{\partial \beta} \tag{A.14}$$

Since β as a parameter vector contains the elements of A and B, we can obtain expressions $\partial E/\partial A$ and $\partial E/\partial B$ for A and B vectors, with $\sum(.) = \sum_{k=1}^{N}(.)$

$$\begin{aligned} \frac{\partial E}{\partial A} &= \sum(\dot{x} - Ax - Bu)\,(-x^T) = A\sum x\,x^T + B\sum u\,x^T - \sum \dot{x}\,x^T \\ \frac{\partial E}{\partial B} &= \sum(\dot{x} - Ax - Bu)\,(-u) = A\sum x\,u + B\sum u^2 - \sum \dot{x}\,u \end{aligned} \tag{A.15}$$

Expanding, we get, for $A(2,2)$ and $B(2,1)$,

$$\begin{bmatrix} \frac{\partial E}{\partial a_{11}} & \frac{\partial E}{\partial a_{12}} \\ \frac{\partial E}{\partial a_{21}} & \frac{\partial E}{\partial a_{22}} \end{bmatrix} = \begin{bmatrix} a_{11} & a_{12} \\ a_{21} & a_{22} \end{bmatrix}\begin{bmatrix} \sum x_1^2 & \sum x_1 x_2 \\ \sum x_2 x_1 & \sum x_2^2 \end{bmatrix} + \begin{bmatrix} b_1 \\ b_2 \end{bmatrix}\left[\sum u x_1\ \sum u x_2\right] - \begin{bmatrix} \sum \dot{x}_1 x_1 & \sum \dot{x}_1 x_2 \\ \sum \dot{x}_2 x_1 & \sum \dot{x}_2 x_2 \end{bmatrix} \tag{A.16}$$

Simplifying, we get

$$\frac{\partial E}{\partial a_{11}} = a_{11}\sum x_1^2 + a_{12}\sum x_2 x_1 + b_1\sum x_1 u - \sum \dot{x}_1 x_1$$

$$\frac{\partial E}{\partial a_{12}} = a_{11}\sum x_1 x_2 + a_{12}\sum x_2^2 + b_1\sum u x_2 - \sum \dot{x}_1 x_2 \quad \text{(A.17)}$$

$$\frac{\partial E}{\partial a_{21}} = a_{21}\sum x_1^2 + a_{22}\sum x_2 x_1 + b_2\sum u x_1 - \sum \dot{x}_2 x_1$$

$$\frac{\partial E}{\partial a_{22}} = a_{21}\sum x_1 x_2 + a_{22}\sum x_2^2 + b_2\sum u x_2 - \sum \dot{x}_2 x_2$$

In addition, we have

$$\frac{\partial E}{\partial b_1} = a_{11}\Sigma x_1 u + a_{12}\Sigma x_2 u + b\Sigma u^2 - \Sigma \dot{x}_1 u$$

$$\frac{\partial E}{\partial b_2} = a_{21}\Sigma x_1 u + a_{22}\Sigma x_2 u + b_2\Sigma u^2 - \Sigma \dot{x}_2 u \quad \text{(A.18)}$$

Next, assuming that the impedance R is very high, we describe the dynamics of RNNs as

$$\dot{x}_i = \sum_{j=1}^{n} w_{ij}\beta_j + b_i \quad \text{(A.19)}$$

We also have $E = -1/2 \sum_i \sum_j W_{ij}\beta_i\beta_j - \sum_i b_i\beta_i$ as the energy landscape of the recurrent neural network. Then, we get

$$\frac{\partial E}{\partial \beta_i} = -\sum_{j=1}^{n} w_{ij}\beta_j - b_i \quad \text{(A.20)}$$

or

$$\frac{\partial E}{\partial \beta_i} = -\left[\sum_{j=1}^{n} w_{ij}\beta_j + b_i\right] = -\dot{x}_i \quad \text{(A.21)}$$

or

$$\dot{x}_i = -\frac{\partial E}{\partial \beta_i}$$

Since

$$\beta_i = f(x_i),\ \dot{x}_i = \left(f^{-1}\right)' \dot{\beta}_i \tag{A.22}$$

Thus, $\left(f^{-1}\right) \dot{\beta}_i = -\dfrac{\partial E}{\partial \beta_i}$

Here, $'$ denotes derivative w.r.t β.
Hence,

$$\dot{\beta}_i = -\frac{1}{\left(f^{-1}\right)'(\beta_i)} \frac{\partial E}{\partial \beta_i} = \frac{1}{\left(f^{-1}\right)'(\beta_i)} \left[\sum_{j=1}^{n} w_{ij}\beta_j + b_i\right] \tag{A.23}$$

We get expressions for the weight matrix W and the bias vector b as

$$W = -\begin{bmatrix} \sum x_1^2 & \sum x_2x_1 & 0 & 0 & \sum ux_1 & 0 \\ \sum x_1x_2 & \sum x_2^2 & 0 & 0 & \sum ux_2 & 0 \\ 0 & 0 & \sum x_1^2 & \sum x_2x_1 & 0 & \sum ux_1 \\ 0 & 0 & \sum x_1x_2 & \sum x_2^2 & 0 & \sum ux_2 \\ \sum x_1u & \sum x_2u & 0 & 0 & \sum u^2 & 0 \\ 0 & 0 & \sum x_1u & \sum x_2u & 0 & \sum u^2 \end{bmatrix} \tag{A.24}$$

$$b = -\begin{bmatrix} \sum \dot{x}_1x_1 \\ \sum \dot{x}_1x_2 \\ \sum \dot{x}_2x_1 \\ \sum \dot{x}_2x_2 \\ \sum \dot{x}_1u \\ \sum \dot{x}_2u \end{bmatrix} \tag{A.25}$$

Thus, the algorithm for parameter estimation of the dynamic system can be given as

i. compute W matrix, since the measurements of x, $\dot{x}$, and u are available (equation error formulation) for a certain time interval T,
ii. compute bias vector in the similar way,
iii. choose the initial values of β_i randomly,
iv. then solve the following differential equation.

Since $\beta_i = f(x_i)$ and since the sigmoid nonlinearity is a known function f, by differentiating and simplifying, we get

$$\frac{d\beta_i}{dt} = \frac{\lambda\,(\rho^2 - \beta_i^2)}{2\rho}\left[\sum_{j=1}^{n} w_{ij}\beta_j + b_i\right] \tag{A.26}$$

Here,

$$f(x_i) = \rho\left(\frac{1-e^{-\lambda x_i}}{1+e^{-\lambda x_i}}\right) \tag{A.27}$$

Integration of eqn. (A.26) yields the solution to parameter estimation problem. For good convergence of the estimates to the true parameters, proper tuning of λ and ρ is very essential. Often, λ is chosen small, i.e. less than 1.0. The ρ is chosen such that when x_i (of recurrent neural network) approaches $\pm\infty$, the function f approaches $\pm\rho$. The eqn. (A.26) can be discretized to obtain the estimates by recursion:

$$\beta_i(k+1) = \beta_i(k) + \frac{\lambda(\rho^2 - \beta_i^2(k))}{2\rho}\left[\sum_{j=1}^{n} w_{ij}\beta_j(k) + b_j\right] \tag{A.28}$$

The Δt can be absorbed in the constants of the 2nd term of eqn. (A.28).

A.3 GENETIC ALGORITHM–BASED MODELLING

GAs are search methods inspired by nature's evolutionary systems. They can be used to obtain global and robust solutions to many optimization problems in science, engineering, economics, psychology, and biology. The natural systems have evolved over millions of years. They have gone through iterations over many generations and in the process have become very robust, especially to their many different environments. Due to its strong evolutionary 'experience', the nature offers good solutions whenever robustness is called for. Biological systems are generally more robust, efficient, and flexible compared to the most sophisticated artificial systems. Artificial systems have to learn from biological systems to improve their performance and carry out their daily required functions for longer period of time and with greater efficiency.

TABLE A.2
GA and Natural Genetic System

Natural Genetic System	Genetic Algorithm
Chromosomes	String of numbers
Gene	Feature or detection
Allele	Feature value
Locus	String position
Genotype	Structure
Phenotype	Parameter set, alternative form, a decoded structure

GAs are based on some of the principles that govern the natural systems. GAs are computational optimization schemes with an approach that seems rather unconventional. The algorithms solve optimization problems imitating the nature in the way it has been working for millions of years on the evolution of life-forms. Inspired by the biological systems, GAs adopt the rules of natural selection and genetics to attain robustness. Acting on the premise of survival of the fittest, a population or samples of feasible solutions are combined in a manner similar to the combination of chromosomes in a natural genetic system. The fitter population members pass on their structures as genes in far greater measure than their less fit members do. As the generations evolve, the net effect is evolution of the population towards an optimum (species, solution, etc.). GAs operate by combining the information present in different possible solutions in a way that a better solution is obtained in next/future generations. The terms used in the study of GAs are given in Table A.2.

A.3.1 Operations in a Typical Genetic Algorithm

A.3.1.1 Chromosomes

They represent encoding of information in a string of finite length, and each chromosome consists of a string of bits (binary digit, 0 or 1). Or it could be a symbol from a set of more than two elements. Generally, for function optimization, chromosomes are constructed from binary strings as seen from the following table:

Parameter value	String
6	000110
34	100010

The long stretches of DNA that carry the genetic information needed to build an organism are called chromosomes. The chromosomes consist of genes. Each gene represents a unit of information, and it takes different values. These values are called alleles at different locations called loci. The strings, composed of features or detectors, assume values like 0 or 1, which are located at different positions in the string. The total package or system is called the genotype or structure. The phenotype results when interaction of genotype with environment takes place.

A.3.1.2 Population and Fitness

GAs operate on population of possible solutions with chromosomes. The population members are known as individuals. Each individual is assigned a fitness value based on the objective function or cost function. Better individuals (solutions) have higher fitness values, and weaker ones have lower fitness values.

A.3.1.3 Initialization and Reproduction

By randomly selecting information from the search space and encoding it, a population of possible initial solutions is created. Reproduction is a process in which individual strings are copied as per their fitness values. Thus, the strings with a greater fitness value have higher probability of contributing one or more offsprings to the next generation.

A.3.1.4 Crossover

In a crossover, a site is selected randomly along the length of the chromosomes, and each chromosome is split into two pieces at the crossover site. The new ones are formed by joining the top piece of one chromosome with the tailpiece of the other.

A.3.1.5 Mutation

It is a small operation in which a bit in a string is changed at a random location. The main idea is to break monotony and add a bit of novelty. This operation would help gain information not available to the rest of the population. It lends diversity to the population.

A.3.1.6 Generation

Each iteration in the optimization procedure is called a generation. In each generation, pairs are chosen for crossover operation, the fitness is determined, and mutation is carried out during crossover operation (during or after has a subtle distinction). With these operations performed, a new population evolves which is carried forward.

A.3.1.7 Survival of the Fittest

The individuals may be fitter or weaker than some other population members. So the members must be ranked as per their fitness value. In each generation, the weaker members are allowed to wither out and the ones with good fitness values take part in the genetic operations. The net result is the evolution of the population towards the global optimum.

A.3.1.8 Cost Function, Decision Variables, and Search Space

In most practical optimization problems, the goal is to find optimal parameters to increase the production and/or to reduce the expenditure/loss, i.e. to get maximum profit by reorganizing the system and its parameters that affect the cost function. Since this, in effect, reflects on the cost, it is represented by the cost function. A carefully devised and convergent computational algorithm would eventually find an optimum solution to the problem. The parameters of the system that decide the cost are termed decision variables. The search space is a Euclidean space in which parameters take different values, and each point in the space is a probable solution.

A.3.2 Stopping Strategies for Genetic Algorithms

One needs to know as to where and when to stop the GA iterations. If the population size is fixed, then more generations might be needed for the convergence of a GA to an optimal solution. One way is to track the fitness value for no further improvement. As the algorithmic steps progress, a situation would occur wherein we need a large number of generations to bring about a small improvement in the fitness value. One can define a predetermined number of generation/iterations to solve the problem. Also, insignificant change in the norm of estimated parameters can be tracked for a few consecutive iterations before stopping the search. It must be possible to do an effective search if one exploits some important similarities in the coding used in GAs. Another way is to evaluate gradient of the cost function and use the conventional approaches for assessing the quality of the estimates for their convergence to true values. An increase in the number of samples would generally increase the success rate.

A.3.3 Genetic Algorithms without Coding of Parameters

GAs become more complex because of coding the chromosomes, especially for more complex problems. In the problems of science and engineering, we come across real numbers. Thus, we need to use real numbers and still use GAs on these numbers for solving optimization problems. A major change is in the crossover and mutation operations. Averaging the two samples, i.e. the two sets of parameter values, can perform the crossover operation. After the crossover, the best individual is mutated. In mutation, a small noise is added. Assume that two individuals have β_1 and β_2 as numerical values of the parameters. Then after crossover, we obtain the new individual as $(\beta_1 + \beta_2)/2$. For mutation, we have $\beta_3 = \beta_1 + \varepsilon^* v$, where d is a constant, and v is a number chosen randomly between −1 and 1. Thus, all the GA operations can be performed by using real numbers like 4.8904, etc., without coding the samples. This feature is extremely well suited for several engineering applications: parameter estimation, control, optimization, and signal processing.

A.3.4 Parameter Estimation using Genetic Algorithm

As we have seen in previous chapters, most of the parameter estimation methods are based on the minimization of the cost function resulting in utilization of the gradient of the cost function. The application of GA to parameter estimation problem does not need utilization of the gradient of the cost function.

Consider the problem of parameter estimation as follows:

$$z = H\beta + v;\ \hat{z} = H\hat{\beta} \tag{A.29}$$

The cost function is formulated as

$$E = \frac{1}{2}\sum (z - \hat{z})^T (z - \hat{z}) = \frac{1}{2}\sum \left(z - H\hat{\beta}\right)^T \left(z - H\hat{\beta}\right) \tag{A.30}$$

Then, procedure outlined in A.3.2 and A.3.3 is followed.

A.4 FUZZY LOGIC–BASED MODELLING

A conventional control/system + logic can be just regarded as intelligent control/system: If say $y >$ some number, then use one controller else use other controller. A FL-based controller is suited to keep (i) output variables between the specified limits and (ii) control actuation (i.e. control inputs or related variables) between the limits. FL is one form of logic that can be used in intelligent control; to avoid contradictions in rules, human intervention is needed at every stage to tune various adjustable parameters. FL is used: (i) when system dynamics are slow and/or nonlinear, (ii) when models are not available, and (iii) when competent human operators are available; this signifies that FL can be used as an alternative modelling type compared to the traditional methods of mathematical modelling. However, it is still feasible to use FL in the conventional procedure for parameter estimation.

A.4.1 Fuzzy Logic

A comparison of the classical crisp set and the fuzzy set is shown in Figure A.3 (Figure A.4).

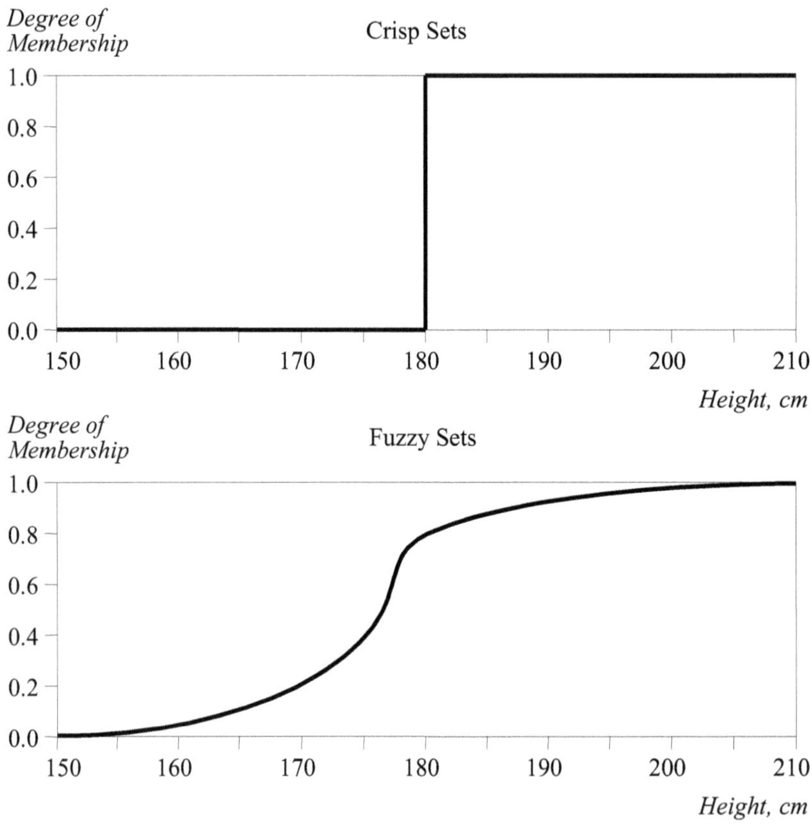

FIGURE A.3 Comparison of crisp and fuzzy membership functions/sets.

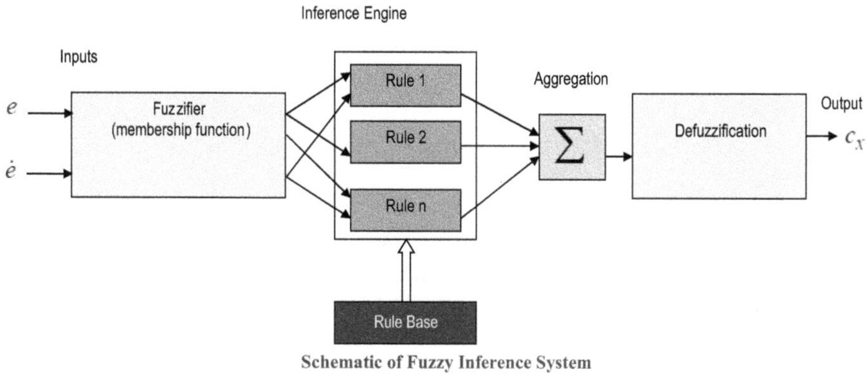

FIGURE A.4 FL inference engine/system.

The x-axis represents the universe of discourse (UOD) – the range of all possible values applicable to a chosen variable. In the present case, the variable is the man's height. According to this representation, the universe of men's heights consists of all tall men. The y-axis represents the membership value of the fuzzy set; the fuzzy set of 'tall men' maps height values into corresponding membership values.

For modelling and control purposes, a fuzzy inference system (FIS) is used.

FL is a rule-based logic. A fuzzy variable is one whose values can be considered labels of fuzzy sets: (i) temperature → fuzzy variable → linguistic values such as low, medium, normal, high, and very high; (ii) membership values (on the UOD – degree C); and (iii) the dependence of a linguistic variable on another which can be described by means of a fuzzy conditional statement; these statements are called the rules/rule base:

R: If s1 (is true), then S2 (is true) Or S1 → S2

More specifically,

i. If the load is small, then torque is very high
ii. If the error is negative large, then output is –ve large.

Composite conditional statement/rule:

1. R1: if S1, then (if S2 then S3) is equivalent to:
2. R1: if S1, then R2 AND
3. R2: if S2, then S3

No. of rules could be as large as 30; for complex process control plant, the rules could be 60–80 rules; for small task, 5–10 rules might be sufficient like washing m/c.

A4.2 Fuzzy Identification/Parameter Estimation Process

Steps are as follows: (i) I/O data collection, (ii) model structure selection (confined to the choice of the system's dynamics order and complexity of the fuzzy model),

(iii) model estimation, and (iv) model validation. Selection of system dynamics can be using nonlinear autoregressive with exogenous input (NARX), where the relation between past I/O data and predicted out has the form:

$$y(k+1) = F(y(k),..., y(k-n_y+1), u(k),..., u(k-n_u+1)) \tag{A.31}$$

where n_y & n_u are related to the system's order. Complexity of the fuzzy model is mainly connected to the number of fuzzy rules that are used to define the model itself. The number of fuzzy rules is essentially related to the choice of the number of clusters that are used to partitioned I/O data. Data clustering is a technique to process and classify a set of data according to similarity by minimizing the following objective function:

$$J(X;U;V) = \sum_{i=1}^{c}\sum_{k=1}^{n}(\mu_{ik})^m d^2(v_i, x_k) \tag{A.32}$$

Here, c is the number of fuzzy clusters, and $X = [x_1,...,x_n] \in R^{pxn}$ is the matrix containing data set composed of n observations with each consisting of p measured variables. $U = [\mu_{ik}]$ is the fuzzy partition matrix, where $\mu_{ik} \in [0,1]$ is the membership function of datum x_k to cluster i, and $d(v_i, v_k)$ is the distance. The parameter $m > 1$ is the fuzziness index: when m -> 1, the cluster tends to be crisp, whereas when m -> infinity, the partition becomes maximum fuzzy and $\mu_{ik} - > 1/c$.

Subsequently, a rule base is formulated for parameter estimation of the system. Also, adaptive neural network fuzzy inference system (ANFIS) can be used for parameter estimation using the procedure outlined in Section A.2.2 for FFNN.

REFERENCES

1. Raol, J. R., Girija, G., and Singh J. *Modelling and Parameter Estimation of Dynamic Systems.* IET/IEE Control Series Book, Vol. 65, IEEE, London, UK, 2004.
2. Raol, J. R. and Singh J. *Flight Mechanics Modelling and Analysis.* Revised edition. CRC Press, Boca Raton, FL, 2023.
3. Raol, J. R., Girija, G., and Twala, B. *Nonlinear Filtering: Concepts and Engineering Applications.* CRC Press, Boca Raton, FL, 2017.
4. Raol, J. R. *Data Fusion Mathematics-Theory and Practice.* CRC Press, Boca Raton, FL, 2016.

Appendix B
Obstacle Avoidance in Robot Path Planning

Sudesh K. Kashyap and Jitendra R. Raol

B.1 INTRODUCTION

A quadrotor is a type of unmanned/uninhibited aerial vehicle (UAV) that is controlled through its four motors. It can perform roll, pitch, and yaw motions by varying the motor speed. Quadrotors are popular test beds for UAV's experiment because of their versatility [1–5]. The real-life applications of quadrotors are search and rescue (SAR) operation, monitoring of crops, surveillance, and military. In such cases, a quadrotor may encounter collisions with its neighbours or obstacles in its path. To prevent this occurrence, the quadrotor must have an obstacle avoidance controller that will avoid certain obstacles.

Artificial potential field/force (APF)-based algorithm is one of the most popular algorithms when it comes to mobile robot path planning [1]. In this method, the robot is assumed to experience artificial potential force from its surroundings. These potential forces are attractive force and repulsive force. Through this, the robot is expected to avoid obstacles based on the total potential forces.

B.2 ARTIFICIAL POTENTIAL FIELD

In an APF approach, the position to be reached would induce attractive forces and the obstacles from the surroundings will introduce repulsive forces [1]. APF consists of creation of potential fields in which goal has minimum potential field; the obstacles are assumed to have high potential. Thus, the vehicle robot will be attracted to the target and at the same time repelled by the obstacles. Figure B.1 shows the target point (small circle) and obstacle (big circle); all points that are far from the target have higher force, while points that are near the target have lower force, denoted by the length of the arrows [1]. On the other hand, repulsion is high when the robot is near the obstacle and low when the robot is far from the obstacle. As the robot approaches to the obstacle, it will experience high repulsive force so the robot will move towards low potential which is the target.

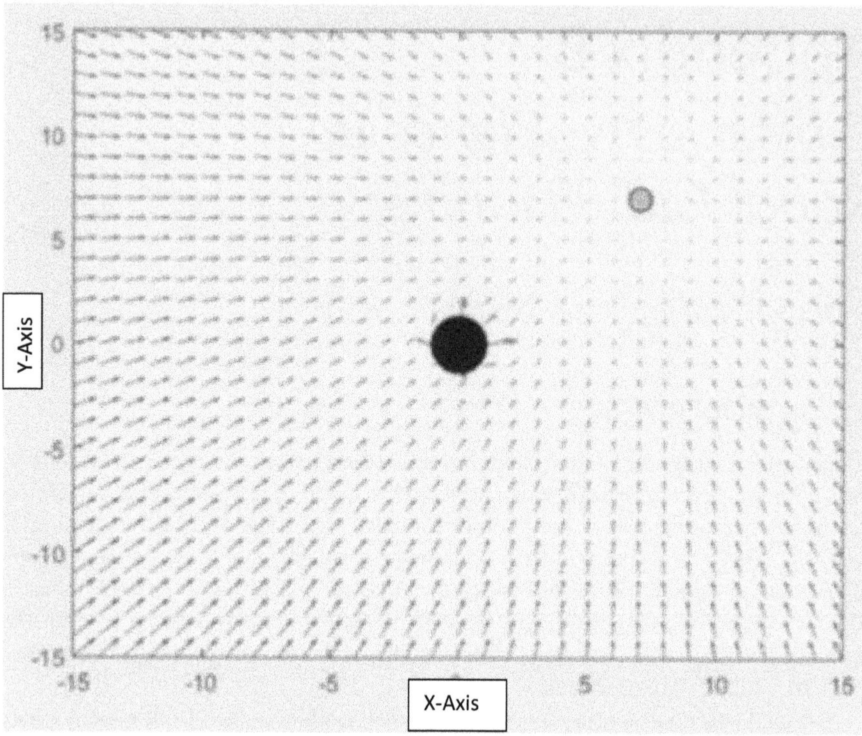

FIGURE B.1 Potential fields: attractive and repulsive.

B.2.1 Attractive Potential Field

This is given by

$$U_{\text{at}(q)} = \frac{1}{2}\xi d^k(q, q_m) \tag{B.1}$$

The symbol ξ is a positive scaling factor, q is the current position of robot, q_m is the goal position, and $d(q, q_m)$ is the distance between the current position of robot and the goal position; $k = 1, 2$; and for $k = 1$, the attractive potential field has a conic shape as in Figure B.2 [1]. The attractive force $F_{\text{at}}(q)$ can be represented by the negative gradient of the potential field using $k = 1$ in (B.2).

$$F_{\text{at}}(q) = -\nabla U_{\text{at}(q)} \tag{B.2}$$

The operator ∇ denotes the vector differential operator which is defined as $\left(\frac{\partial}{\partial x}, \frac{\partial}{\partial y}\right)$

$$F_{\text{at}}(q) = -\nabla U_{\text{at}(q)} = -\frac{1}{2}\xi\frac{(q - q_m)}{d(q, q_m)} \tag{B.3}$$

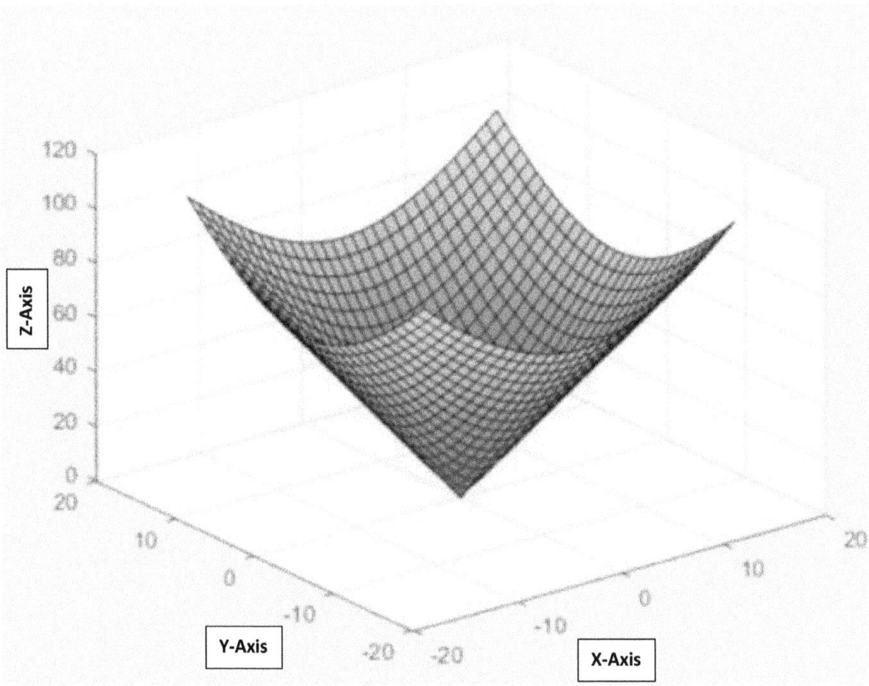

FIGURE B.2 Convex cup of the attractive potential field for $k=1$.

In (B.3), the symbol ξ is a positive scaling factor, $(q-q_m)$ is the difference between the robot current position and goal position, and $d(q,q_m)$ is the distance between the current position of robot and the goal position. For $k=2$, the attractive potential field has a parabolic shape as shown in Figure B.3 [1].

Again, attractive force $F_{at}(q)$ can be represented by the negative gradient of the potential field using $k=2$ in (B.4).

$$F_{at}(q) = -\nabla U_{at(q)} = -\xi(q-q_m) \tag{B.4}$$

B.2.2 Repulsive Potential Field

The repulsive force is inversely proportional to the distance from the obstacle. The common repulsive potential function formula is determined by

$$U_{rep(q)} = \frac{1}{2}\eta\left[\frac{1}{d(q,q_{obs})} - \frac{1}{\rho_o}\right]^2, \text{ for } d(q,q_{obs}) \le \rho_o \tag{B.5}$$

$$U_{rep(q)} = 0 \text{ , for } d(q,q_{obs}) > \rho_o \tag{B.6}$$

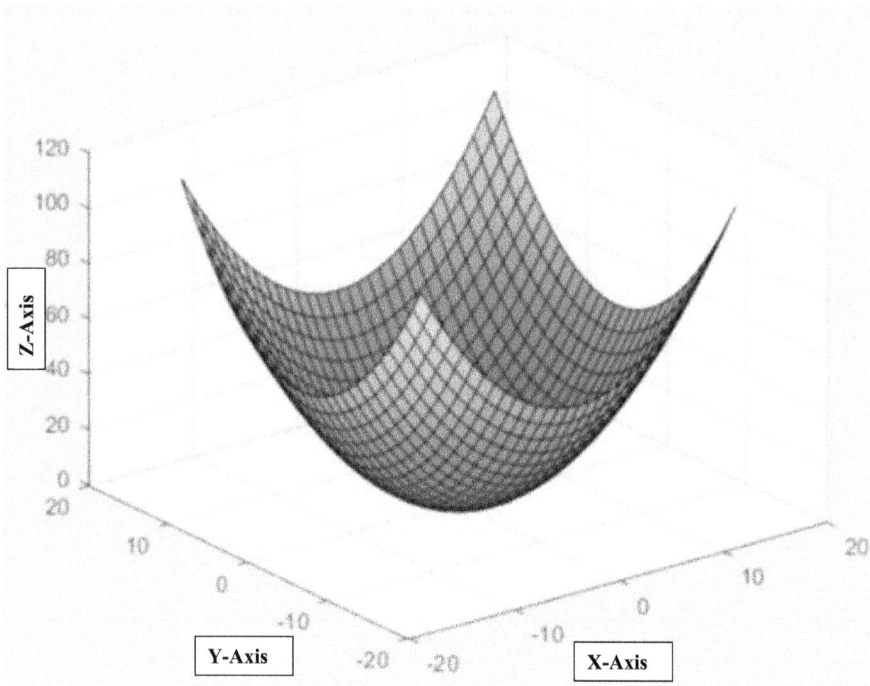

FIGURE B.3 Convex cup of the attractive potential field for $k=2$.

where η is a positive scaling factor, $d(q,q_{\text{obs}})$ is the shortest distance between the robot's current position and obstacle, and ρ_o is a positive constant that represents the distance of obstacle's influence. Taking the negative gradient of this potential function, repulsive force is determined by

$$F_{\text{rep}(q)} = -\nabla U_{\text{rep}}(q)$$

$$F_{\text{rep}(q)} = \eta\left[\frac{1}{d(q,q_{\text{obs}})} - \frac{1}{\rho_o}\right]^2 \frac{(q - q_{\text{obs}})}{d^3(q,q_{\text{obs}})}, \text{ for } d(q,q_{\text{obs}}) \leq \rho_o \tag{B.7}$$

$$F_{\text{rep}(q)} = 0, \text{ for } d(q,q_{\text{obs}}) > \rho_o \tag{B.8}$$

The graphical representation of this potential field is shown in Figure B.4 [1]. The obstacle is located at the highest point and has high repulsive force. Note that one must have knowledge about the coordinates of goal, obstacle, and current position of the robot while implementing this method in both 2D and 3D space.

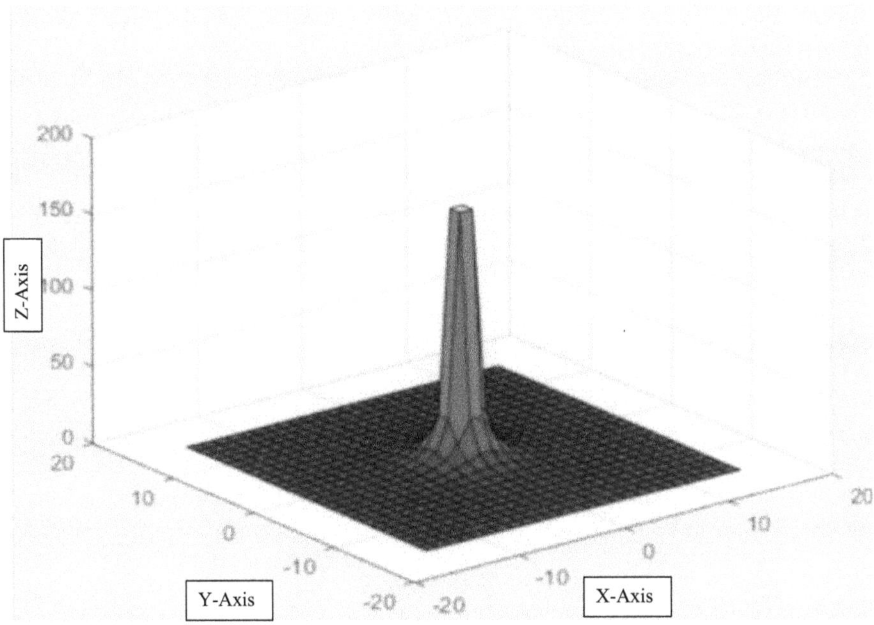

FIGURE B.4 Shape of the repulsive potential field (obstacle function).

B.3 OBSTACLE AVOIDANCE

Obstacle avoidance systems are majorly used in transportation systems such as aircraft traffic control, autonomous cars, and underwater vehicles. The importance of obstacle avoidance is to obtain a collision-free trajectory from the starting point to a target point. Obstacles can either be static or dynamic; static obstacle is the obstacle which is predefined and has a fixed position, whereas the dynamic obstacle is the obstacle which is not pre-known and has indefinite movement.

B.4 VEHICLE GUIDANCE

In this section, a simple planning strategy for control of the position of an autonomous vehicle to move it towards a goal position (i.e., to guide the vehicle) while avoiding an obstacle is described [6]. It would give intuitive insights into how planning strategies operate.

B.4.1 Obstacle Course and Vehicle Characteristics

We seek to guide a vehicle from an initial position to a goal position while avoiding collisions with obstacles. We assume that (i) we have perfect information about where obstacles are, (ii) the vehicle is in a rectangular room, and (iii) the obstacles are poles with known (x, y) positions. We consider a room such that the x-coordinate, $x \in [0, 30]$, and the y-coordinate, $y \in [0, 30]$. We assume that the initial vehicle position is (5, 5) and that the goal position is (25, 25). We assume that each side of the cubical vehicle measures 2.5 units so that it cannot fit in between the three poles that are at positions (8, 10), (10, 10), and (10, 12).

We do not consider vehicle dynamics and assume that when it decides to move from one position to another position, it can approximately do so in one time step (but we do not put explicit units on distance or length of the time step). The feedback control is used for guidance: the current position is sensed, and the command is made to move the vehicle to the new position. The vehicle may not end up where it was commanded to go, but at the next time step, we will sense the vehicle's position and the adjustments would be made.

B.4.2 Path Planning Strategy

The assumption that where all the obstacles are greatly simplifies the planning problem and allows us to focus only on some basic features of planning strategies. In a certain sense, we have a perfect model of a part of our environment. We do not have a perfect model of the entire environment due to the uncertainty in reaching a desired commanded position.

B.4.3 Obstacle and Goal Functions

Since we are using a planning strategy, it is critical to realize that we need to formulate the path-finding problem as an optimization problem. To do this, we take the simple approach of constructing a surface (sometimes called a 'potential field') that represents where the obstacles are. To represent the obstacles, we take Gaussian functions of unity height and centre them at each of the obstacles and compute an 'obstacle function' Jo(x, y) that is the maximum value of each of those functions at each point (x, y). Note that if you moved the vehicle about the environment in a way that the vehicle position is moved to points that try to minimize Jo(x, y) (e.g., via hill climbing), then the vehicle will avoid the obstacles, due to the tails of the Gaussian functions. For many initial vehicle positions, the vehicle would move to the edge of the region, and when it arrives there, we always keep it on the edge.

A goal function is also defined, by giving penalty for not reaching the goal.

B.4.4 Plan Generation and Selection

By now, it must be clear that if we commanded the vehicle to move a distance of λ in a direction θ that is chosen by simply moving in the 'direction of steepest descent' on the function $J_o(x, y)$ (obstacle function), then the vehicle would avoid obstacles but not reach the goal position and stay there. Similarly, if the direction was chosen to be the one with steepest descent for the goal function $J_g(x, y)$, then it would move to the goal position but may collide with some obstacle for some initial vehicle positions. Hence, a multi-objective cost function is used:

$$J(x, y) = w_1 J_o(x, y) + w_2 J_g(x, y) \tag{B.9}$$

This is a Pareto optimal cost function. It also means that the problem is solved that gives conditionally optimal solution. In (B.9), the weights w_1 and w_2 specify the relative importance of achieving obstacle avoidance and reaching the goal: the choices of the weight values represent that obstacle avoidance is important, but the vehicle must also keep moving towards the goal position. The choice of the weights will affect the shape of the Pareto cost function.

B.5 ILLUSTRATIVE EXAMPLE

This example illustrates the vehicle guidance as discussed in Section B.4.

Run the code in the MATLAB command window: pathplan.m; it calls the scripts 'goalfuntion.m' and 'obstaclefuntion.m'. The code can be found in the book's website. Here, only the two submodules are shown for displaying the formulae of the goal function and the obstacle function [6]:

```
% Goal function:
%
function Jg=goalfunction(x,xgoal,w2)
% An example goal function:
Jg=w2*(x-xgoal)'*(x-xgoal);
% Jg=0.01*(x-xgoal)'*(x-xgoal)/(1-0.01*(x-xgoal)'*(x-xgoal));
% END of the file
% Obstacle function:
%
function J=obstaclefunction(x,w1)
% An example function to represent sensed obstacles:
J=...
w1*max([exp(-0.8*((x(1,1)-20)^2+(x(2,1)-15)^2)),...
exp(-0.8*((x(1,1)-8)^2+(x(2,1)-10)^2)),...
exp(-0.8*((x(1,1)-10)^2+(x(2,1)-12)^2)),...
exp(-0.8*((x(1,1)-12)^2+(x(2,1)-14)^2)),...
exp(-0.8*((x(1,1)-24)^2+(x(2,1)-20)^2)),...
exp(-0.8*((x(1,1)-18)^2+(x(2,1)-20)^2))]);
% END OF FILE
```

The results generated by running the 'pathplan' are plotted in Figures B.5–B.12.

Figure B.5 shows the positions of the obstacles, initial vehicle position, and the goal position. Figure B.6 shows several obstacle function values. Figure B.7 depicts the contours (map) of the obstacle function, initial position, and the goal position. Figure B.8 shows the scaled goal function. Figure B.9 shows the contours of the goal function, initial position, and the goal position. Figure B.10 depicts the Pareto (optimal) cost function (multi-objective cost), initial position, and the goal position. The Pareto optimal cost function is also known as the conditionally optimal cost function, since the optimality of the joint, or the multi-objective cost function depends on the choices of the weights used in the cost function.

Once the vehicle has reached the goal from its initial position, the achieved trajectories are the ones as shown in Figure B.11. Finally, Figure B.12 depicts the path traversed by the vehicle avoiding the obstacles and reaching the goal.

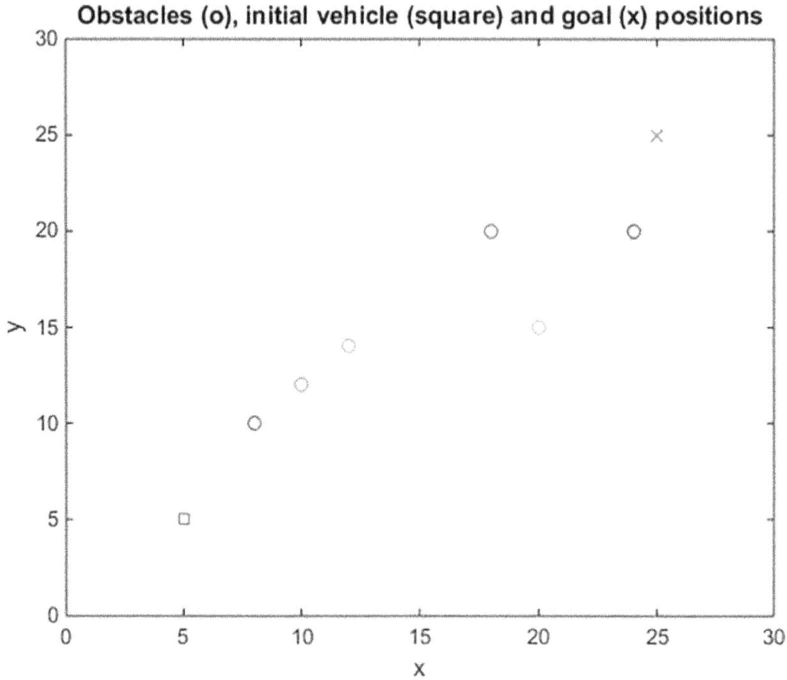

FIGURE B.5 Positions of the obstacles, initial vehicle, and the goal.

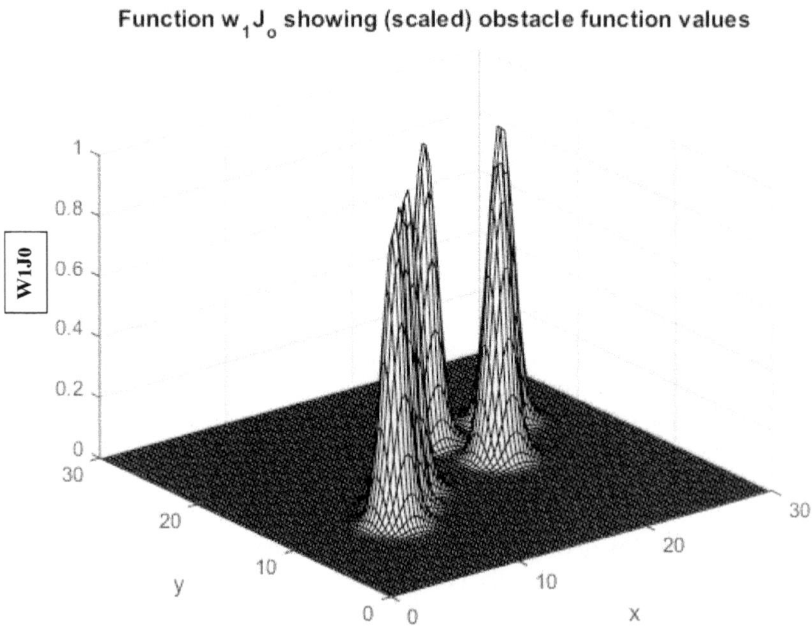

FIGURE B.6 Obstacle function values.

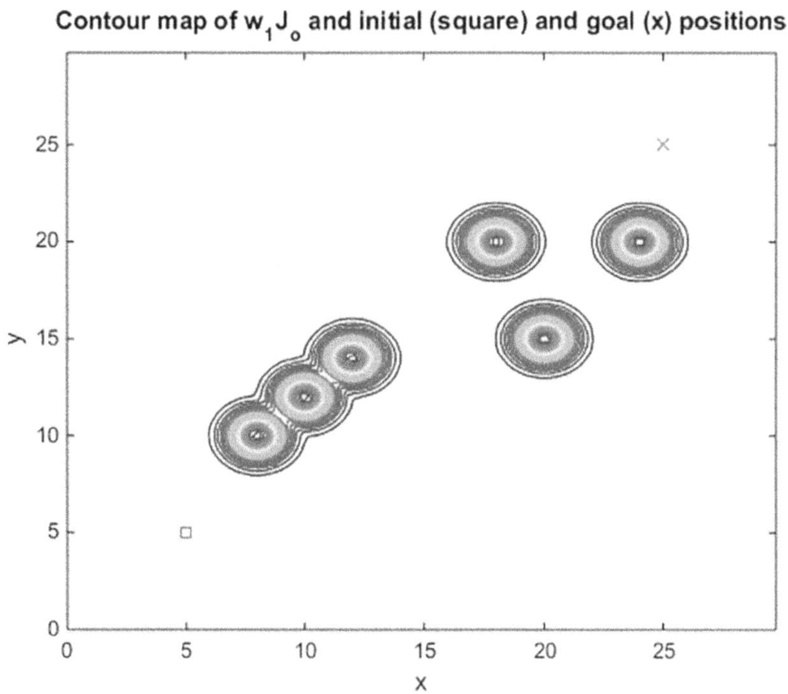

FIGURE B.7 Contour map of the obstacle function, initial position, and the goal position.

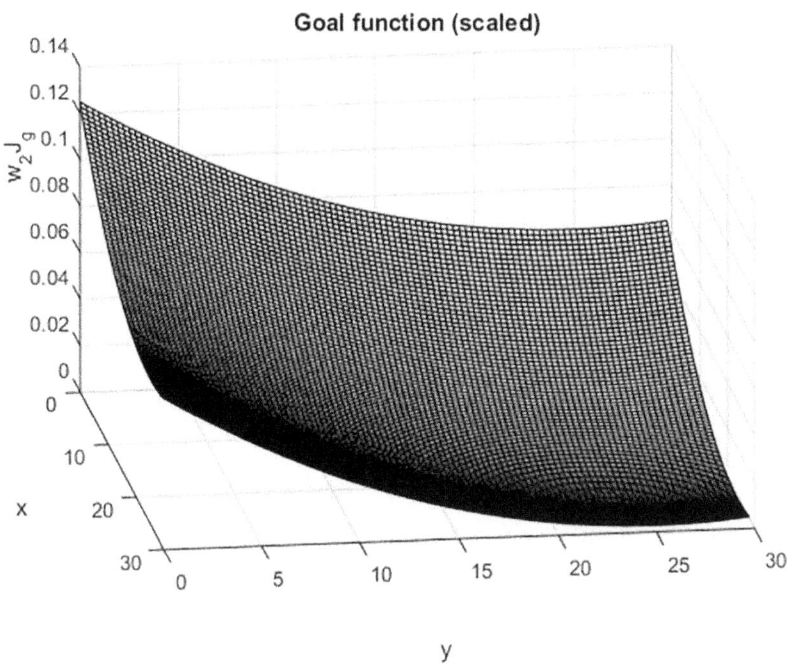

FIGURE B.8 Scaled goal function.

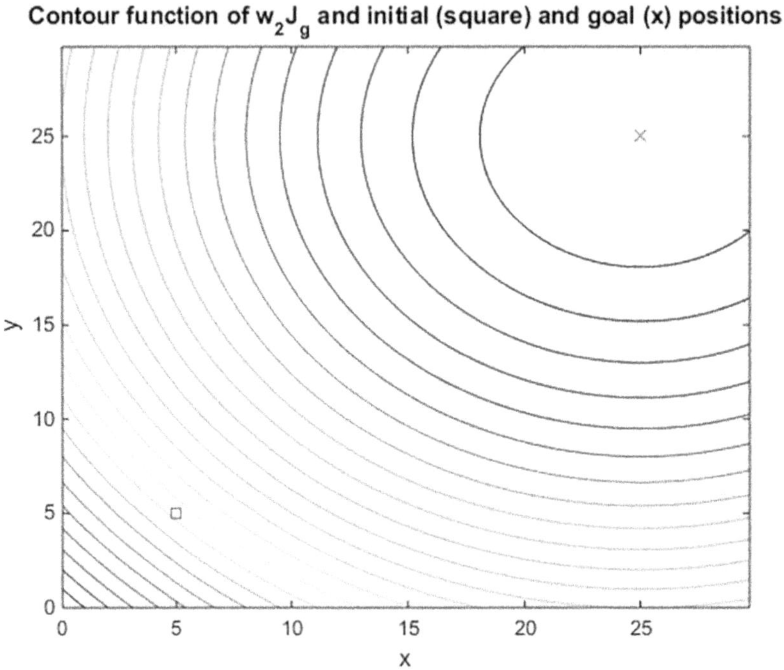

FIGURE B.9 Contours of the goal function, initial position, and the goal position.

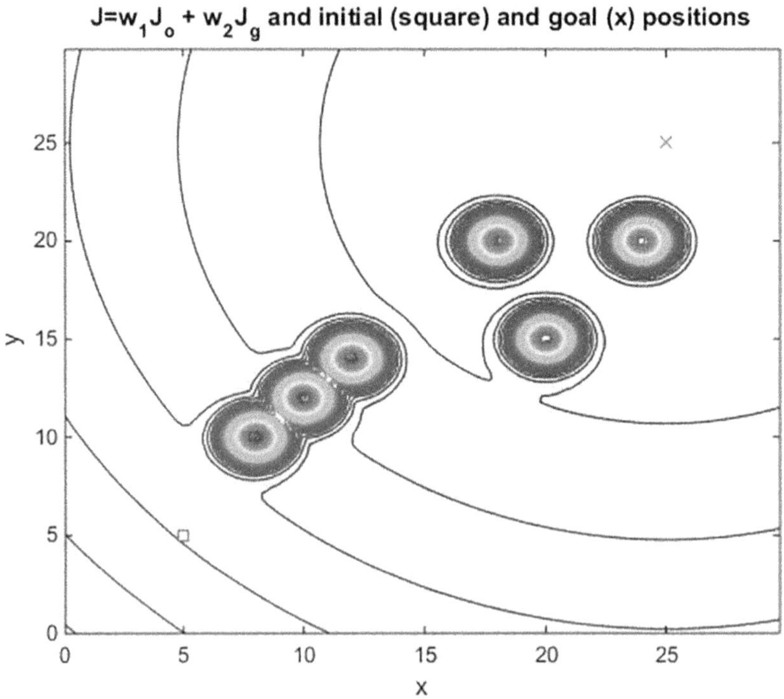

FIGURE B.10 Pareto cost function (multi-objective cost), initial position, and the goal position.

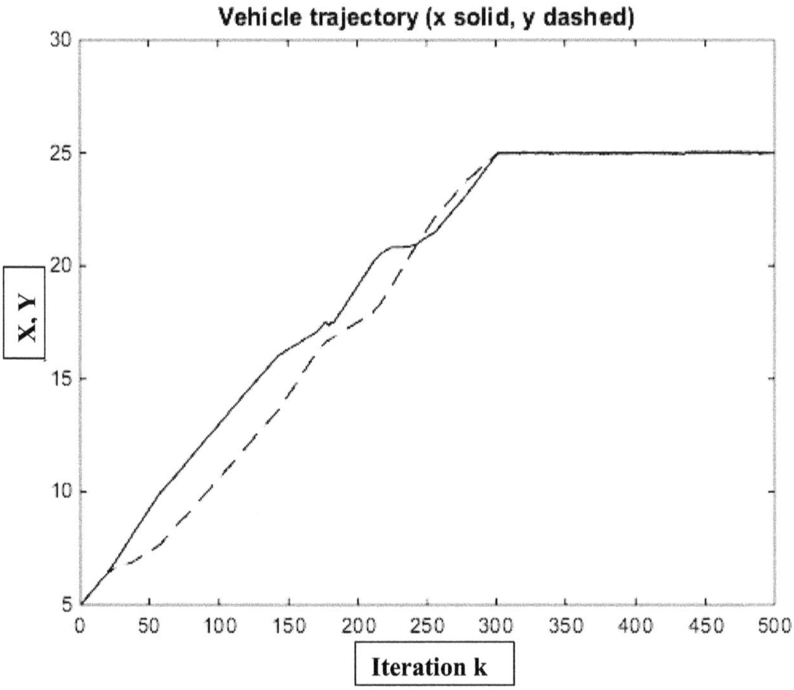

FIGURE B.11 Trajectories of the vehicle.

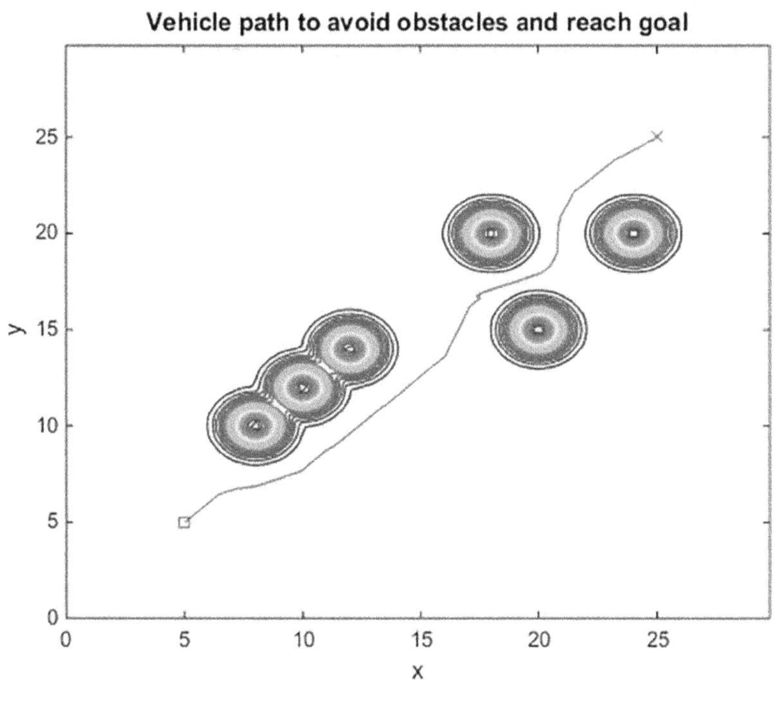

FIGURE B.12 Path traversed by the vehicle avoiding the obstacles and reaching the goal.

REFERENCES

1. Galvez, R. L., Faelden, G. E. U., Maningo, J. M. Z., Reiichiro Christian, S., Dadios, N. E. P., Bandala, A. A., Vicerra, R. R. P., and Fernando, A. H. Obstacle avoidance algorithm for swarm of quadrotor unmanned aerial vehicle using artificial potential fields, *Proceedings of the 2017 IEEE Region 10 Conference (TENCON) on,* Malaysia, 5–8 November 2017, Malaysia.
2. Faelden, G., Maningo, J., Nakano, R., Bandala, A., and Dadios, E. Blind localization method for quadrotor-unmanned aerial vehicle (QUAV) utilizing genetic algorithm. *2014 International Conference on Humanoid, Nanotechnology, Information Technology, Communication and Control, Environment and Management (HNICEM),* Palawan, 2014, Palawan.
3. Faelden, G., Maningo, J., Nakano, R., Bandala, A., and Dadios, E. A neural network approach to a cooperative balancing problem in quadrotor-unmanned aerial vehicles (QUAVs). *2015 International Conference on Humanoid, Nanotechnology, Information Technology, Communication and Control, Environment and Management (HNICEM),* Cebu City, Philippines.
4. Galvez, R. L., Dadios, E. P. and Bandala, A. A. Path planning of quadrotor UAV using genetic algorithm. In *7th IEEE International Conference on Humanoid, Nanotechnology, Information Technology Communication and Control, Environment and Management (HNICEM)*, 2014, Palawan, Philippines.
5. Nakano, R., Bandala, A., Faelden, G., Maningo, J. and Dadios, E. A genetic algorithm approach to swarm centroid tracking in quadrotor unmanned aerial vehicles. *2014 International Conference on Humanoid, Nanotechnology, Information Technology, Communication and Control, Environment and Management (HNICEM),* Palawan, 2014, Palawan.
6. Passino, K. M. *Biomimicry for Optimization, Control, and Automation.* Springer-Verlag, 2005, London, UK.

Appendix C
Enhanced Occlusion Detection Algorithm for Surveillance Cameras

C.1 INTRODUCTION

In the dynamic landscape of the internet of things (IoT) and evolving communication technologies, ranging from smart homes to smart cities, the scope of IoT applications continues to expand exponentially. Surveillance cameras, a pivotal component of IoT, are increasingly pervasive, deeply intertwined with various aspects of our daily lives and professional endeavors. Offering functionalities such as live monitoring, video playback, and anomaly detection, these cameras play a crucial role in safeguarding both individual and societal security. However, they are susceptible to occlusion due to diverse incidents or deliberate human interference, such as criminals intentionally obstructing cameras to evade detection [1–2]. Thus, ensuring uninterrupted surveillance holds paramount importance in practical contexts.

Presently, occlusion detection methods primarily rely on frame differentiation, aiming to identify significant changes in monitoring images during occlusion incidents. However, these methods exhibit high rates of false positives and false negatives, rendering them limited in their applicability. Recognizing that the background scene monitored by a camera typically remains unchanged, scholars have proposed techniques that distinguish between the stationary background and the altered foreground in the monitoring image. Despite their potential, these methods may be affected by the presence of foreground elements in the current frame, thereby compromising detection accuracy.

Addressing the limitations of existing approaches, this book proposes an enhanced background modelling method that integrates background difference. This refined method facilitates the rapid and accurate extraction of the video's background image. Leveraging local binary patterns (LBPs) as image features, which are relatively straightforward to extract and compute while offering high accuracy, this study constructs a feature vector to gauge the similarity between the background image and a reference image. Subsequently, the determination of camera occlusion hinges on this similarity metric. This approach simplifies occlusion detection through image similarity comparison, mitigating the impact of foreground elements and exhibiting robustness to varying illumination conditions.

By enhancing the background modelling technique and employing a robust feature descriptor, this study offers a straightforward yet effective approach to camera occlusion detection, thereby contributing to the reliability and resilience of surveillance systems in diverse environments.

C.2 METHODS

C.2.1 Adaptive Background Modelling Method

The technique of adaptive background subtraction is fundamental and essential in the field of computer vision for the process of isolating or extracting the background elements from a sequence of frames in a video. Its significance lies in its ability to effectively handle dynamic scenes where the background undergoes gradual changes due to factors like lighting variations, moving objects, or camera motion. This method is fundamental for a range of applications such as object detection, motion tracking, and anomaly detection, where accurately delineating foreground objects from the background is paramount.

At the core of the algorithm lies the concept of learning and adaptation. Instead of assuming a static background, the algorithm dynamically adjusts the background model to accommodate gradual changes in lighting, shadows, and other environmental factors. This adaptability makes it robust in scenarios where the background undergoes variations over time. The process begins by initializing parameters, including the learning rate ('alpha') and the video file to be processed. The algorithm then selects a reference frame from the video, usually one that best represents the background scene. This reference frame serves as the initial estimate of the background. As the algorithm iterates through each frame of the video, it updates the background estimate by blending the current frame with the previous background using the learning rate ('alpha'). This blending ensures that the background model gradually adapts to changes in the scene while preserving its stability. Once all frames have been processed, the algorithm compares the reference frame with the final background estimate to identify pixels that have remained unchanged. These pixels are considered part of the background. By calculating the percentage of unchanged pixels relative to the total number of pixels in the reference frame, the algorithm quantifies the success of the background adaptation process. Finally, the algorithm visualizes the reference frame and the adapted background image for inspection and further analysis.

This adaptive background modelling technique finds widespread application in various domains such as surveillance, traffic monitoring, and object tracking. Its ability to adapt to changing environments makes it particularly suitable for real-world scenarios where the background is not static. Additionally, by providing a reliable estimate of the background scene, the algorithm enables accurate detection and tracking of moving objects, thus facilitating tasks like anomaly detection and event recognition.

C.2.2 Mean-Median Modelling Method

This algorithm implements a background modelling technique utilizing both mean and median filtering methods. It's important to understand the differences between mean and median filtering.

Mean filtering is a common image processing technique used to smooth images by replacing each pixel value with the average value of its neighboring pixels. In the context of background modelling, mean filtering is applied to a sequence of video frames to compute the average pixel intensity across the frames at each spatial location. This results in a smoothed background image that represents the average appearance of the scene over time. Mean filtering is sensitive to outliers and can blur sharp edges or details in the image.

Median filtering is another image processing technique used for noise reduction and preserving edge details. Instead of computing the average pixel value, median filtering replaces each pixel with the median value of its neighboring pixels. This makes it robust to outliers and preserves edge information better compared to mean filtering. In background modelling, median filtering is often employed to compute a background image by taking the median pixel intensity across a sequence of video frames. This results in a background image that is less affected by sudden changes or outliers in the scene.

Initially, this method determines the dimensions and extracts a reference frame. A sliding window approach is then employed to process video frames incrementally, updating a buffer with each frame for subsequent background modelling. Here, the sliding window approach enables the algorithm to analyze the video sequence iteratively, updating the background model based on a continuous stream of frames, and facilitating accurate background subtraction and scene understanding in computer vision applications. This method calculates the background model using both mean and median filtering techniques on the buffered frames, resulting in smoothed representations of the background scene.

Once the background models are computed, the algorithm compares the reference frame with the median-filtered background image to identify unchanged pixels, quantifying the success of the background modelling process. Similarly, the success percentage is computed for the mean-filtered background image. These comparisons provide insights into the effectiveness of each filtering method in accurately capturing the background scene's dynamics.

Overall, this method utilizes a sliding window approach and mean and median filtering techniques to the background model. By comparing the reference frame with the background images, the algorithm accurately captures the scene's dynamics and facilitates reliable background subtraction for various computer vision applications.

C.2.3 LBP Algorithm

In image analysis, various features such as gray histogram, edge histogram, color histogram, corner features, and scale-invariant features can be employed to quantify differences between images. After careful analysis, it has been observed that occluded and unoccluded videos exhibit distinct texture features in the background. Consequently, the LBP algorithm has been chosen to measure these differences in background images due to its efficacy in numerous applications including face recognition [3], facial expression recognition [4], image retrieval [5], and image classification [6]. Unlike other feature extraction methods, the LBP algorithm offers higher recognition accuracy while maintaining simplicity in calculation and ease of feature extraction. Leveraging these characteristics, the LBP algorithm is well suited to meet the requirements of real-time detection accuracy in surveillance camera occlusion scenarios.

p_7	p_6	p_5
p_0	p_c	p_4
p_1	p_2	p_3

FIGURE C.1 Window of the LBP algorithm.

The LBP algorithm operates as a nonparametric method for describing differences in gray values between a central pixel and its neighboring pixels within an image. It efficiently captures local texture features, with the original LBP operator defining a 3×3 window around the central pixel as shown in Figure C.1. Here, the gray values of the eight neighboring pixels are compared against the threshold of the central pixel. If a neighbor's gray value exceeds the threshold, its coding value is set to 1; otherwise, it's set to 0. Each neighbor's coding value is then assigned a weight of 2^i, where i ranges from 0 to 7, forming an eight-bit binary pattern. The resulting binary pattern represents the LBP value, which effectively encapsulates texture information within the window.

This LBP value replaces the gray value of the original central pixel. The calculation of the LBP value, denoted as $LBP_{P,R}$, can be expressed mathematically as

$$LBP_{P,R} = \sum_{i=0}^{p-1} s\left(p_i - p_c\right) 2^i \tag{C.1}$$

In this equation, P represents the number of neighboring pixel points ($P = 8$), and $s(x)$ is a symbolic function defined as

$$s(x) = \begin{cases} 1, \ x \geq 0 \\ 0, \ x < 0 \end{cases} \tag{C.2}$$

The LBP algorithm simplifies the process of extracting local texture features by only considering differences in gray values between central and neighboring pixels within the specified window. Its straightforward calculation and wide applicability make it a versatile tool in image analysis tasks.

C.2.4 Occlusion Detection Algorithm for the Surveillance Camera

The occlusion detection algorithm for surveillance cameras begins with the utilization of the improved frame difference method, as described in [7], to extract a reference image from an unoccluded surveillance video. Subsequently, the background modelling method based on background difference and frame difference is employed to extract the background image from the video under surveillance. These images are then processed using the LBP algorithm, resulting in what is termed as a 'mapping' of the images. During practical application, the histogram of these mappings is computed to construct feature vectors, which serve as representations of the images.

To measure the difference between these feature vectors, the chi-square statistic method [8] is employed. In this approach, two histograms, denoted as *S* and *M*, are compared using the formula:

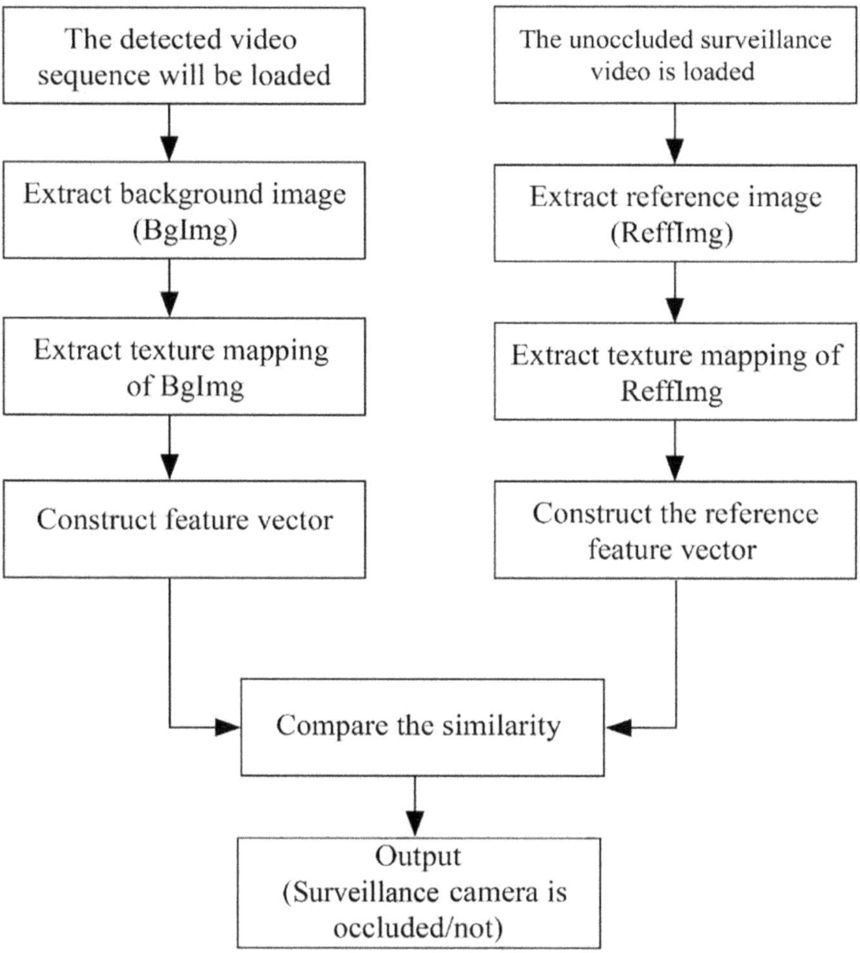

FIGURE C.2 Process of occlusion detection for the surveillance camera.

$$X^2\ (S,M) = \sqrt{\sum_i \frac{(S_i - M_i)^2}{S_i + M_i}} \tag{C.3}$$

Here, S and M represent two different feature vectors, and S_i and M_i denote the values at the same location within the respective vectors.

The specific process of the occlusion detection algorithm unfolds as follows:

Initially, a reference image is extracted from an unoccluded surveillance video, followed by processing with the LBP algorithm to generate a texture mapping. Subsequently, the histogram of the mapping is computed to construct feature vectors, which are stored as reference feature vectors.

1. For the video sequence under scrutiny, the background image is extracted using the background modelling method, and further processed using the LBP algorithm to obtain a mapping. The histogram of this mapping is computed to derive feature vectors. Subsequently, the chi-square statistic is applied to measure the similarity between these feature vectors and the reference feature vectors. Based on this similarity, the presence of occlusion within the video is determined. Higher similarity indicates less likelihood of occlusion, whereas lower similarity suggests a higher degree of occlusion.
2. The algorithm continues to process subsequent frames of the video, repeating Step 2, until the last frame is read. The continuous detection of occlusion is tracked, and based on the frequency of occlusion detection, the severity of occlusion for the surveillance camera is inferred.

This algorithm provides a systematic approach to detecting occlusion in surveillance videos, offering valuable insights into the integrity and functionality of surveillance camera systems.

There are other algorithms for extracting background information from video sequences. This background modelling process is implemented using two distinct techniques: the median and mean techniques. These methods aim to compute an accurate representation of the background scene from a sequence of video frames. In the median technique, the algorithm calculates the median pixel value for each spatial location across multiple frames stored in a buffer. This approach is known for its robustness to outliers, as it is less influenced by extreme pixel values. Conversely, the mean technique computes the average pixel value for each spatial location across the buffered frames. While this method provides a smoother background model, this simple yet effective strategy enables the extraction of background information from video sequences with minimal computational complexity, facilitating tasks such as background subtraction in surveillance and monitoring applications.

C.3 RESULTS

C.3.1 Adaptive Background Modelling Method

The analysis of the adaptive background method is supplemented with visual evidence in the form of images depicting the reference frame and the corresponding computed

FIGURE C.3 Reference image.

background image. The reference frame utilized in the analysis is extracted from frame number 445 of the video sequence, providing a static representation of the scene without any foreground objects. Using the algorithm, a computed background image is generated through a process of mean filtering applied to a sliding window of frames from the video sequence. This filtering methodology ensures a smoothed representation of the background scene, facilitating accurate background subtraction. Upon visual inspection, the comparison between the reference frame and the computed background image reveals a high degree of congruence. Each pixel in the reference frame aligns precisely with its counterpart in the computed background image, indicating a successful adaptation process. This alignment signifies the algorithm's efficacy in accurately capturing and modelling the background scene, leading to a 100% success percentage in background adaptation.

Figure C.3 is designated as the reference image, depicting a static snapshot of the scene extracted from frame number 445 of the video sequence. This reference image serves as the basis for comparison in evaluating the algorithm's performance in background adaptation.

Figure C.4 represents the background-extracted image generated by the adaptive background method algorithm. This computed background image showcases the algorithm's ability to accurately model the background scene, achieving a 100% success rate in background adaptation.

C.3.2 Mean-Median Modelling Method

The analysis of the mean-median background method is similarly supplemented with visual evidence, providing a comprehensive evaluation of the algorithm's performance in background adaptation. In this context, *Figure C.5* serves as the reference image, depicting a static snapshot of the scene extracted from frame

FIGURE C.4 Background image by the adaptive background method.

FIGURE C.5 Reference image.

number 445 of the video sequence. The reference frame provides a basis for evaluating how well the algorithm captures the background scene accurately. The sliding window approach, with a frame difference of 20 frames, is utilized to capture temporal dynamics and enhance the robustness of background modelling. Using the mean-median background method algorithm, computed background images are generated through both median and mean filtering techniques. Figure C.6 represents the background-extracted image generated by median filtering, while Figure C.7 depicts the background image generated by mean filtering. These computed background

FIGURE C.6 Background image extracted by median filtering.

FIGURE C.7 Background image extracted by mean filtering.

images showcase the algorithm's ability to accurately model the background scene, achieving a remarkable 93.83% success rate in background adaptation.

C.4 CONCLUSION

Both the adaptive background algorithm and the mean-median background method demonstrate effectiveness in accurately modelling and adapting to the background scene in video sequences. The adaptive algorithm, with its incremental approach and utilization of a learning rate parameter, excels in capturing subtle changes in the background over time. It achieves a 100% success rate in background adaptation, as evidenced by the precise alignment between the reference frame and the computed background image. Similarly, the mean-median background method, employing mean and median filtering techniques, proves successful in generating background images that closely resemble the reference frame. With both methods, the reference frame serves as a benchmark for evaluating the accuracy of the background adaptation process. Overall, these findings underscore the algorithms' capabilities in facilitating reliable background subtraction and scene understanding in various computer vision applications.

REFERENCES

1. Kuang, C. *Research on Detection and Classification of Camera Tampering Events*, Harbin Institute of Technology, Harbin, China, 2016 (in Chinese).
2. Qiu, M. *Research and Achievement of the Diagnoses of the Surveillance Video Image Quality*, East China University of Science and Technology, Shanghai, China, 2014 (in Chinese).
3. Shi, L., Wang, X. and Shen, Y. Research on 3D face recognition method based on LBP and SVM. *Optik*, 220, 165157, 2020.
4. Bejaoui, H., Ghazouani, H. and Barhoumi, W. Sparse coding-based representation of LBP difference for 3D/4D facial expression recognition. *Multim. Tools Appl.*, 78(16), 22773–22796, 2019.
5. Garg, M. and Dhiman, G. A novel content-based image retrieval approach for classification using GLCM features and texture fused LBP variants. *Neural Comput. Appl.*, 33, 1311–1328, 2020.
6. Ye, Z., Dong, R., Lin, B., Jin, C. and Nian, Y. Hyperspectral image classification based on segmented local binary patterns. *Sensi. Imag. J.*, 21(2), 556–567, 2020.
7. Kong, X. *Research on Video Anomaly Detection Method Based on Image Analysis*. Beijing University of Posts and Tele communications, Beijing, China, 2017 (in Chinese).
8. Shi, P., Hou, B., Chen, J. and Zu, Y. *An Algorithm of Occlusion Detection for the Surveillance Camera*. Beijing University of Posts and Telecommunications, Beijing, China, 2021.

ABBREVIATIONS

IoT	Internet of Things
LBP	Local Binary Pattern

QUESTIONS AND ANSWERS

1. What technique is utilized for image feature extraction?
2. What is the advantage of employing LBP as an image feature?
3. What are some widely utilized background modelling methods in machine vision?
4. What distinguishes the frame difference approach from other background modelling methods?
5. What limitation does the enhanced frame difference method still face?
6. What is the significance of Thread1 in the algorithm?

7. What is the role of Thread1 in Step 3 of the background modelling algorithm?
8. How does the iterative process conclude in Step 5 of the algorithm?
9. What are some of the features commonly used in image analysis to quantify differences between images?
10. How does the LBP algorithm capture local texture features within an image?
11. How does the LBP algorithm simplify the process of extracting local texture features?
12. How does the algorithm track the presence and severity of occlusion throughout the video sequence?
13. What role does the chi-square statistic play in the occlusion detection algorithm? Occlusion presence.

Index